Student Solutions Manual for Turner/McKeague
Elementary Algebra

Prepared by

Ross Rueger

Department of Mathematics

College of the Sequoias

Visalia, California

Student Solutions Manual for Turner/McKeague Elementary Algebra

Ross Rueger

Publisher: XYZ Textbooks

Sales: Amy Jacobs, Richard Jones, Bruce Spears, Rachael Hillman

Cover Design: Rachel Hintz

ISBN-13: 978-1-630980-67-2 / ISBN-10: 1-63098-067-6

For product information and technology assistance, contact us at
XYZ Textbooks, 1-877-745-3499

For permission to use material from this text or product,
e-mail: **info@mathtv.com**

XYZ Textbooks
1339 Marsh Street
San Luis Obispo, CA 93401
USA

Printed in the United States of America

For your course and learning solutions, visit **www.xyztextbooks.com**

Student Solutions Manual to Accompany

Elementary Algebra

By

Mark D. Turner
Charles P. McKeague

Prepared by
Ross Rueger
Department of Mathematics
College of the Sequoias
Visalia, California

XYZ Textbooks

Contents

Preface

This *Student Solutions Manual* contains complete solutions to all odd-numbered exercises, and complete solutions to all chapter tests, of *Elementary Algebra* by Charles P. McKeague and Mark D. Turner. I have attempted to format solutions for readability and accuracy, and apologize to you for any errors that you may encounter. If you have any comments, suggestions, error corrections, or alternative solutions please feel free to send me an email (address below).

Please use this manual with some degree of caution. Be sure that you have attempted a solution, and re-attempted it, before you look it up in this manual. Mathematics can only be learned by *doing*, and not by observing! As you use this manual, do not just read the solution but work it along with the manual, using my solution to check your work. If you use this manual in that fashion then it should be helpful to you in your studying.

I would like to thank Katherine Heistand Shields, Amy Jacobs and Mark Turner at XYZ Textbooks for their help with this project and for getting back to me with corrections so quickly. Producing a manual such as this is a team effort, and this is an excellent team to work with.

I wish to express my appreciation to Pat McKeague for asking me to be involved with this textbook. His books continue to refine the subject of intermediate algebra, and you will find the text very easy to read and understand. I especially appreciate his efforts through XYZ Textbooks to make textbooks affordable for students to purchase.

Good luck!

Ross Rueger
College of the Sequoias
rossrueger@gmail.com

March, 2016

Chapter 1
The Basics

1.1 Variables, Symbols, and the Order of Operations

1. The equivalent expression is $x+5$.

3. The equivalent expression is $5y$.

5. The equivalent expression is $5(y-16)$.

7. The equivalent expression is $\frac{x}{3}$.

9. Evaluating the exponent: $3^2=3\cdot 3=9$

11. Evaluating the exponent: $7^2=7\cdot 7=49$

13. Evaluating the exponent: $2^3=2\cdot 2\cdot 2=8$

15. Evaluating the exponent: $4^3=4\cdot 4\cdot 4=64$

17. Evaluating the exponent: $2^4=2\cdot 2\cdot 2\cdot 2=16$

19. Evaluating the exponent: $10^2=10\cdot 10=100$

21. Evaluating the exponent: $11^2=11\cdot 11=121$

23. Using the order of operations: $2\cdot 3+5=6+5=11$

25. Using the order of operations: $2(3+5)=2(8)=16$

27. Using the order of operations: $5+2\cdot 6=5+12=17$

29. Using the order of operations: $(5+2)\cdot 6=7\cdot 6=42$

31. Using the order of operations: $5\cdot 4+5\cdot 2=20+10=30$

33. Using the order of operations: $5(4+2)=5(6)=30$

35. Using the order of operations: $8+2(5+3)=8+2(8)=8+16=24$

37. Using the order of operations: $(8+2)(5+3)=(10)(8)=80$

39. Using the order of operations: $20+2(8-5)+1=20+2(3)+1=20+6+1=27$

41. Using the order of operations: $5+2(3\cdot 4-1)+8=5+2(12-1)+8=5+2(11)+8=5+22+8=35$

43. Using the order of operations: $8+10\div 2=8+5=13$

45. Using the order of operations: $4+8\div 4-2=4+2-2=4$

47. Using the order of operations: $3+12\div 3+6\cdot 5=3+4+30=37$

49. Using the order of operations: $3\cdot 8+10\div 2+4\cdot 2=24+5+8=37$

51. Using the order of operations: $(5+3)(5-3)=(8)(2)=16$

53. Using the order of operations: $5^2-3^2=5\cdot 5-3\cdot 3=25-9=16$

55. Using the order of operations: $(4+5)^2=9^2=9\cdot 9=81$

57. Using the order of operations: $4^2+5^2=4\cdot 4+5\cdot 5=16+25=41$

59. Using the order of operations: $3\cdot 10^2+4\cdot 10+5=300+40+5=345$

61. Using the order of operations: $2\cdot 10^3+3\cdot 10^2+4\cdot 10+5=2{,}000+300+40+5=2{,}345$

63. Using the order of operations: $10-2(4\cdot 5-16)=10-2(20-16)=10-2(4)=10-8=2$

65. Using the order of operations: $4[7+3(2\cdot 9-8)]=4[7+3(18-8)]=4[7+3(10)]=4(7+30)=4(37)=148$

67. Using the order of operations: $5(7-3)+8(6-4)=5(4)+8(2)=20+16=36$

69. Using the order of operations: $3(4\cdot 5-12)+6(7\cdot 6-40)=3(20-12)+6(42-40)=3(8)+6(2)=24+12=36$

71. Using the order of operations: $3^4+4^2\div 2^3-5^2=81+16\div 8-25=81+2-25=58$

73. Using the order of operations: $5^2+3^4\div 9^2+6^2=25+81\div 81+36=25+1+36=62$

75. Using the order of operations: $20\div 2\cdot 10=10\cdot 10=100$

77. Using the order of operations: $24 \div 8 \cdot 3 = 3 \cdot 3 = 9$
79. Using the order of operations: $36 \div 6 \cdot 3 = 6 \cdot 3 = 18$
81. Using the order of operations: $48 \div 12 \cdot 2 = 4 \cdot 2 = 8$
83. Using the order of operations: $16 - 8 + 4 = 8 + 4 = 12$
85. Using the order of operations: $24 - 14 + 8 = 10 + 8 = 18$
87. Using the order of operations: $36 - 6 + 12 = 30 + 12 = 42$
89. Using the order of operations: $48 - 12 + 17 = 36 + 17 = 53$
91. The perimeter is $4(1\text{ in.}) = 4\text{ in.}$, and the area is $(1\text{ in.})^2 = 1\text{ in.}^2$.
93. The perimeter is $2(1.5\text{ in.}) + 2(0.75\text{ in.}) = 3.0\text{ in.} + 1.5\text{ in.} = 4.5\text{ in.}$, and the area is $(1.5\text{ in.})(0.75\text{ in.}) = 1.125\text{ in.}^2$.
95. The perimeter is $2.75\text{ cm} + 4\text{ cm} + 3.5\text{ cm} = 10.25\text{ cm}$, and the area is $\frac{1}{2}(4\text{ cm})(2.5\text{ cm}) = 5.0\text{ cm}^2$.
97. There are $5 \cdot 2 = 10$ cookies in the package.
99. The total number of calories is $210 \cdot 2 = 420$ calories.
101. There are $7 \cdot 32 = 224$ chips in the bag.
103. For a person eating 3,000 calories per day, the recommended amount of fat would be $80 + 15 = 95$ grams.
105. a. The amount of caffeine is: $6(280) = 1{,}680$ mg

b. The amount of caffeine is: $2(34) + 3(47) = 68 + 141 = 209$ mg

107. Completing the table:

Activity	Calories burned in 1 hour
Bicycling	374
Bowling	265
Handball	680
Jogging	680
Skiing	544

109. The area is given by: $(8.5)(11) = 93.5$ square inches

The perimeter is given by: $2(8.5) + 2(11) = 17 + 22 = 39$ inches

111. Translating into symbols: $4(x - 9)$. The correct answer is d.
113. Using the order of operations: $6 + 4(3 \cdot 5 - 8) - 10 = 6 + 4(15 - 8) - 10 = 6 + 4(7) - 10 = 6 + 28 - 10 = 24$

The correct answer is a.

115. Finding the perimeter: $2(8) + 2(10) = 16 + 20 = 36$ inches. The correct answer is b.

1.2 The Real Numbers

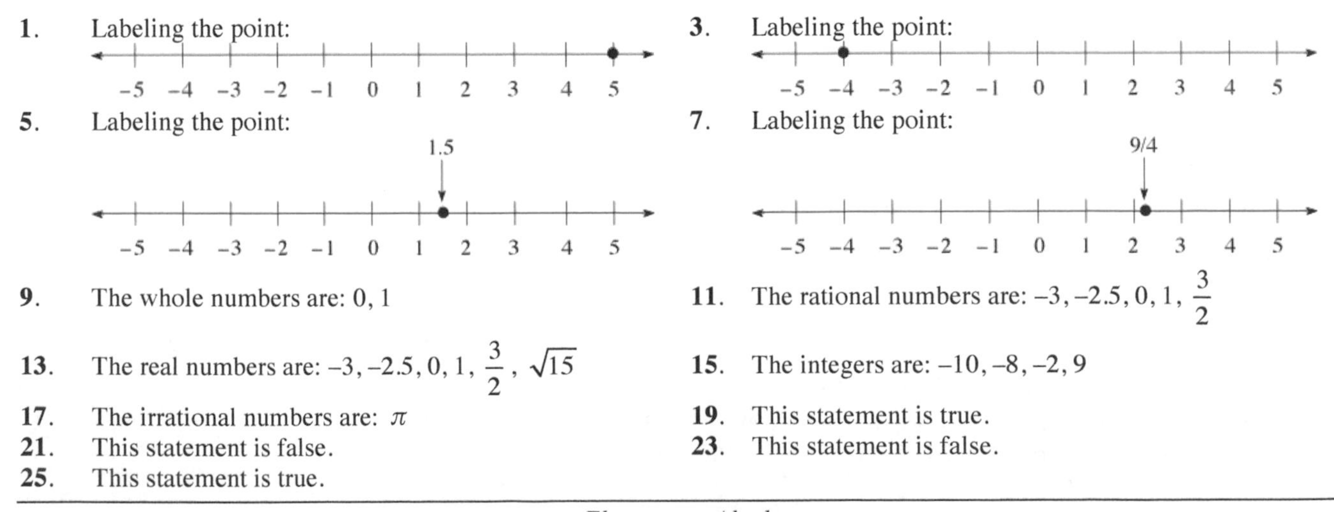

27. The opposite of 10 is –10 and the absolute value is $|10| = 10$.

29. The opposite of $\frac{3}{4}$ is $-\frac{3}{4}$ and the absolute value is $\left|\frac{3}{4}\right| = \frac{3}{4}$.

31. The opposite of $\frac{11}{2}$ is $-\frac{11}{2}$ and the absolute value is $\left|\frac{11}{2}\right| = \frac{11}{2}$.

33. The opposite of –3 is 3 and the absolute value is $|-3| = 3$.

35. The opposite of $-\frac{2}{5}$ is $\frac{2}{5}$ and the absolute value is $\left|-\frac{2}{5}\right| = \frac{2}{5}$.

37. The opposite of x is $-x$ and the absolute value is $|x|$.

39. The correct symbol is <: $-5 < -3$

41. The correct symbol is >: $-3 > -7$

43. Since $|-4| = 4$ and $-|-4| = -4$, the correct symbol is >: $|-4| > -|-4|$

45. Since $-|-7| = -7$, the correct symbol is >: $7 > -|-7|$

47. Simplifying the expression: $|8-2| = |6| = 6$

49. Simplifying the expression: $\left|5 \cdot 2^3 - 2 \cdot 3^2\right| = |5 \cdot 8 - 2 \cdot 9| = |40-18| = |22| = 22$

51. Simplifying the expression: $|7-2| - |4-2| = |5| - |2| = 5 - 2 = 3$

53. Simplifying the expression: $10 - |7 - 2(5-3)| = 10 - |7 - 2(2)| = 10 - |7-4| = 10 - |3| = 10 - 3 = 7$

55. Simplifying the expression:

$$\begin{aligned} 15 - |8 - 2(3 \cdot 4 - 9)| - 10 &= 15 - |8 - 2(12-9)| - 10 \\ &= 15 - |8 - 2(3)| - 10 \\ &= 15 - |8-6| - 10 \\ &= 15 - |2| - 10 \\ &= 15 - 2 - 10 \\ &= 3 \end{aligned}$$

57. A loss of 8 yards corresponds to –8 on a number line. The total yards gained corresponds to –2 yards.

59. The temperature can be represented as –64°. The new (warmer) temperature corresponds to –54°.

61. The wind chill temperature is –15°.

63. His depth can be represented by –100 feet. After he descends another 5 feet, his depth is –105 feet.

65. The calories burned is: $2(544) + 1(299) = 1{,}088 + 299 = 1{,}387$ calories

67. The difference in calories burned is: $3(653) - 3(435) = 1{,}959 - 1{,}305 = 654$ more calories

69. **a.** This is $3 \cdot 31 = 93$ million phones.
    b. The chart shows $4.5 \cdot 31 = 139.5$ million phones, so this statement is false.
    c. The chart shows $9.5 \cdot 31 = 294.5$ million phones, so this statement is true.

71. The point is $-1\frac{3}{4}$. The correct answer is c.

73. Evaluating: $|-9| = 9$. The correct answer is a.

1.3 Addition of Real Numbers

1. Adding all positive and negative combinations of 3 and 5:

$3+5=8$ $\quad$ $3+(-5)=-2$

$-3+5=2$ $\quad$ $(-3)+(-5)=-8$

3. Adding all positive and negative combinations of 15 and 20:

$15+20=35$ $\quad$ $15+(-20)=-5$

$-15+20=5$ $\quad$ $(-15)+(-20)=-35$

5. Adding the numbers: $6+(-3)=3$
7. Adding the numbers: $13+(-20)=-7$
9. Adding the numbers: $18+(-32)=-14$
11. Adding the numbers: $-6+3=-3$
13. Adding the numbers: $-30+5=-25$
15. Adding the numbers: $-6+(-6)=-12$
17. Adding the numbers: $-9+(-10)=-19$
19. Adding the numbers: $-10+(-15)=-25$
21. Performing the additions: $5+(-6)+(-7)=5+(-13)=-8$
23. Performing the additions: $-7+8+(-5)=-12+8=-4$
25. Performing the additions: $5+[6+(-2)]+(-3)=5+4+(-3)=9+(-3)=6$
27. Performing the additions: $[6+(-2)]+[3+(-1)]=4+2=6$
29. Performing the additions: $20+(-6)+[3+(-9)]=14+(-6)=8$
31. Performing the additions: $-3+(-2)+[5+(-4)]=-3+(-2)+1=-5+1=-4$
33. Performing the additions: $(-9+2)+[5+(-8)]+(-4)=-7+(-3)+(-4)=-14$
35. Performing the additions: $[-6+(-4)]+[7+(-5)]+(-9)=-10+2+(-9)=-19+2=-17$
37. Performing the additions: $(-6+9)+(-5)+(-4+3)+7=3+(-5)+(-1)+7=10+(-6)=4$
39. Using order of operations: $-5+2(-3+7)=-5+2(4)=-5+8=3$
41. Using order of operations: $9+3(-8+10)=9+3(2)=9+6=15$
43. Using order of operations: $-10+2(-6+8)+(-2)=-10+2(2)+(-2)=-10+4+(-2)=-12+4=-8$
45. Using order of operations: $2(-4+7)+3(-6+8)=2(3)+3(2)=6+6=12$
47. Writing the expression and simplifying: $5+9=14$
49. Writing the expression and simplifying: $[-7+(-5)]+4=(-12)+4=-8$
51. The expression is: $[-2+(-3)]+10=-5+10=5$
53. The expression is: $4+x$
55. The expression is: $-8+x$
57. The expression is: $[x+(-2)]+3$
59. The number is 3, since $-8+3=-5$.
61. The number is –3, since $-3+(-6)=-9$.
63. The pattern is to add 5, so the next two terms are $18+5=23$ and $23+5=28$.
65. The pattern is to add 5, so the next two terms are $25+5=30$ and $30+5=35$.
67. The pattern is to add –5, so the next two terms are $5+(-5)=0$ and $0+(-5)=-5$.
69. The pattern is to add –6, so the next two terms are $-6+(-6)=-12$ and $-12+(-6)=-18$.
71. The pattern is to add –4, so the next two terms are $0+(-4)=-4$ and $-4+(-4)=-8$.
73. Yes, since each successive odd number is 2 added to the previous one.
75. The expression is $-12°+4°=-8°$.
77. The expression is $\$10+(-\$6)+(-\$8)=\$10+(-\$14)=-\$4$.
79. The new balance is $-\$30+\$40=\$10$.

81. The profit was: $\$11{,}500-\$9{,}500=\$2{,}000$.

83. There was a loss in the year 2006: $\$7{,}750-\$8{,}250=-\$500$

85. Translating the phrase: $-3+4$. The correct answer is c.

1.4 Subtraction of Real Numbers

1. Subtracting the numbers: $5-8=5+(-8)=-3$

3. Subtracting the numbers: $3-9=3+(-9)=-6$

5. Subtracting the numbers: $5-5=5+(-5)=0$

7. Subtracting the numbers: $-8-2=-8+(-2)=-10$

9. Subtracting the numbers: $-4-12=-4+(-12)=-16$

11. Subtracting the numbers: $-6-6=-6+(-6)=-12$

13. Subtracting the numbers: $-8-(-1)=-8+1=-7$

15. Subtracting the numbers: $15-(-20)=15+20=35$

17. Subtracting the numbers: $-4-(-4)=-4+4=0$

19. Using order of operations: $3-2-5=3+(-2)+(-5)=3+(-7)=-4$

21. Using order of operations: $9-2-3=9+(-2)+(-3)=9+(-5)=4$

23. Using order of operations: $-6-8-10=-6+(-8)+(-10)=-24$

25. Performing the additions: $-22+4-10=-22+4+(-10)=-32+4=-28$

27. Using order of operations: $10-(-20)-5=10+20+(-5)=30+(-5)=25$

29. Using order of operations: $8-(2-3)-5=8-(-1)-5=8+1+(-5)=9+(-5)=4$

31. Using order of operations: $7-(3-9)-6=7-(-6)-6=7+6+(-6)=13+(-6)=7$

33. Using order of operations: $5-(-8-6)-2=5-(-14)-2=5+14+(-2)=19+(-2)=17$

35. Using order of operations: $-(5-7)-(2-8)=-(-2)-(-6)=2+6=8$

37. Using order of operations: $-(3-10)-(6-3)=-(-7)-3=7+(-3)=4$

39. Performing the additions: $16-[(4-5)-1]=16-(-1-1)=16-(-2)=16+2=18$

41. Using order of operations: $5-[(2-3)-4]=5-(-1-4)=5-(-5)=5+5=10$

43. Using order of operations:

$$21-[-(3-4)-2]-5=21-[-(-1)-2]-5=21-(1-2)-5=21-(-1)-5=21+1+(-5)=22+(-5)=17$$

45. Using order of operations: $2\cdot 8-3\cdot 5=16-15=1$

47. Using order of operations: $3\cdot 5-2\cdot 7=15-14=15+(-14)=1$

49. Using order of operations: $5\cdot 9-2\cdot 3-6\cdot 2=45-6-12=45+(-6)+(-12)=45+(-18)=27$

51. Using order of operations: $3\cdot 8-2\cdot 4-6\cdot 7=24-8-42=24+(-8)+(-42)=24+(-50)=-26$

53. Using order of operations: $2\cdot 3^2-5\cdot 2^2=2\cdot 9-5\cdot 4=18-20=18+(-20)=-2$

55. Using order of operations: $4\cdot 3^3-5\cdot 2^3=4\cdot 27-5\cdot 8=108-40=108+(-40)=68$

57. Writing the expression and simplifying: $-7-4=-7+(-4)=-11$

59. Writing the expression and simplifying: $12-(-8)=12+8=20$

61. Writing the expression and simplifying: $-5-(-7)=-5+7=2$

63. Writing the expression and simplifying: $[4+(-5)]-17=(-1)+(-17)=-18$

65. Writing the expression and simplifying: $8-5=3$

67. Writing the expression and simplifying: $-8-5=-8+(-5)=-13$

69. Writing the expression and simplifying: $8-(-5)=8+5=13$

71. Writing the expression: $x-6$

73. Writing the expression: $-4-x$

75. Writing the expression: $(x+12)-5$

77. The number is 10, since $8-10=-2$.

79. The number is –2, since $8-(-2)=8+2=10$.

81. The expression is $\$1{,}500-\$730=\$770$.

83. The expression is $-\$35+\$15-\$20=-\$35+(-\$20)+\$15=-\$55+\$15=-\$40$.

85. The expression is $73°+10°-8°=83°-8°=75°$ F.

87. The sequence of values is \$4,500, \$3,950, \$3,400, \$2,850, and \$23,00. This is an arithmetic sequence, since –\$550 is added to each value to obtain the new value.

89. a. Completing the table:

Day	0	2	4	6	8	10
Plant Height (inches)	0	1	3	6	13	23

b. Subtracting: $13-1=12$. The grass is 12 inches higher after 8 days than after 2 days.

91. a. Yes, the numbers appear to be rounded to the hundreds place.

b. Subtracting: $21{,}400-13{,}000=8{,}400$. There were 8,400 more participants in 2004 than in 2000.

c. If the amount increases by another 8,400, there will be 21,400 + 8,400 = 29,800 participants in 2008.

93. Writing the expression: $-9-4$. The correct answer is a.

1.5 Multiplication and Division of Real Numbers

1. Finding the product: $7(-6)=-42$

3. Finding the product: $-8(2)=-16$

5. Finding the product: $-3(-1)=3$

7. Finding the product: $-11(-11)=121$

9. Using order of operations: $-3(2)(-1)=6$

11. Using order of operations: $-3(-4)(-5)=-60$

13. Using order of operations: $-2(-4)(-3)(-1)=24$

15. Using order of operations: $(-7)^2=(-7)(-7)=49$

17. Using order of operations: $(-3)^3=(-3)(-3)(-3)=-27$

19. Finding the quotient: $\frac{8}{-4}=-2$

21. Finding the quotient: $\frac{-48}{16}=-3$

23. Finding the quotient: $\frac{-7}{21}=-\frac{1}{3}$

25. Finding the quotient: $\frac{-39}{-13}=3$

27. Finding the quotient: $\frac{-6}{-42}=\frac{1}{7}$

29. Finding the quotient: $\frac{0}{-32}=0$

31. Performing the operations: $-3+12=9$

33. Performing the operations: $-3-12=-3+(-12)=-15$

35. Performing the operations: $-3(12)=-36$

37. Performing the operations: $-3\div 12=\frac{-3}{12}=-\frac{1}{4}$

39. Using order of operations: $\frac{3(-2)}{-10}=\frac{-6}{-10}=\frac{3}{5}$

41. Using order of operations: $\frac{-5(-5)}{-15}=\frac{25}{-15}=-\frac{5}{3}$

43. Using order of operations: $\frac{-8(-7)}{-28}=\frac{56}{-28}=-2$

45. Using order of operations: $-2(2-5)=-2(-3)=6$

47. Using order of operations: $-5(8-10)=-5(-2)=10$

49. Using order of operations: $(4-7)(6-9)=(-3)(-3)=9$

51. Using order of operations: $(-3-2)(-5-4)=(-5)(-9)=45$

53. Using order of operations: $-3(-6)+4(-1)=18+(-4)=14$

55. Using order of operations: $2(3)-3(-4)+4(-5)=6+12+(-20)=18+(-20)=-2$

57. Using order of operations: $\frac{27}{4-13}=\frac{27}{-9}=-3$

59. Using order of operations: $\frac{20-6}{5-5}=\frac{14}{0}$, which is undefined

61. Using order of operations: $\frac{-3+9}{2\cdot 5-10}=\frac{6}{10-10}=\frac{6}{0}$, which is undefined

63. Using order of operations: $\frac{15(-5)-25}{2(-10)}=\frac{-75-25}{-20}=\frac{-100}{-20}=5$

65. Using order of operations: $\frac{27-2(-4)}{-3(5)}=\frac{27+8}{-15}=\frac{35}{-15}=-\frac{7}{3}$

67. Using order of operations: $\frac{12-6(-2)}{12(-2)}=\frac{12+12}{-24}=\frac{24}{-24}=-1$

69. Using order of operations: $4(-3)^2+5(-6)^2=4(9)+5(36)=36+180=216$

71. Using order of operations: $7(-2)^3-2(-3)^3=7(-8)-2(-27)=-56+54=-2$

73. Using order of operations: $6-4(8-2)=6-4(6)=6-24=6+(-24)=-18$

75. Using order of operations: $9-4(3-8)=9-4(-5)=9+20=29$

77. Using order of operations: $-4(3-8)-6(2-5)=-4(-5)-6(-3)=20+18=38$

79. Using order of operations: $\frac{5^2-2^2}{-5+2}=\frac{25-4}{-3}=\frac{21}{-3}=-7$

81. Using order of operations: $\frac{8^2-2^2}{8^2+2^2}=\frac{64-4}{64+4}=\frac{60}{68}=\frac{15}{17}$

83. Using order of operations: $\frac{(5+3)^2}{-5^2-3^2}=\frac{8^2}{-25-9}=\frac{64}{-34}=-\frac{32}{17}$

85. Using order of operations: $\frac{(8-4)^2}{8^2-4^2}=\frac{4^2}{64-16}=\frac{16}{48}=\frac{1}{3}$

87. Using order of operations: $7-2[-6-4(-3)]=7-2(-6+12)=7-2(6)=7-12=7+(-12)=-5$

89. Using order of operations:

$$7-3[2(-4-4)-3(-1-1)]=7-3[2(-8)-3(-2)]=7-3(-16+6)=7-3(-10)=7+30=37$$

91. Using order of operations:

$$8-6[-2(-3-1)+4(-2-3)]=8-6[-2(-4)+4(-5)]=8-6(8-20)=8-6(-12)=8+72=80$$

93. Using order of operations: $\frac{-4\cdot 3^2-5\cdot 2^2}{-8(7)}=\frac{-4\cdot 9-5\cdot 4}{-56}=\frac{-36-20}{-56}=\frac{-56}{-56}=1$

95. Using order of operations: $\frac{3\cdot 10^2+4\cdot 10+5}{345}=\frac{300+40+5}{345}=\frac{345}{345}=1$

97. Using order of operations: $\frac{7-[(2-3)-4]}{-1-2-3}=\frac{7-(-1-4)}{-6}=\frac{7-(-5)}{-6}=\frac{7+5}{-6}=\frac{12}{-6}=-2$

99. Using order of operations: $\frac{6(-4)-2(5-8)}{-6-3-5}=\frac{-24-2(-3)}{-14}=\frac{-24+6}{-14}=\frac{-18}{-14}=\frac{9}{7}$

101. Using order of operations: $\frac{3(-5-3)+4(7-9)}{5(-2)+3(-4)}=\frac{3(-8)+4(-2)}{-10+(-12)}=\frac{-24+(-8)}{-22}=\frac{-32}{-22}=\frac{16}{11}$

103. Using order of operations: $\frac{|3-9|}{3-9}=\frac{|-6|}{-6}=\frac{6}{-6}=-1$

105. **a.** Simplifying: $20 \div 4 \cdot 5 = 5 \cdot 5 = 25$

b. Simplifying: $-20 \div 4 \cdot 5 = -5 \cdot 5 = -25$

c. Simplifying: $20 \div (-4) \cdot 5 = -5 \cdot 5 = -25$

d. Simplifying: $20 \div 4(-5) = 5(-5) = -25$

e. Simplifying: $-20 \div 4(-5) = -5(-5) = 25$

107. Writing the expression: $3(-10) + 5 = -30 + 5 = -25$

109. Evaluating: $(-9)(2) - 8 = -18 - 8 = -18 + (-8) = -26$

111. The quotient is $\frac{-12}{-4} = 3$.

113. The number is –10, since $\frac{-10}{-5} = 2$.

115. The number is –3, since $\frac{27}{-3} = -9$.

117. The expression is: $\frac{-20}{4} - 3 = -5 - 3 = -8$

119. Simplifying the expression: $3(x-5) + 4 = 3x - 15 + 4 = 3x - 11$

121. Simplifying the expression: $2(3) - 4 - 3(-4) = 6 - 4 + 12 = 2 + 12 = 14$

123. The pattern is to multiply by 2, so the next number is $4 \cdot 2 = 8$.

125. The pattern is to multiply by –2, so the next number is $40 \cdot (-2) = -80$.

127. The pattern is to multiply by –2, so the next number is $12 \cdot (-2) = -24$.

129. The temperature is: $25° - 4(6°) = 25° - 24° = 1°$ F.

131. Each person would lose: $\frac{13{,}600 - 15{,}000}{4} = \frac{-1{,}400}{4} = -350 = \$350$ loss

133. The change per hour is: $\frac{61° - 75°}{4} = \frac{-14°}{4} = -3.5°$ per hour

135. **a.** Since they predict \$50 revenue for every 25 people, their projected revenue is:

$$\frac{\$50}{25 \text{ people}} \cdot 10{,}000 \text{ people} = \$20{,}000$$

b. Since they predict \$50 revenue for every 25 people, their projected revenue is:

$$\frac{\$50}{25 \text{ people}} \cdot 25{,}000 \text{ people} = \$50{,}000$$

c. Since they predict \$50 revenue for every 25 people, their projected revenue is:

$$\frac{\$50}{25 \text{ people}} \cdot 5{,}000 \text{ people} = \$10{,}000$$

Since this projected revenue is more than the \$5,000 cost for the list, it is a wise purchase.

137. The net change in calories is: $2(630) - 3(265) = 1260 - 795 = 465$ calories

139. Dividing: $\frac{-18}{0}$ is undefined. The correct answer is a.

141. The pattern is to multiply by –3, so the next number is $-54 \cdot (-3) = 162$. The correct answer is b.

1.6 Fractions

1. This number is composite: $48 = 6 \cdot 8 = (2 \cdot 3) \cdot (2 \cdot 2 \cdot 2) = 2^4 \cdot 3$

3. This number is prime.

5. This number is composite: $1{,}023 = 3 \cdot 341 = 3 \cdot 11 \cdot 31$

7. Factoring the number: $144 = 12 \cdot 12 = (3 \cdot 4) \cdot (3 \cdot 4) = (3 \cdot 2 \cdot 2) \cdot (3 \cdot 2 \cdot 2) = 2^4 \cdot 3^2$

9. Factoring the number: $38 = 2 \cdot 19$

11. Factoring the number: $105 = 5 \cdot 21 = 5 \cdot (3 \cdot 7) = 3 \cdot 5 \cdot 7$

13. Factoring the number: $180 = 10 \cdot 18 = (2 \cdot 5) \cdot (3 \cdot 6) = (2 \cdot 5) \cdot (3 \cdot 2 \cdot 3) = 2^2 \cdot 3^2 \cdot 5$

15. Factoring the number: $385 = 5 \cdot 77 = 5 \cdot (7 \cdot 11) = 5 \cdot 7 \cdot 11$

17. Factoring the number: $121 = 11 \cdot 11 = 11^2$

19. Factoring the number: $420 = 10 \cdot 42 = (2 \cdot 5) \cdot (7 \cdot 6) = (2 \cdot 5) \cdot (7 \cdot 2 \cdot 3) = 2^2 \cdot 3 \cdot 5 \cdot 7$

21. Factoring the number: $620 = 10 \cdot 62 = (2 \cdot 5) \cdot (2 \cdot 31) = 2^2 \cdot 5 \cdot 31$

23. Reducing the fraction: $\frac{105}{165} = \frac{3 \cdot 5 \cdot 7}{3 \cdot 5 \cdot 11} = \frac{7}{11}$

25. Reducing the fraction: $\frac{525}{735} = \frac{3 \cdot 5 \cdot 5 \cdot 7}{3 \cdot 5 \cdot 7 \cdot 7} = \frac{5}{7}$

27. Reducing the fraction: $\frac{385}{455} = \frac{5 \cdot 7 \cdot 11}{5 \cdot 7 \cdot 13} = \frac{11}{13}$

29. Reducing the fraction: $\frac{322}{345} = \frac{2 \cdot 7 \cdot 23}{3 \cdot 5 \cdot 23} = \frac{2 \cdot 7}{3 \cdot 5} = \frac{14}{15}$

31. Reducing the fraction: $\frac{205}{369} = \frac{5 \cdot 41}{3 \cdot 3 \cdot 41} = \frac{5}{3 \cdot 3} = \frac{5}{9}$

33. Reducing the fraction: $\frac{215}{344} = \frac{5 \cdot 43}{2 \cdot 2 \cdot 2 \cdot 43} = \frac{5}{2 \cdot 2 \cdot 2} = \frac{5}{8}$

35. Simplifying and factoring: $3 \cdot 8 + 3 \cdot 7 + 3 \cdot 5 = 24 + 21 + 15 = 60 = 6 \cdot 10 = (2 \cdot 3) \cdot (2 \cdot 5) = 2^2 \cdot 3 \cdot 5$

37. Building the fraction: $\frac{3}{4} = \frac{3}{4} \cdot \frac{6}{6} = \frac{18}{24}$

39. Building the fraction: $\frac{1}{2} = \frac{1}{2} \cdot \frac{12}{12} = \frac{12}{24}$

41. Building the fraction: $\frac{5}{8} = \frac{5}{8} \cdot \frac{3}{3} = \frac{15}{24}$

43. Building the fraction: $\frac{3}{5} = \frac{3}{5} \cdot \frac{12}{12} = \frac{36}{60}$

45. Building the fraction: $\frac{11}{30} = \frac{11}{30} \cdot \frac{2}{2} = \frac{22}{60}$

47. The correct symbol is $<$: $-\frac{3}{4} < -\frac{1}{4}$

49. The correct symbol is $<$: $-\frac{3}{2} < -\frac{3}{4}$

51. Multiplying the fractions: $\frac{2}{3} \cdot \frac{4}{5} = \frac{8}{15}$

53. Multiplying the fractions: $\frac{1}{2}(3) = \frac{1}{2} \cdot \frac{3}{1} = \frac{3}{2}$

55. Multiplying the fractions: $\frac{1}{4}(5) = \frac{1}{4} \cdot \frac{5}{1} = \frac{5}{4}$

57. Multiplying the fractions: $\frac{4}{3} \cdot \frac{3}{4} = \frac{12}{12} = 1$

59. Multiplying the fractions: $6\left(\frac{1}{6}\right) = \frac{6}{1} \cdot \frac{1}{6} = \frac{6}{6} = 1$

61. Multiplying the fractions: $3 \cdot \frac{1}{3} = \frac{3}{1} \cdot \frac{1}{3} = \frac{3}{3} = 1$

63. Expanding the exponent: $\left(\frac{3}{4}\right)^2 = \frac{3}{4} \cdot \frac{3}{4} = \frac{9}{16}$

65. Expanding the exponent: $\left(\frac{2}{3}\right)^3 = \frac{2}{3} \cdot \frac{2}{3} \cdot \frac{2}{3} = \frac{8}{27}$

67. Expanding the exponent: $\left(\frac{1}{10}\right)^4 = \frac{1}{10} \cdot \frac{1}{10} \cdot \frac{1}{10} \cdot \frac{1}{10} = \frac{1}{10{,}000}$

69. Multiplying the fractions: $-\frac{2}{3} \cdot \frac{5}{7} = -\frac{2 \cdot 5}{3 \cdot 7} = -\frac{10}{21}$

71. Multiplying the fractions: $-8\left(\frac{1}{2}\right) = -\frac{8}{1} \cdot \frac{1}{2} = -\frac{8}{2} = -4$

73. Multiplying the fractions: $-\frac{3}{4}\left(-\frac{4}{3}\right) = \frac{12}{12} = 1$

75. Multiplying the fractions: $\left(-\frac{3}{4}\right)^2=\left(-\frac{3}{4}\right)\left(-\frac{3}{4}\right)=\frac{9}{16}$

77. Multiplying the expressions: $-\frac{1}{3}(-3x)=\left[-\frac{1}{3}\bullet(-3)\right]x=1x=x$

79. Dividing and reducing: $\frac{4}{5}\div\frac{3}{4}=\frac{4}{5}\bullet\frac{4}{3}=\frac{16}{15}$

81. Dividing and reducing: $-\frac{5}{6}\div\left(-\frac{5}{8}\right)=-\frac{5}{6}\bullet\left(-\frac{8}{5}\right)=\frac{40}{30}=\frac{4}{3}$

83. Dividing and reducing: $\frac{10}{13}\div\left(-\frac{5}{4}\right)=\frac{10}{13}\bullet\left(-\frac{4}{5}\right)=-\frac{40}{65}=-\frac{8}{13}$

85. Dividing and reducing: $-\frac{5}{6}\div\frac{5}{6}=-\frac{5}{6}\bullet\frac{6}{5}=-\frac{30}{30}=-1$

87. Dividing and reducing: $-\frac{3}{4}\div\left(-\frac{3}{4}\right)=-\frac{3}{4}\bullet\left(-\frac{4}{3}\right)=\frac{12}{12}=1$

89. **a.** Simplifying: $8\div\frac{4}{5}=8\bullet\frac{5}{4}=10$

b. Simplifying: $8\div\frac{4}{5}-10=8\bullet\frac{5}{4}-10=10-10=0$

c. Simplifying: $8\div\frac{4}{5}(-10)=8\bullet\frac{5}{4}(-10)=10(-10)=-100$

d. Simplifying: $8\div\left(-\frac{4}{5}\right)-10=8\bullet\left(-\frac{5}{4}\right)-10=-10-10=-20$

91. Combining the fractions: $\frac{3}{6}+\frac{1}{6}=\frac{4}{6}=\frac{2}{3}$

93. Combining the fractions: $\frac{3}{8}-\frac{5}{8}=-\frac{2}{8}=-\frac{1}{4}$

95. Combining the fractions: $-\frac{1}{4}+\frac{3}{4}=\frac{2}{4}=\frac{1}{2}$

97. Combining the fractions: $\frac{x}{3}-\frac{1}{3}=\frac{x-1}{3}$

99. Combining the fractions: $\frac{1}{4}+\frac{2}{4}+\frac{3}{4}=\frac{6}{4}=\frac{3}{2}$

101. Combining the fractions: $\frac{x+7}{2}-\frac{1}{2}=\frac{x+7-1}{2}=\frac{x+6}{2}$

103. Combining the fractions: $\frac{1}{10}-\frac{3}{10}-\frac{4}{10}=-\frac{6}{10}=-\frac{3}{5}$

105. Combining the fractions: $\frac{1}{a}+\frac{4}{a}+\frac{5}{a}=\frac{10}{a}$

107. Completing the table:

First Number a	Second Number b	The Sum of a and b $a+b$
$\frac{1}{2}$	$\frac{1}{3}$	$\frac{5}{6}$
$\frac{1}{3}$	$\frac{1}{4}$	$\frac{7}{12}$
$\frac{1}{4}$	$\frac{1}{5}$	$\frac{9}{20}$
$\frac{1}{5}$	$\frac{1}{6}$	$\frac{11}{30}$

109. Completing the table:

First Number a	Second Number b	The Sum of a and b $a+b$
$\frac{1}{12}$	$\frac{1}{2}$	$\frac{7}{12}$
$\frac{1}{12}$	$\frac{1}{3}$	$\frac{5}{12}$
$\frac{1}{12}$	$\frac{1}{4}$	$\frac{1}{3}$
$\frac{1}{12}$	$\frac{1}{6}$	$\frac{1}{4}$

111. Combining the fractions: $\frac{4}{9}+\frac{1}{3}=\frac{4}{9}+\frac{1\cdot 3}{3\cdot 3}=\frac{4}{9}+\frac{3}{9}=\frac{7}{9}$

113. Combining the fractions: $2+\frac{1}{3}=\frac{2\cdot 3}{1\cdot 3}+\frac{1}{3}=\frac{6}{3}+\frac{1}{3}=\frac{7}{3}$

115. Combining the fractions: $-\frac{3}{4}+1=-\frac{3}{4}+\frac{1\cdot 4}{1\cdot 4}=-\frac{3}{4}+\frac{4}{4}=\frac{1}{4}$

117. Combining the fractions: $\frac{1}{2}+\frac{2}{3}=\frac{1\cdot 3}{2\cdot 3}+\frac{2\cdot 2}{3\cdot 2}=\frac{3}{6}+\frac{4}{6}=\frac{7}{6}$

119. Combining the fractions: $\frac{5}{12}-\left(-\frac{3}{8}\right)=\frac{5}{12}+\frac{3}{8}=\frac{5\cdot 2}{12\cdot 2}+\frac{3\cdot 3}{8\cdot 3}=\frac{10}{24}+\frac{9}{24}=\frac{19}{24}$

121. Combining the fractions: $-\frac{1}{20}+\frac{8}{30}=-\frac{1\cdot 3}{20\cdot 3}+\frac{8\cdot 2}{30\cdot 2}=-\frac{3}{60}+\frac{16}{60}=\frac{13}{60}$

123. First factor the denominators to find the LCM:

$30=2\cdot 3\cdot 5$

$42=2\cdot 3\cdot 7$

$\text{LCM}=2\cdot 3\cdot 5\cdot 7=210$

Combining the fractions: $\frac{17}{30}+\frac{11}{42}=\frac{17\cdot 7}{30\cdot 7}+\frac{11\cdot 5}{42\cdot 5}=\frac{119}{210}+\frac{55}{210}=\frac{174}{210}=\frac{2\cdot 3\cdot 29}{2\cdot 3\cdot 5\cdot 7}=\frac{29}{5\cdot 7}=\frac{29}{35}$

125. First factor the denominators to find the LCM:

$84=2\cdot 2\cdot 3\cdot 7$

$90=2\cdot 3\cdot 3\cdot 5$

$\text{LCM}=2\cdot 2\cdot 3\cdot 3\cdot 5\cdot 7=1{,}260$

Combining the fractions: $\frac{25}{84}+\frac{41}{90}=\frac{25\cdot 15}{84\cdot 15}+\frac{41\cdot 14}{90\cdot 14}=\frac{375}{1{,}260}+\frac{574}{1{,}260}=\frac{949}{1{,}260}$

127. First factor the denominators to find the LCM:

$126=2\cdot 3\cdot 3\cdot 7$

$180=2\cdot 2\cdot 3\cdot 3\cdot 5$

$\text{LCM}=2\cdot 2\cdot 3\cdot 3\cdot 5\cdot 7=1{,}260$

Combining the fractions:

$$\frac{13}{126}-\frac{13}{180}=\frac{13\cdot 10}{126\cdot 10}-\frac{13\cdot 7}{180\cdot 7}=\frac{130}{1{,}260}-\frac{91}{1{,}260}=\frac{39}{1{,}260}=\frac{3\cdot 13}{2\cdot 2\cdot 3\cdot 3\cdot 5\cdot 7}=\frac{13}{2\cdot 2\cdot 3\cdot 5\cdot 7}=\frac{13}{420}$$

129. Combining the fractions: $\frac{3}{4}+\frac{1}{8}+\frac{5}{6}=\frac{3\cdot 6}{4\cdot 6}+\frac{1\cdot 3}{8\cdot 3}+\frac{5\cdot 4}{6\cdot 4}=\frac{18}{24}+\frac{3}{24}+\frac{20}{24}=\frac{41}{24}$

131. Combining the fractions: $\frac{1}{2}+\frac{1}{3}+\frac{1}{4}+\frac{1}{6}=\frac{1\cdot 6}{2\cdot 6}+\frac{1\cdot 4}{3\cdot 4}+\frac{1\cdot 3}{4\cdot 3}+\frac{1\cdot 2}{6\cdot 2}=\frac{6}{12}+\frac{4}{12}+\frac{3}{12}+\frac{2}{12}=\frac{15}{12}=\frac{5}{4}$

133. Combining the fractions: $1-\frac{5}{2}=1\cdot\frac{2}{2}-\frac{5}{2}=\frac{2}{2}-\frac{5}{2}=-\frac{3}{2}$

135. Combining the fractions: $1+\frac{1}{2}=1\cdot\frac{2}{2}+\frac{1}{2}=\frac{2}{2}+\frac{1}{2}=\frac{3}{2}$

137. The sum is given by: $\frac{3}{7}+2+\frac{1}{9}=\frac{3\cdot 9}{7\cdot 9}+\frac{2\cdot 63}{1\cdot 63}+\frac{1\cdot 7}{9\cdot 7}=\frac{27}{63}+\frac{126}{63}+\frac{7}{63}=\frac{160}{63}$

139. The difference is given by: $\frac{7}{8}-\frac{1}{4}=\frac{7}{8}-\frac{1\cdot 2}{4\cdot 2}=\frac{7}{8}-\frac{2}{8}=\frac{5}{8}$

141. The pattern is to add $-\frac{1}{3}$, so the fourth term is: $-\frac{1}{3}+\left(-\frac{1}{3}\right)=-\frac{2}{3}$

143. The pattern is to add $\frac{2}{3}$, so the fourth term is: $\frac{5}{3}+\frac{2}{3}=\frac{7}{3}$

145. The pattern is to multiply by $\frac{1}{5}$, so the fourth term is: $\frac{1}{25}\cdot\frac{1}{5}=\frac{1}{125}$

147. The perimeter is: $4\left(\frac{3}{8}\right)=\frac{12}{8}=\frac{3}{2}=1\frac{1}{2}$ feet

149. The perimeter is: $2\left(\frac{4}{5}\right)+2\left(\frac{3}{10}\right)=\frac{8}{5}+\frac{3}{5}=\frac{11}{5}=2\frac{1}{5}$ centimeters (cm)

151. The number of blankets is: $12\div\frac{6}{7}=12\cdot\frac{7}{6}=\frac{84}{6}=14$ blankets

153. The number of bags is: $12\div\frac{1}{4}=12\cdot\frac{4}{1}=48$ bags

155. The number of spoons is: $\frac{3}{4}\div\frac{1}{8}=\frac{3}{4}\cdot\frac{8}{1}=\frac{24}{4}=6$ spoons

157. The number of cartons is: $14\div\frac{1}{2}=14\cdot\frac{2}{1}=28$ cartons

159. Adding: $\frac{1}{2}+4=\frac{1}{2}+\frac{8}{2}=\frac{9}{2}=4\frac{1}{2}$ pints

161. Multiplying: $\frac{5}{8}\cdot 2{,}120=\frac{5}{8}\cdot\frac{2{,}120}{1}=\$1{,}325$

163. Adding: $\frac{1}{4}+\frac{3}{20}=\frac{5}{20}+\frac{3}{20}=\frac{8}{20}=\frac{2}{5}$ of the students

165. Completing the table:

Grade	Number of Students	Fraction of Students
A	5	$\frac{1}{8}$
B	8	$\frac{1}{5}$
C	20	$\frac{1}{2}$
below C	7	$\frac{7}{40}$
Total	40	1

167. Dividing: $\frac{6 \text{ acres}}{3/5 \text{ acres/lot}}=\frac{6}{1}\cdot\frac{5}{3}=10$ lots

169. Factoring the number: $132=4\cdot 33=(2\cdot 2)\cdot(3\cdot 11)=2^2\cdot 3\cdot 11$. The correct answer is b.

171. The reciprocal is $-\frac{7}{6}$. The correct answer is c.

173. Adding the fractions: $\frac{2}{5}+\frac{4}{3}=\frac{2\cdot 3}{5\cdot 3}+\frac{4\cdot 5}{3\cdot 5}=\frac{6}{15}+\frac{20}{15}=\frac{26}{15}$. The correct answer is a.

1.7 Properties of Real Numbers

1. commutative property (of addition)
3. multiplicative inverse property
5. commutative property (of addition)
7. distributive property
9. commutative and associative properties (of addition)
11. commutative and associative properties (of addition)
13. commutative property (of addition)
15. commutative and associative properties (of multiplication)
17. commutative property (of multiplication)
19. additive inverse property

21. The expression should read $3(x+2)=3x+6$.

23. The expression should read $9(a+b)=9a+9b$.

25. The expression should read $3(0)=0$.

27. The expression should read $3+(-3)=0$.

29. The expression should read $10(1)=10$.

31. Simplifying the expression: $4+(2+x)=(4+2)+x=6+x$

33. Simplifying the expression: $(x+2)+7=x+(2+7)=x+9$

35. Simplifying the expression: $3(5x)=(3\cdot 5)x=15x$

37. Simplifying the expression: $-9(6y)=(-9\cdot 6)y=-54y$

39. Simplifying the expression: $\frac{1}{2}(3a)=\left(\frac{1}{2}\cdot 3\right)a=\frac{3}{2}a$

41. Simplifying the expression: $-\frac{1}{3}(3x)=\left(-\frac{1}{3}\cdot 3\right)x=-1x=-x$

43. Simplifying the expression: $\frac{1}{2}(2y)=\left(\frac{1}{2}\cdot 2\right)y=1y=y$

45. Simplifying the expression: $-\frac{3}{4}\left(\frac{4}{3}x\right)=\left(-\frac{3}{4}\cdot\frac{4}{3}\right)x=-1x=-x$

47. Simplifying the expression: $-\frac{6}{5}\left(-\frac{5}{6}a\right)=\left(-\frac{6}{5}\cdot -\frac{5}{6}\right)a=1a=a$

49. Applying the distributive property: $8(x+2)=8\cdot x+8\cdot 2=8x+16$

51. Applying the distributive property: $8(x-2)=8\cdot x-8\cdot 2=8x-16$

53. Applying the distributive property: $4(y+1)=4\cdot y+4\cdot 1=4y+4$

55. Applying the distributive property: $3(6x+5)=3\cdot 6x+3\cdot 5=18x+15$

57. Applying the distributive property: $-2(3a+7)=-2\cdot 3a-2\cdot 7=-6a-14$

59. Applying the distributive property: $-9(6y-8)=-9\cdot 6y-9\cdot(-8)=-54y+72$

61. Applying the distributive property: $\frac{1}{3}(3x+6)=\frac{1}{3}\cdot 3x+\frac{1}{3}\cdot 6=x+2$

63. Applying the distributive property: $6(2x+3y)=6\cdot 2x+6\cdot 3y=12x+18y$

65. Applying the distributive property: $4(3a-2b)=4\cdot 3a-4\cdot 2b=12a-8b$

67. Applying the distributive property: $\frac{1}{2}(6x+4y)=\frac{1}{2}\cdot 6x+\frac{1}{2}\cdot 4y=3x+2y$

69. Applying the distributive property: $-4(a+2)=-4a+(-4)(2)=-4a-8$

71. Applying the distributive property: $-\frac{1}{2}(3x-6)=-\frac{3}{2}x-\frac{1}{2}(-6)=-\frac{3}{2}x+3$

73. Applying the distributive property: $10\left(\frac{x}{2}+\frac{3}{5}\right)=10\left(\frac{x}{2}\right)+10\left(\frac{3}{5}\right)=5x+6$

75. Applying the distributive property: $15\left(\frac{x}{5}-\frac{4}{3}\right)=15\left(\frac{x}{5}\right)-15\left(\frac{4}{3}\right)=3x-20$

77. Applying the distributive property: $x\left(\frac{3}{x}+1\right)=x\left(\frac{3}{x}\right)+x(1)=3+x$

79. Applying the distributive property: $-21\left(\frac{x}{7}-\frac{y}{3}\right)=-21\left(\frac{x}{7}\right)-21\left(-\frac{y}{3}\right)=-3x+7y$

81. Applying the distributive property: $a\left(\frac{3}{a}-\frac{2}{a}\right)=a\left(\frac{3}{a}\right)-a\left(\frac{2}{a}\right)=3-2=1$

83. Applying the distributive property: $4(a+4)+9=4a+16+9=4a+25$

85. Applying the distributive property: $2(3x+5)+2=6x+10+2=6x+12$

87. Applying the distributive property: $7(2x+4)+10=14x+28+10=14x+38$

89. Applying the distributive property: $-3(2x-5)-7=-6x+15-7=-6x+8$

91. Applying the distributive property: $-5(3x+4)-10=-15x-20-10=-15x-30$

93. Simplifying: $\left(\frac{1}{2}\cdot 18\right)^2=(9)^2=(9)(9)=81$

95. Simplifying: $\left(\frac{1}{2}\cdot 3\right)^2=\left(\frac{3}{2}\right)^2=\left(\frac{3}{2}\right)\left(\frac{3}{2}\right)=\frac{9}{4}$

97. Applying the distributive property: $\frac{1}{2}(4x+2)=\frac{1}{2}\cdot 4x+\frac{1}{2}\cdot 2=2x+1$

99. Applying the distributive property: $\frac{3}{4}(8x-4)=\frac{3}{4}\cdot 8x-\frac{3}{4}\cdot 4=6x-3$

101. Applying the distributive property: $\frac{5}{6}(6x+12)=\frac{5}{6}\cdot 6x+\frac{5}{6}\cdot 12=5x+10$

103. Applying the distributive property: $10\left(\frac{3}{5}x+\frac{1}{2}\right)=10\cdot\frac{3}{5}x+10\cdot\frac{1}{2}=6x+5$

105. Applying the distributive property: $15\left(\frac{1}{3}x+\frac{2}{5}\right)=15\cdot\frac{1}{3}x+15\cdot\frac{2}{5}=5x+6$

107. Applying the distributive property: $12\left(\frac{1}{2}m-\frac{5}{12}\right)=12\cdot\frac{1}{2}m-12\cdot\frac{5}{12}=6m-5$

109. Applying the distributive property: $21\left(\frac{1}{3}+\frac{1}{7}x\right)=21\cdot\frac{1}{3}+21\cdot\frac{1}{7}x=7+3x$

111. Applying the distributive property: $6\left(\frac{1}{2}x-\frac{1}{3}y\right)=6\cdot\frac{1}{2}x-6\cdot\frac{1}{3}y=3x-2y$

113. Simplifying: $-\frac{1}{3}(-2x+6)=-\frac{1}{3}\cdot(-2x)-\frac{1}{3}\cdot 6=\frac{2}{3}x-2$

115. Simplifying: $8\left(-\frac{1}{4}x+\frac{1}{8}y\right)=8\cdot\left(-\frac{1}{4}x\right)+8\cdot\left(\frac{1}{8}y\right)=-2x+y$

117. Applying the distributive property: $0.09(x+2{,}000)=0.09x+180$

119. Applying the distributive property: $0.12(x+500)=0.12x+60$

121. Applying the distributive property: $a\left(1+\frac{1}{a}\right)=a\cdot 1+a\cdot\frac{1}{a}=a+1$

123. Applying the distributive property: $a\left(\frac{1}{a}-1\right)=a\cdot\frac{1}{a}-a\cdot 1=1-a$

125. No. The man cannot reverse the order of putting on his socks and putting on his shoes.

127. Division is not a commutative operation. For example, $8\div 4=2$ while $4\div 8=\frac{1}{2}$.

129. Computing his hours worked:

$4(2+3)=4(5)=20$ hours $\qquad$ $4\cdot 2+4\cdot 3=8+12=20$ hours

131. Applying the associative property: $-4(8x)=(-4\cdot 8)x=-32x$. The correct answer is a.

Chapter 1 Test

1. Evaluating the exponent: $12^2=12\cdot 12=144$

2. Evaluating the exponent: $4^3=4\cdot 4\cdot 4=64$

3. Simplifying using order of operations: $10+2(7-3)-4^2=10+2(4)-4^2=10+8-16=2$

4. Simplifying using order of operations: $15+24\div 6-3^2=15+24\div 6-9=15+4-9=10$

5. The integers are: $-3, 2$

6. The rational numbers are: $-3,-\frac{1}{2},2$

7. Translating into symbols: $6+(-9)=-3$

8. Translating into symbols: $-5-(-12)=-5+12=7$

9. Translating into symbols: $6(-7)=-42$

10. Translating into symbols: $\frac{32}{-8}=-4$

11. The pattern is to add 4, so the next number is: $9+4=13$

12. The pattern is to multiply by $-\frac{1}{3}$, so the next number is: $-3\cdot\left(-\frac{1}{3}\right)=1$

13. Simplifying: $-2(3)-7=-6-7=-6+(-7)=-13$

14. Simplifying: $2(3)^3-4(-2)^4=2\cdot 27-4\cdot 16=54-64=-10$

15. Simplifying: $9+4(2-6)=9+4(-4)=9-16=-7$

16. Simplifying: $5-3[-2(1+4)+3(-3)]=5-3[-2(5)+(-9)]=5-3(-10-9)=5-3(-19)=5+57=62$

17. Simplifying: $\frac{-4(3)+5(-2)}{-5-6}=\frac{-12-10}{-11}=\frac{-22}{-11}=2$

18. Simplifying: $\frac{4(3-5)-2(-6+8)}{4(-2)+10}=\frac{4(-2)-2(2)}{-8+10}=\frac{-8-4}{2}=\frac{-12}{2}=-6$

19. Factoring the number: $660=10\cdot 66=(2\cdot 5)\cdot(6\cdot 11)=(2\cdot 5)\cdot(2\cdot 3\cdot 11)=2^2\cdot 3\cdot 5\cdot 11$

20. Factoring the number: $4,725=25\cdot 189=(5\cdot 5)\cdot(9\cdot 21)=(5\cdot 5)\cdot(3\cdot 3\cdot 3\cdot 7)=3^3\cdot 5^2\cdot 7$

21. Combining the fractions: $\frac{5}{24}+\frac{9}{36}=\frac{5\cdot 3}{24\cdot 3}+\frac{9\cdot 2}{36\cdot 2}=\frac{15}{72}+\frac{18}{72}=\frac{33}{72}=\frac{11}{24}$

22. Combining the fractions: $\frac{5}{y}+\frac{6}{y}=\frac{11}{y}$

23. associative property of multiplication (d)

24. distributive property (e)

25. commutative property of addition (a)

26. associative property of addition (c)

27. Simplifying: $5+(7+3x)=3x+5+7=3x+12$

28. Simplifying: $3(-5y)=-15y$

29. Simplifying: $-5(2x-3)=-5(2x)+5(3)=-10x+15$

30. Simplifying: $\frac{1}{3}(6x+12)=\frac{1}{3}(6x)+\frac{1}{3}(12)=2x+4$

Chapter 2
Linear Equations and Inequalities

2.1 Simplifying Expressions

1. Simplifying the expression: $3x-6x=(3-6)x=-3x$
3. Simplifying the expression: $-2a+a=(-2+1)a=-a$
5. Simplifying the expression: $7x+3x+2x=(7+3+2)x=12x$
7. Simplifying the expression: $3a-2a+5a=(3-2+5)a=6a$
9. Simplifying the expression: $4x-3+2x=4x+2x-3=6x-3$
11. Simplifying the expression: $3a+4a+5=(3+4)a+5=7a+5$
13. Simplifying the expression: $2x-3+3x-2=2x+3x-3-2=5x-5$
15. Simplifying the expression: $3a-1+a+3=3a+a-1+3=4a+2$
17. Simplifying the expression: $-4x+8-5x-10=-4x-5x+8-10=-9x-2$
19. Simplifying the expression: $7a+3+2a+3a=7a+2a+3a+3=12a+3$
21. Simplifying the expression: $5(2x-1)+4=10x-5+4=10x-1$
23. Simplifying the expression: $7(3y+2)-8=21y+14-8=21y+6$
25. Simplifying the expression: $-3(2x-1)+5=-6x+3+5=-6x+8$
27. Simplifying the expression: $5-2(a+1)=5-2a-2=-2a+3$
29. Simplifying the expression: $6-4(x-5)=6-4x+20=-4x+20+6=-4x+26$
31. Simplifying the expression: $-9-4(2-y)+1=-9-8+4y+1=4y+1-9-8=4y-16$
33. Simplifying the expression: $-6+2(2-3x)+1=-6+4-6x+1=-6x-6+4+1=-6x-1$
35. Simplifying the expression: $(4x-7)-(2x+5)=4x-7-2x-5=4x-2x-7-5=2x-12$
37. Simplifying the expression: $8(2a+4)-(6a-1)=16a+32-6a+1=16a-6a+32+1=10a+33$
39. Simplifying the expression: $3(x-2)+(x-3)=3x-6+x-3=3x+x-6-3=4x-9$
41. Simplifying the expression: $4(2y-8)-(y+7)=8y-32-y-7=8y-y-32-7=7y-39$
43. Simplifying the expression: $-9(2x+1)-(x+5)=-18x-9-x-5=-18x-x-9-5=-19x-14$
45. Evaluating when $x=2$: $3x-1=3(2)-1=6-1=5$
47. Evaluating when $x=2$: $-2x-5=-2(2)-5=-4-5=-9$
49. Evaluating when $x=2$: $x^2-8x+16=(2)^2-8(2)+16=4-16+16=4$
51. Evaluating when $x=2$: $(x-4)^2=(2-4)^2=(-2)^2=4$

53. Evaluating when $x=-5$: $7x-4-x-3=7(-5)-4-(-5)-3=-35-4+5-3=-42+5=-37$

Now simplifying the expression: $7x-4-x-3=7x-x-4-3=6x-7$

Evaluating when $x=-5$: $6x-7=6(-5)-7=-30-7=-37$

Note that the two values are the same.

55. Evaluating when $x=-5$: $5(2x+1)+4=5[2(-5)+1]+4=5(-10+1)+4=5(-9)+4=-45+4=-41$

Now simplifying the expression: $5(2x+1)+4=10x+5+4=10x+9$

Evaluating when $x=-5$: $10x+9=10(-5)+9=-50+9=-41$

Note that the two values are the same.

57. Evaluating when $x=-3$ and $y=5$: $x^2-2xy+y^2=(-3)^2-2(-3)(5)+(5)^2=9+30+25=64$

59. Evaluating when $x=-3$ and $y=5$: $(x-y)^2=(-3-5)^2=(-8)^2=64$

61. Evaluating when $x=-3$ and $y=5$: $x^2+6xy+9y^2=(-3)^2+6(-3)(5)+9(5)^2=9-90+225=144$

63. Evaluating when $x=-3$ and $y=5$: $(x+3y)^2=[-3+3(5)]^2=(-3+15)^2=(12)^2=144$

65. Evaluating when $x=\frac{1}{2}$: $12x-3=12\left(\frac{1}{2}\right)-3=6-3=3$

67. Evaluating when $x=\frac{1}{4}$: $12x-3=12\left(\frac{1}{4}\right)-3=3-3=0$

69. Evaluating when $x=\frac{3}{2}$: $12x-3=12\left(\frac{3}{2}\right)-3=18-3=15$

71. Evaluating when $x=\frac{3}{4}$: $12x-3=12\left(\frac{3}{4}\right)-3=9-3=6$

73. **a.** Substituting the values for n:

n	1	2	3	4
$3n$	3	6	9	12

b. Substituting the values for n:

n	1	2	3	4
n^3	1	8	27	64

75. Substituting $n=1,2,3,4$:

$n=1$: $3(1)-2=3-2=1$

$n=2$: $3(2)-2=6-2=4$

$n=3$: $3(3)-2=9-2=7$

$n=4$: $3(4)-2=12-2=10$

The sequence is 1, 4, 7, 10, ..., which is an arithmetic sequence.

77. Substituting $n=1,2,3,4$:

$n=1$: $(1)^2-2(1)+1=1-2+1=0$

$n=2$: $(2)^2-2(2)+1=4-4+1=1$

$n=3$: $(3)^2-2(3)+1=9-6+1=4$

$n=4$: $(4)^2-2(4)+1=16-8+1=9$

The sequence is 0, 1, 4, 9, ..., which is a sequence of squares.

79. Simplifying the expression: $7-3(2y+1)=7-6y-3=-6y+4$

81. Simplifying the expression: $0.08x+0.09x=0.17x$

83. Simplifying the expression: $(x+y)+(x-y)=x+x+y-y=2x$

85. Simplifying the expression: $3x+2(x-2)=3x+2x-4=5x-4$

87. Simplifying the expression: $4(x+1)+3(x-3)=4x+4+3x-9=7x-5$

89. Simplifying the expression: $x+(x+3)(-3)=x-3x-9=-2x-9$

91. Simplifying the expression: $3(4x-2)-(5x-8)=12x-6-5x+8=7x+2$

93. Simplifying the expression: $-(3x+1)-(4x-7)=-3x-1-4x+7=-7x+6$

95. Simplifying the expression: $(x+3y)+3(2x-y)=x+3y+6x-3y=7x$

97. Simplifying the expression: $3(2x+3y)-2(3x+5y)=6x+9y-6x-10y=-y$

99. Simplifying the expression: $-6\left(\frac{1}{2}x-\frac{1}{3}y\right)+12\left(\frac{1}{4}x+\frac{2}{3}y\right)=-3x+2y+3x+8y=10y$

101. Simplifying the expression: $0.08x+0.09(x+2{,}000)=0.08x+0.09x+180=0.17x+180$

103. Simplifying the expression: $0.10x+0.12(x+500)=0.10x+0.12x+60=0.22x+60$

105. Evaluating the expression: $b^2-4ac=(-5)^2-4(1)(-6)=25-(-24)=25+24=49$

107. Evaluating the expression: $b^2-4ac=(4)^2-4(2)(-3)=16-(-24)=16+24=40$

109. **a**. Substituting $x = 8{,}000$: $-0.0035(8000)+70=42°\text{F}$

b. Substituting $x = 12{,}000$: $-0.0035(12000)+70=28°\text{F}$

c. Substituting $x = 24{,}000$: $-0.0035(24000)+70=-14°\text{F}$

111. **a**. Substituting $t = 10$: $35+0.25(10)=\$37.50$

b. Substituting $t = 20$: $35+0.25(20)=\$40.00$

c. Substituting $t = 30$: $35+0.25(30)=\$42.50$

113. Simplifying: $17-5=12$

115. Simplifying: $2-5=-3$

117. Simplifying: $-2.4+(-7.3)=-9.7$

119. Simplifying: $-\frac{1}{2}+\left(-\frac{3}{4}\right)=-\frac{1}{2}\bullet\frac{2}{2}-\frac{3}{4}=-\frac{2}{4}-\frac{3}{4}=-\frac{5}{4}$

121. Simplifying: $4(2\bullet 9-3)-7=4(18-3)-7=4(15)-7=60-7=53$

123. Simplifying: $4(2a-3)-7a=8a-12-7a=a-12$

125. Evaluating when $x = 5$: $2(5)-3=10-3=7$

127. Simplifying the expression: $7x+4-2x-1=7x-2x+4-1=5x+3$. The correct answer is c.

129. Evaluating when $x = -2$: $(-2)^2-5(-2)-4=4+10-4=10$. The correct answer is c.

2.2 Addition Property of Equality

1. Substituting $x=4$: $3(4)-5=12-5=7$. Yes, it is a solution to the equation.

3. Substituting $y=-2$: $3(-2)-4(-2+6)+2=3(-2)-4(4)+2=-6-16+2=-20\neq 8$. No, it is not a solution to the equation.

5. a. Substituting $m=-1$:

$$2(-1)+3=-2+3=1$$
$$-1-5=-6$$

No, it is not a solution to the equation.

b. Substituting $m=-8$:

$$2(-8)+3=-16+3=-13$$
$$-8-5=-13$$

Yes, it is a solution to the equation.

7. a. Substituting $x=-\frac{13}{6}$:

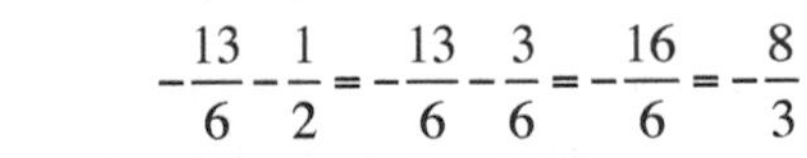

$$2\left(-\frac{13}{6}\right)+\frac{5}{3}=-\frac{13}{3}+\frac{5}{3}=-\frac{8}{3}$$
$$-\frac{13}{6}-\frac{1}{2}=-\frac{13}{6}-\frac{3}{6}=-\frac{16}{6}=-\frac{8}{3}$$

Yes, it is a solution to the equation.

b. Substituting $x=\frac{7}{6}$:

$$2\left(\frac{7}{6}\right)+\frac{5}{3}=\frac{7}{3}+\frac{5}{3}=\frac{12}{3}=4$$
$$\frac{7}{6}-\frac{1}{2}=\frac{7}{6}-\frac{3}{6}=\frac{4}{6}=\frac{2}{3}$$

No, it is not a solution to the equation.

9. Solving the equation:

$$x-3=8$$
$$x-3+3=8+3$$
$$x=11$$

11. Solving the equation:

$$x+2=6$$
$$x+2+(-2)=6+(-2)$$
$$x=4$$

13. Solving the equation:

$$a+\frac{1}{2}=-\frac{1}{4}$$
$$a+\frac{1}{2}+\left(-\frac{1}{2}\right)=-\frac{1}{4}+\left(-\frac{1}{2}\right)$$
$$a=-\frac{1}{4}+\left(-\frac{2}{4}\right)$$
$$a=-\frac{3}{4}$$

15. Solving the equation:

$$x+2.3=-3.5$$
$$x+2.3+(-2.3)=-3.5+(-2.3)$$
$$x=-5.8$$

17. Solving the equation:

$$y+11=-6$$
$$y+11+(-11)=-6+(-11)$$
$$y=-17$$

19. Solving the equation:

$$x-\frac{5}{8}=-\frac{3}{4}$$
$$x-\frac{5}{8}+\frac{5}{8}=-\frac{3}{4}+\frac{5}{8}$$
$$x=-\frac{6}{8}+\frac{5}{8}$$
$$x=-\frac{1}{8}$$

21. Solving the equation:

$$m-6=-10$$
$$m-6+6=-10+6$$
$$m=-4$$

23. Solving the equation:

$$6.9+x=3.3$$
$$-6.9+6.9+x=-6.9+3.3$$
$$x=-3.6$$

25. Solving the equation:

$$\begin{aligned} 5 &= a+4 \\ 5-4 &= a+4-4 \\ a &= 1 \end{aligned}$$

27. Solving the equation:

$$\begin{aligned} -\frac{5}{9} &= x-\frac{2}{5} \\ -\frac{5}{9}+\frac{2}{5} &= x-\frac{2}{5}+\frac{2}{5} \\ -\frac{25}{45}+\frac{18}{45} &= x \\ x &= -\frac{7}{45} \end{aligned}$$

29. Solving the equation:

$$\begin{aligned} 4x+2-3x &= 4+1 \\ x+2 &= 5 \\ x+2+(-2) &= 5+(-2) \\ x &= 3 \end{aligned}$$

31. Solving the equation:

$$\begin{aligned} 8a-\frac{1}{2}-7a &= \frac{3}{4}+\frac{1}{8} \\ a-\frac{1}{2} &= \frac{6}{8}+\frac{1}{8} \\ a-\frac{1}{2} &= \frac{7}{8} \\ a-\frac{1}{2}+\frac{1}{2} &= \frac{7}{8}+\frac{1}{2} \\ a &= \frac{7}{8}+\frac{4}{8} \\ a &= \frac{11}{8} \end{aligned}$$

33. Solving the equation:

$$\begin{aligned} -3-4x+5x &= 18 \\ -3+x &= 18 \\ 3-3+x &= 3+18 \\ x &= 21 \end{aligned}$$

35. Solving the equation:

$$\begin{aligned} -11x+2+10x+2x &= 9 \\ x+2 &= 9 \\ x+2+(-2) &= 9+(-2) \\ x &= 7 \end{aligned}$$

37. Solving the equation:

$$\begin{aligned} -2.5+4.8 &= 8x-1.2-7x \\ 2.3 &= x-1.2 \\ 2.3+1.2 &= x-1.2+1.2 \\ x &= 3.5 \end{aligned}$$

39. Solving the equation:

$$\begin{aligned} 2y-10+3y-4y &= 18-6 \\ y-10 &= 12 \\ y-10+10 &= 12+10 \\ y &= 22 \end{aligned}$$

41. Solving the equation:

$$\begin{aligned} 2(x+3)-x &= 4 \\ 2x+6-x &= 4 \\ x+6 &= 4 \\ x+6+(-6) &= 4+(-6) \\ x &= -2 \end{aligned}$$

43. Solving the equation:

$$\begin{aligned} -3(x-4)+4x &= 3-7 \\ -3x+12+4x &= -4 \\ x+12 &= -4 \\ x+12+(-12) &= -4+(-12) \\ x &= -16 \end{aligned}$$

45. Solving the equation:

$$\begin{aligned} 5(2a+1)-9a &= 8-6 \\ 10a+5-9a &= 2 \\ a+5 &= 2 \\ a+5+(-5) &= 2+(-5) \\ a &= -3 \end{aligned}$$

47. Solving the equation:

$$\begin{aligned} -(x+3)+2x-1 &= 6 \\ -x-3+2x-1 &= 6 \\ x-4 &= 6 \\ x-4+4 &= 6+4 \\ x &= 10 \end{aligned}$$

49. Solving the equation:

$$\begin{aligned} 4y-3(y-6)+2&=8\\ 4y-3y+18+2&=8\\ y+20&=8\\ y+20+(-20)&=8+(-20)\\ y&=-12 \end{aligned}$$

51. Solving the equation:

$$\begin{aligned} -3(2m-9)+7(m-4)&=12-9\\ -6m+27+7m-28&=3\\ m-1&=3\\ m-1+1&=3+1\\ m&=4 \end{aligned}$$

53. Solving the equation:

$$\begin{aligned} 4x&=3x+2\\ 4x+(-3x)&=3x+(-3x)+2\\ x&=2 \end{aligned}$$

55. Solving the equation:

$$\begin{aligned} 8a&=7a-5\\ 8a+(-7a)&=7a+(-7a)-5\\ a&=-5 \end{aligned}$$

57. Solving the equation:

$$\begin{aligned} 2x&=3x+1\\ (-2x)+2x&=(-2x)+3x+1\\ 0&=x+1\\ 0+(-1)&=x+1+(-1)\\ x&=-1 \end{aligned}$$

59. Solving the equation:

$$\begin{aligned} 3y+4&=2y+1\\ 3y+(-2y)+4&=2y+(-2y)+1\\ y+4&=1\\ y+4+(-4)&=1+(-4)\\ y&=-3 \end{aligned}$$

61. Solving the equation:

$$\begin{aligned} 2m-3&=m+5\\ 2m+(-m)-3&=m+(-m)+5\\ m-3&=5\\ m-3+3&=5+3\\ m&=8 \end{aligned}$$

63. Solving the equation:

$$\begin{aligned} 4x-7&=5x+1\\ 4x+(-4x)-7&=5x+(-4x)+1\\ -7&=x+1\\ -7+(-1)&=x+1+(-1)\\ x&=-8 \end{aligned}$$

65. Solving the equation:

$$\begin{aligned} 5x-\frac{2}{3}&=4x+\frac{4}{3}\\ 5x+(-4x)-\frac{2}{3}&=4x+(-4x)+\frac{4}{3}\\ x-\frac{2}{3}&=\frac{4}{3}\\ x-\frac{2}{3}+\frac{2}{3}&=\frac{4}{3}+\frac{2}{3}\\ x&=\frac{6}{3}=2 \end{aligned}$$

67. Solving the equation:

$$\begin{aligned} 8a-7.1&=7a+3.9\\ 8a+(-7a)-7.1&=7a+(-7a)+3.9\\ a-7.1&=3.9\\ a-7.1+7.1&=3.9+7.1\\ a&=11 \end{aligned}$$

69. Solving the equation:

$$\begin{aligned} 11y-2.9&=12y+2.9\\ 11y+(-11y)-2.9&=12y+(-11y)+2.9\\ -2.9&=y+2.9\\ -2.9-2.9&=y+2.9-2.9\\ y&=-5.8 \end{aligned}$$

71. **a.** Solving for R:

$$\begin{aligned} T+R+A&=100\\ 88+R+6&=100\\ 94+R&=100\\ R&=6\% \end{aligned}$$

b. Solving for R:

$$\begin{aligned} T+R+A&=100\\ 0+R+95&=100\\ 95+R&=100\\ R&=5\% \end{aligned}$$

c. Solving for A:

$$\begin{aligned} T+R+A&=100\\ 0+98+A&=100\\ 98+A&=100\\ A&=2\% \end{aligned}$$

d. Solving for R:

$$\begin{aligned} T+R+A&=100\\ 0+R+25&=100\\ 25+R&=100\\ R&=75\% \end{aligned}$$

73. Simplifying: $\frac{3}{2}\left(\frac{2}{3}y\right)=y$

75. Simplifying: $\frac{1}{5}(5x)=x$

77. Simplifying: $\frac{1}{5}(30)=6$

79. Simplifying: $\frac{3}{2}(4)=\frac{12}{2}=6$

81. Simplifying: $12\left(-\frac{3}{4}\right)=-\frac{36}{4}=-9$

83. Simplifying: $\frac{3}{2}\left(-\frac{5}{4}\right)=-\frac{15}{8}$

85. Simplifying: $13+(-5)=8$

87. Simplifying: $-\frac{3}{4}+\left(-\frac{1}{2}\right)=-\frac{3}{4}-\frac{1}{2}\cdot\frac{2}{2}=-\frac{3}{4}-\frac{2}{4}=-\frac{5}{4}$

89. Simplifying: $7x+(-4x)=3x$

91. **a.** Substituting $x=9$:

$$\begin{aligned} &5(9)-6=45-6=39\\ &4(9)+3=36+3=39 \end{aligned}$$

Yes, it is a solution to the equation.

b. Substituting $x=-3$:

$$\begin{aligned} &5(-3)-6=-15-6=-21\\ &4(-3)+3=-12+3=-9 \end{aligned}$$

No, it is not a solution to the equation.

c. Substituting $x=3$:

$$\begin{aligned} &5(3)-6=15-6=9\\ &4(3)+3=12+3=15 \end{aligned}$$

No, it is not a solution to the equation.

d. Substituting $x=-9$:

$$\begin{aligned} &5(-9)-6=-45-6=-51\\ &4(-9)+3=-36+3=-33 \end{aligned}$$

No, it is not a solution to the equation.

The correct answer is a.

93. Solving the equation:

$$\begin{aligned} -4(x-2)+5x&=3-9\\ -4x+8+5x&=-6\\ x+8&=-6\\ x+8+(-8)&=-6+(-8)\\ x&=-14 \end{aligned}$$

The correct answer is c.

2.3 Multiplication Property of Equality

1. Solving the equation:

$$5x = 10$$
$$\frac{1}{5}(5x) = \frac{1}{5}(10)$$
$$x = 2$$

3. Solving the equation:

$$7a = 28$$
$$\frac{1}{7}(7a) = \frac{1}{7}(28)$$
$$a = 4$$

5. Solving the equation:

$$-8x = 4$$
$$-\frac{1}{8}(-8x) = -\frac{1}{8}(4)$$
$$x = -\frac{1}{2}$$

7. Solving the equation:

$$8m = -16$$
$$\frac{1}{8}(8m) = \frac{1}{8}(-16)$$
$$m = -2$$

9. Solving the equation:

$$-3x = -9$$
$$-\frac{1}{3}(-3x) = -\frac{1}{3}(-9)$$
$$x = 3$$

11. Solving the equation:

$$-7y = -28$$
$$-\frac{1}{7}(-7y) = -\frac{1}{7}(-28)$$
$$y = 4$$

13. Solving the equation:

$$2x = 0$$
$$\frac{1}{2}(2x) = \frac{1}{2}(0)$$
$$x = 0$$

15. Solving the equation:

$$-5x = 0$$
$$-\frac{1}{5}(-5x) = -\frac{1}{5}(0)$$
$$x = 0$$

17. Solving the equation:

$$\frac{x}{3} = 2$$
$$3\left(\frac{x}{3}\right) = 3(2)$$
$$x = 6$$

19. Solving the equation:

$$-\frac{m}{5} = 10$$
$$-5\left(-\frac{m}{5}\right) = -5(10)$$
$$m = -50$$

21. Solving the equation:

$$-\frac{x}{2} = -\frac{3}{4}$$
$$-2\left(-\frac{x}{2}\right) = -2\left(-\frac{3}{4}\right)$$
$$x = \frac{3}{2}$$

23. Solving the equation:

$$\frac{2}{3}a = 8$$
$$\frac{3}{2}\left(\frac{2}{3}a\right) = \frac{3}{2}(8)$$
$$a = 12$$

25. Solving the equation:

$$-\frac{3}{5}x = \frac{9}{5}$$
$$-\frac{5}{3}\left(-\frac{3}{5}x\right) = -\frac{5}{3}\left(\frac{9}{5}\right)$$
$$x = -3$$

27. Solving the equation:

$$-\frac{5}{8}y = -20$$
$$-\frac{8}{5}\left(-\frac{5}{8}y\right) = -\frac{8}{5}(-20)$$
$$y = 32$$

29. Simplifying and then solving the equation:

$$\begin{aligned} -4x-2x+3x&=24 \\ -3x&=24 \\ -\frac{1}{3}(-3x)&=-\frac{1}{3}(24) \\ x&=-8 \end{aligned}$$

31. Simplifying and then solving the equation:

$$\begin{aligned} 4x+8x-2x&=15-10 \\ 10x&=5 \\ \frac{1}{10}(10x)&=\frac{1}{10}(5) \\ x&=\frac{1}{2} \end{aligned}$$

33. Simplifying and then solving the equation:

$$\begin{aligned} -3-5&=3x+5x-10x \\ -8&=-2x \\ -\frac{1}{2}(-8)&=-\frac{1}{2}(-2x) \\ x&=4 \end{aligned}$$

35. Simplifying and then solving the equation:

$$\begin{aligned} 18-13&=\frac{1}{2}a+\frac{3}{4}a-\frac{5}{8}a \\ 8(5)&=8\left(\frac{1}{2}a+\frac{3}{4}a-\frac{5}{8}a\right) \\ 40&=4a+6a-5a \\ 40&=5a \\ \frac{1}{5}(40)&=\frac{1}{5}(5a) \\ a&=8 \end{aligned}$$

37. Solving by multiplying both sides of the equation by –1:

$$\begin{aligned} -x&=4 \\ -1(-x)&=-1(4) \\ x&=-4 \end{aligned}$$

39. Solving by multiplying both sides of the equation by –1:

$$\begin{aligned} -x&=-4 \\ -1(-x)&=-1(-4) \\ x&=4 \end{aligned}$$

41. Solving by multiplying both sides of the equation by –1:

$$\begin{aligned} 15&=-a \\ -1(15)&=-1(-a) \\ a&=-15 \end{aligned}$$

43. Solving by multiplying both sides of the equation by –1:

$$\begin{aligned} -y&=\frac{1}{2} \\ -1(-y)&=-1\left(\frac{1}{2}\right) \\ y&=-\frac{1}{2} \end{aligned}$$

45. Solving the equation:

$$\begin{aligned} 3x-2&=7 \\ 3x-2+2&=7+2 \\ 3x&=9 \\ \frac{1}{3}(3x)&=\frac{1}{3}(9) \\ x&=3 \end{aligned}$$

47. Solving the equation:

$$\begin{aligned} 2a+1&=3 \\ 2a+1+(-1)&=3+(-1) \\ 2a&=2 \\ \frac{1}{2}(2a)&=\frac{1}{2}(2) \\ a&=1 \end{aligned}$$

49. Multiplying by 8 to eliminate fractions:

$$\frac{1}{8}+\frac{1}{2}x=\frac{1}{4}$$
$$8\left(\frac{1}{8}+\frac{1}{2}x\right)=8\left(\frac{1}{4}\right)$$
$$1+4x=2$$
$$(-1)+1+4x=(-1)+2$$
$$4x=1$$
$$\frac{1}{4}(4x)=\frac{1}{4}(1)$$
$$x=\frac{1}{4}$$

51. Solving the equation:

$$6x=2x-12$$
$$6x+(-2x)=2x+(-2x)-12$$
$$4x=-12$$
$$\frac{1}{4}(4x)=\frac{1}{4}(-12)$$
$$x=-3$$

53. Solving the equation:

$$2y=-4y+18$$
$$2y+4y=-4y+4y+18$$
$$6y=18$$
$$\frac{1}{6}(6y)=\frac{1}{6}(18)$$
$$y=3$$

55. Solving the equation:

$$-7x=-3x-8$$
$$-7x+3x=-3x+3x-8$$
$$-4x=-8$$
$$-\frac{1}{4}(-4x)=-\frac{1}{4}(-8)$$
$$x=2$$

57. Solving the equation:

$$8x+4=2x-5$$
$$8x+(-8x)+4=2x+(-8x)-5$$
$$-6x-5=4$$
$$-6x-5+5=4+5$$
$$-6x=9$$
$$-\frac{1}{6}(-6x)=-\frac{1}{6}(9)$$
$$x=-\frac{3}{2}$$

59. Solving the equation:

$$6m-3=m+2$$
$$6m+(-6m)-3=m+(-6m)+2$$
$$-5m+2=-3$$
$$-5m+2+(-2)=-3+(-2)$$
$$-5m=-5$$
$$-\frac{1}{5}(-5m)=-\frac{1}{5}(-5)$$
$$m=1$$

61. Solving the equation:

$$9y+2=6y-4$$
$$9y+(-9y)+2=6y+(-9y)-4$$
$$-3y-4=2$$
$$-3y-4+4=2+4$$
$$-3y=6$$
$$-\frac{1}{3}(-3y)=-\frac{1}{3}(6)$$
$$y=-2$$

63. Solving the equation:

$$2(x+3)=12$$
$$2x+6=12$$
$$2x+6+(-6)=12+(-6)$$
$$2x=6$$
$$\frac{1}{2}(2x)=\frac{1}{2}(6)$$
$$x=3$$

65. Solving the equation:

$$\begin{aligned} 6(x-1) &= -18 \\ 6x-6 &= -18 \\ 6x-6+6 &= -18+6 \\ 6x &= -12 \\ \frac{1}{6}(6x) &= \frac{1}{6}(-12) \\ x &= -2 \end{aligned}$$

67. Solving the equation:

$$\begin{aligned} 2(4a+1) &= -6 \\ 8a+2 &= -6 \\ 8a+2+(-2) &= -6+(-2) \\ 8a &= -8 \\ \frac{1}{8}(8a) &= \frac{1}{8}(-8) \\ a &= -1 \end{aligned}$$

69. Solving the equation:

$$\begin{aligned} 14 &= 2(5x-3) \\ 14 &= 10x-6 \\ 14+6 &= 10x-6+6 \\ 20 &= 10x \\ \frac{1}{10}(20) &= \frac{1}{10}(10x) \\ x &= 2 \end{aligned}$$

71. Solving the equation:

$$\begin{aligned} -2(3y+5) &= 14 \\ -6y-10 &= 14 \\ -6y-10+10 &= 14+10 \\ -6y &= 24 \\ -\frac{1}{6}(-6y) &= -\frac{1}{6}(24) \\ y &= -4 \end{aligned}$$

73. Solving the equation:

$$\begin{aligned} -5(2a+4) &= 0 \\ -10a-20 &= 0 \\ -10a-20+20 &= 0+20 \\ -10a &= 20 \\ -\frac{1}{10}(-10a) &= -\frac{1}{10}(20) \\ a &= -2 \end{aligned}$$

75. Solving the equation:

$$\begin{aligned} 1 &= \frac{1}{2}(4x+2) \\ 1 &= 2x+1 \\ 1+(-1) &= 2x+1+(-1) \\ 0 &= 2x \\ \frac{1}{2}(0) &= \frac{1}{2}(2x) \\ x &= 0 \end{aligned}$$

77. Solving the equation:

$$\begin{aligned} 3(t-4)+5 &= -4 \\ 3t-12+5 &= -4 \\ 3t-7 &= -4 \\ 3t-7+7 &= -4+7 \\ 3t &= 3 \\ \frac{1}{3}(3t) &= \frac{1}{3}(3) \\ t &= 1 \end{aligned}$$

79. Solving the equation:

$$\begin{aligned} 4(2x+1)-7 &= 1 \\ 8x+4-7 &= 1 \\ 8x-3 &= 1 \\ 8x-3+3 &= 1+3 \\ 8x &= 4 \\ \frac{1}{8}(8x) &= \frac{1}{8}(4) \\ x &= \frac{1}{2} \end{aligned}$$

81. **a.** Solving the equation:

$$\begin{aligned} 2x &= 3 \\ \frac{1}{2}(2x) &= \frac{1}{2}(3) \\ x &= \frac{3}{2} \end{aligned}$$

b. Solving the equation:

$$\begin{aligned} 2+x &= 3 \\ 2+(-2)+x &= 3+(-2) \\ x &= 1 \end{aligned}$$

c. Solving the equation:

$$\begin{aligned} 2x+3&=0 \\ 2x+3+(-3)&=0+(-3) \\ 2x&=-3 \\ \frac{1}{2}(2x)&=\frac{1}{2}(-3) \\ x&=-\frac{3}{2} \end{aligned}$$

d. Solving the equation:

$$\begin{aligned} 2x+3&=-5 \\ 2x+3+(-3)&=-5+(-3) \\ 2x&=-8 \\ \frac{1}{2}(2x)&=\frac{1}{2}(-8) \\ x&=-4 \end{aligned}$$

e. Solving the equation:

$$\begin{aligned} 2x+3&=7x-5 \\ 2x+(-7x)+3&=7x+(-7x)-5 \\ -5x+3&=-5 \\ -5x+3+(-3)&=-5+(-3) \\ -5x&=-8 \\ -\frac{1}{5}(-5x)&=-\frac{1}{5}(-8) \\ x&=\frac{8}{5} \end{aligned}$$

83. Solving the equation:

$$\begin{aligned} 7.5x&=1500 \\ \frac{7.5x}{7.5}&=\frac{1500}{7.5} \\ x&=200 \end{aligned}$$

The break-even point is 200 tickets.

85. Solving the equation:

$$\begin{aligned} G-0.21G-0.08G&=987.5 \\ 0.71G&=987.5 \\ G&\approx 1{,}390.85 \end{aligned}$$

Your gross income is approximately $1,390.85.

87. Solving the equation:

$$\begin{aligned} 2x&=4 \\ \frac{1}{2}(2x)&=\frac{1}{2}(4) \\ x&=2 \end{aligned}$$

89. Solving the equation:

$$\begin{aligned} 30&=5x \\ 5x&=30 \\ \frac{1}{5}(5x)&=\frac{1}{5}(30) \\ x&=6 \end{aligned}$$

91. Solving the equation:

$$\begin{aligned} 0.17x&=510 \\ x&=\frac{510}{0.17}=3{,}000 \end{aligned}$$

93. Simplifying: $3(x-5)+4=3x-15+4=3x-11$

95. Simplifying: $0.09(x+2{,}000)=0.09x+180$

97. Simplifying: $7-3(2y+1)=7-6y-3=4-6y=-6y+4$

99. Simplifying: $3(2x-5)-(2x-4)=6x-15-2x+4=4x-11$

101. Simplifying: $10x+(-5x)=5x$

103. Simplifying: $0.08x+0.09x=0.17x$

105. Solving the equation:

$$\begin{aligned} 3x-8x&=20 \\ -5x&=20 \\ -\frac{1}{5}(-5x)&=-\frac{1}{5}(20) \\ x&=-4 \end{aligned}$$

The correct answer is d.

2.4 Solving Linear Equations

1. Solving the equation:

$$\begin{aligned} x+(2x-1)&=2 \\ 3x-1&=2 \\ 3x-1+1&=2+1 \\ 3x&=3 \\ \frac{1}{3}(3x)&=\frac{1}{3}(3) \\ x&=1 \end{aligned}$$

3. Solving the equation:

$$\begin{aligned} 15&=3(x-1) \\ 15&=3x-3 \\ 15+3&=3x-3+3 \\ 18&=3x \\ \frac{1}{3}(18)&=\frac{1}{3}(3x) \\ x&=6 \end{aligned}$$

5. Solving the equation:

$$\begin{aligned} 6-5(2a-3)&=1 \\ 6-10a+15&=1 \\ -10a+21&=1 \\ -10a+21+(-21)&=1+(-21) \\ -10a&=-20 \\ -\frac{1}{10}(-10a)&=-\frac{1}{10}(-20) \\ a&=2 \end{aligned}$$

7. Solving the equation:

$$\begin{aligned} x-(3x+5)&=-3 \\ x-3x-5&=-3 \\ -2x-5&=-3 \\ -2x-5+5&=-3+5 \\ -2x&=2 \\ -\frac{1}{2}(-2x)&=-\frac{1}{2}(2) \\ x&=-1 \end{aligned}$$

9. Solving the equation:

$$\begin{aligned} 7(2y-1)-6y&=-1 \\ 14y-7-6y&=-1 \\ 8y-7&=-1 \\ 8y-7+7&=-1+7 \\ 8y&=6 \\ \frac{1}{8}(8y)&=\frac{1}{8}(6) \\ y&=\frac{3}{4} \end{aligned}$$

11. Solving the equation:

$$\begin{aligned} 5x-8(2x-5)&=7 \\ 5x-16x+40&=7 \\ -11x+40&=7 \\ -11x+40-40&=7-40 \\ -11x&=-33 \\ -\frac{1}{11}(-11x)&=-\frac{1}{11}(-33) \\ x&=3 \end{aligned}$$

13. Solving the equation:

$$\begin{aligned} 4x-(-4x+1)&=5 \\ 4x+4x-1&=5 \\ 8x-1&=5 \\ 8x-1+1&=5+1 \\ 8x&=6 \\ \frac{1}{8}(8x)&=\frac{1}{8}(6) \\ x&=\frac{3}{4} \end{aligned}$$

15. Solving the equation:

$$\begin{aligned} 3(x-3)+2(2x)&=5 \\ 3x-9+4x&=5 \\ 7x-9&=5 \\ 7x-9+9&=5+9 \\ 7x&=14 \\ \frac{1}{7}(7x)&=\frac{1}{7}(14) \\ x&=2 \end{aligned}$$

17. Solving the equation:

$$\begin{aligned}3x+2(x-2)&=6\\3x+2x-4&=6\\5x-4&=6\\5x-4+4&=6+4\\5x&=10\\\frac{1}{5}(5x)&=\frac{1}{5}(10)\\x&=2\end{aligned}$$

19. Solving the equation:

$$\begin{aligned}5x+10(x+8)&=245\\5x+10x+80&=245\\15x+80&=245\\15x+80-80&=245-80\\15x&=165\\x&=\frac{165}{15}=11\end{aligned}$$

21. Solving the equation:

$$\begin{aligned}x+(x+3)(-3)&=x-3\\x-3x-9&=x-3\\-2x-9&=x-3\\-2x-x-9&=x-x-3\\-3x-9&=-3\\-3x-9+9&=-3+9\\-3x&=6\\-\frac{1}{3}(-3x)&=-\frac{1}{3}(6)\\x&=-2\end{aligned}$$

23. Solving the equation:

$$\begin{aligned}5(y+2)&=4(y+1)\\5y+10&=4y+4\\-4y+5y+10&=-4y+4y+4\\y+10&=4\\y+10-10&=4-10\\y&=-6\end{aligned}$$

25. Solving the equation:

$$\begin{aligned}50(x-5)&=30(x+5)\\50x-250&=30x+150\\50x-30x-250&=30x-30x+150\\20x-250&=150\\20x-250+250&=150+250\\20x&=400\\\frac{1}{20}(20x)&=\frac{1}{20}(400)\\x&=20\end{aligned}$$

27. Solving the equation:

$$\begin{aligned}5(x+2)+3(x-1)&=-9\\5x+10+3x-3&=-9\\8x+7&=-9\\8x+7-7&=-9-7\\8x&=-16\\\frac{1}{8}(8x)&=\frac{1}{8}(-16)\\x&=-2\end{aligned}$$

29. Solving the equation:

$$\begin{aligned}-2(3y+1)&=3(1-6y)-9\\-6y-2&=3-18y-9\\-6y-2&=-18y-6\\-6y+18y-2&=-18y+18y-6\\12y-2&=-6\\12y-2+2&=-6+2\\12y&=-4\\\frac{1}{12}(12y)&=\frac{1}{12}(-4)\\y&=-\frac{1}{3}\end{aligned}$$

31. Solving the equation:

$$\begin{aligned}2(t-3)+3(t-2)&=28\\2t-6+3t-6&=28\\5t-12&=28\\5t-12+12&=28+12\\5t&=40\\\frac{1}{5}(5t)&=\frac{1}{5}(40)\\t&=8\end{aligned}$$

33. Solving the equation:

$$\begin{aligned} 5x+10(x+3)+25(x+5) &= 435 \\ 5x+10x+30+25x+125 &= 435 \\ 40x+155 &= 435 \\ 40x+155-155 &= 435-155 \\ 40x &= 280 \\ x &= \frac{280}{40} = 7 \end{aligned}$$

35. Solving the equation:

$$\begin{aligned} 5(x-2)-(3x+4) &= 3(6x-8)+10 \\ 5x-10-3x-4 &= 18x-24+10 \\ 2x-14 &= 18x-14 \\ 2x+(-18x)-14 &= 18x+(-18x)-14 \\ -16x-14 &= -14 \\ -16x-14+14 &= -14+14 \\ -16x &= 0 \\ -\frac{1}{16}(-16x) &= -\frac{1}{16}(0) \\ x &= 0 \end{aligned}$$

37. Solving the equation:

$$\begin{aligned} 2(5x-3)-(2x-4) &= 5-(6x+1) \\ 10x-6-2x+4 &= 5-6x-1 \\ 8x-2 &= -6x+4 \\ 8x+6x-2 &= -6x+6x+4 \\ 14x-2 &= 4 \\ 14x-2+2 &= 4+2 \\ 14x &= 6 \\ \frac{1}{14}(14x) &= \frac{1}{14}(6) \\ x &= \frac{3}{7} \end{aligned}$$

39. Solving the equation:

$$\begin{aligned} -(3x+1)-(4x-7) &= 4-(3x+2) \\ -3x-1-4x+7 &= 4-3x-2 \\ -7x+6 &= -3x+2 \\ -7x+3x+6 &= -3x+3x+2 \\ -4x+6 &= 2 \\ -4x+6+(-6) &= 2+(-6) \\ -4x &= -4 \\ -\frac{1}{4}(-4x) &= -\frac{1}{4}(-4) \\ x &= 1 \end{aligned}$$

41. Multiply by 10 to eliminate fractions:

$$\begin{aligned} 10\left(\frac{x}{2}+4\right) &= 10\left(\frac{x}{5}-1\right) \\ 5x+40 &= 2x-10 \\ 5x+(-2x)+40 &= 2x+(-2x)-10 \\ 3x+40 &= -10 \\ 3x+40-40 &= -10-40 \\ 3x &= -50 \\ x &= -\frac{50}{3} \end{aligned}$$

43. Multiply by 21 to eliminate fractions:

$$\begin{aligned} 21\left(\frac{x}{7}-\frac{5}{21}\right) &= 21\left(\frac{x}{3}+\frac{2}{7}\right) \\ 3x-5 &= 7x+6 \\ 3x+(-7x)-5 &= 7x+(-7x)+6 \\ -4x-5 &= 6 \\ -4x-5+5 &= 6+5 \\ -4x &= 11 \\ -\frac{1}{4}(-4x) &= -\frac{1}{4}(11) \\ x &= -\frac{11}{4} \end{aligned}$$

45. Multiply by 9 to eliminate fractions:

$$9\left(\frac{4}{9}x+\frac{5}{3}\right)=9\left(2+\frac{2}{3}x\right)$$
$$4x+15=18+6x$$
$$4x+(-6x)+15=6x+(-6x)+18$$
$$-2x+15=18$$
$$-2x+15-15=18-15$$
$$-2x=3$$
$$-\frac{1}{2}(-2x)=-\frac{1}{2}(3)$$
$$x=-\frac{3}{2}$$

47. Solving the equation:

$$\frac{1}{2}(x-3)=\frac{1}{4}(x+1)$$
$$\frac{1}{2}x-\frac{3}{2}=\frac{1}{4}x+\frac{1}{4}$$
$$4\left(\frac{1}{2}x-\frac{3}{2}\right)=4\left(\frac{1}{4}x+\frac{1}{4}\right)$$
$$2x-6=x+1$$
$$2x+(-x)-6=x+(-x)+1$$
$$x-6=1$$
$$x-6+6=1+6$$
$$x=7$$

49. Solving the equation:

$$\frac{3}{4}(8x-4)+3=\frac{2}{5}(5x+10)-1$$
$$6x-3+3=2x+4-1$$
$$6x=2x+3$$
$$6x+(-2x)=2x+(-2x)+3$$
$$4x=3$$
$$\frac{1}{4}(4x)=\frac{1}{4}(3)$$
$$x=\frac{3}{4}$$

51. Solving the equation:

$$0.5x+0.2(18-x)=5.4$$
$$0.5x+3.6-0.2x=5.4$$
$$0.3x+3.6=5.4$$
$$0.3x+3.6-3.6=5.4-3.6$$
$$0.3x=1.8$$
$$x=\frac{1.8}{0.3}=6$$

53. Solving the equation:

$$0.06x+0.08(100-x)=6.5$$
$$0.06x+8-0.08x=6.5$$
$$-0.02x+8=6.5$$
$$-0.02x+8+(-8)=6.5+(-8)$$
$$-0.02x=-1.5$$
$$\frac{-0.02x}{-0.02}=\frac{-1.5}{-0.02}$$
$$x=75$$

55. Solving the equation:

$$0.2x-0.5=0.5-0.2(2x-13)$$
$$0.2x-0.5=0.5-0.4x+2.6$$
$$0.2x-0.5=-0.4x+3.1$$
$$0.2x+0.4x-0.5=-0.4x+0.4x+3.1$$
$$0.6x-0.5=3.1$$
$$0.6x-0.5+0.5=3.1+0.5$$
$$0.6x=3.6$$
$$\frac{0.6x}{0.6}=\frac{3.6}{0.6}$$
$$x=6$$

57. Solving the equation:

$$0.08x+0.09(x+2{,}000)=860$$
$$0.08x+0.09x+180=860$$
$$0.17x+180=860$$
$$0.17x+180-180=860-180$$
$$0.17x=680$$
$$x=\frac{680}{0.17}=4{,}000$$

59. Solving the equation:

$$0.10x+0.12(x+500)=214$$
$$0.10x+0.12x+60=214$$
$$0.22x+60=214$$
$$0.22x+60-60=214-60$$
$$0.22x=154$$
$$x=\frac{154}{0.22}=700$$

61. Solving the equation:

$$\begin{aligned}
-0.7(2x-7)&=0.3(11-4x)\\
-1.4x+4.9&=3.3-1.2x\\
-1.4x+1.2x+4.9&=3.3-1.2x+1.2x\\
-0.2x+4.9&=3.3\\
-0.2x+4.9+(-4.9)&=3.3+(-4.9)\\
-0.2x&=-1.6\\
\frac{-0.2x}{-0.2}&=\frac{-1.6}{-0.2}\\
x&=8
\end{aligned}$$

63. Solving the equation:

$$\begin{aligned}
0.2x+0.5(12-x)&=3.6\\
0.2x+6-0.5x&=3.6\\
-0.3x+6&=3.6\\
-0.3x+6-6&=3.6-6\\
-0.3x&=-2.4\\
x&=\frac{-2.4}{-0.3}=8
\end{aligned}$$

65. Since this statement is false, there is no solution. This is a contradiction.

67. Solving the equation:

$$\begin{aligned}
3x-5&=5-3x\\
3x+3x-5&=5-3x+3x\\
6x-5&=5\\
6x-5+5&=5+5\\
6x&=10\\
\frac{1}{6}(6x)&=\frac{1}{6}(10)\\
x&=\frac{5}{3}
\end{aligned}$$

69. Solving the equation:

$$\begin{aligned}
4(3x+2)-x&=7x+2(2x+4)\\
12x+8-x&=7x+4x+8\\
11x+8&=11x+8
\end{aligned}$$

Since this statement is true, the solution set is all real numbers. This is an identity.

71. Multiply by 6 to eliminate fractions:

$$\begin{aligned}
6\left(\frac{4}{3}+\frac{1}{2}x-\frac{1}{6}\right)&=6\left(\frac{1}{3}x+1+\frac{1}{6}x\right)\\
8+3x-1&=2x+6+x\\
3x+7&=3x+6
\end{aligned}$$

Since this statement is false, there is no solution. This is a contradiction.

73. **a.** Solving the equation:

$$\begin{aligned}
4x-5&=0\\
4x-5+5&=0+5\\
4x&=5\\
\frac{1}{4}(4x)&=\frac{1}{4}(5)\\
x&=\frac{5}{4}=1.25
\end{aligned}$$

b. Solving the equation:

$$\begin{aligned}
4x-5&=25\\
4x-5+5&=25+5\\
4x&=30\\
\frac{1}{4}(4x)&=\frac{1}{4}(30)\\
x&=\frac{15}{2}=7.5
\end{aligned}$$

c. Adding: $(4x-5)+(2x+25)=6x+20$

d. Solving the equation:

$$\begin{aligned} 4x-5&=2x+25 \\ 4x-2x-5&=2x-2x+25 \\ 2x-5&=25 \\ 2x-5+5&=25+5 \\ 2x&=30 \\ \frac{1}{2}(2x)&=\frac{1}{2}(30) \\ x&=15 \end{aligned}$$

e. Multiplying: $4(x-5)=4x-20$

f. Solving the equation:

$$\begin{aligned} 4(x-5)&=2x+25 \\ 4x-20&=2x+25 \\ 4x-2x-20&=2x-2x+25 \\ 2x-20&=25 \\ 2x-20+20&=25+20 \\ 2x&=45 \\ \frac{1}{2}(2x)&=\frac{1}{2}(45) \\ x&=\frac{45}{2}=22.5 \end{aligned}$$

75. Solving the equation:

$$\begin{aligned} 40&=2x+12 \\ 2x+12&=40 \\ 2x&=28 \\ x&=14 \end{aligned}$$

77. Solving the equation:

$$\begin{aligned} 12+2y&=6 \\ 2y&=-6 \\ y&=-3 \end{aligned}$$

79. Solving the equation:

$$\begin{aligned} 24x&=6 \\ x&=\frac{6}{24}=\frac{1}{4} \end{aligned}$$

81. Solving the equation:

$$\begin{aligned} 70&=x\cdot 210 \\ x&=\frac{70}{210}=\frac{1}{3} \end{aligned}$$

83. Simplifying: $\frac{1}{2}(-3x+6)=\frac{1}{2}(-3x)+\frac{1}{2}(6)=-\frac{3}{2}x+3$

85. Solving the equation:

$$\begin{aligned} 4(2x-1)+9&=7-(5x+3) \\ 8x-4+9&=7-5x-3 \\ 8x+5&=-5x+4 \\ 8x+5x+5&=-5x+5x+4 \\ 13x+5&=4 \\ 13x+5-5&=4-5 \\ 13x&=-1 \\ \frac{1}{13}(13x)&=\frac{1}{13}(-1) \\ x&=-\frac{1}{13} \end{aligned}$$

The correct answer is a.

87. Solving the equation:

$$\begin{aligned} 0.05x+0.04(x-300)&=312 \\ 0.05x+0.04x-12&=312 \\ 0.09x-12&=312 \\ 0.09x-12+12&=312+12 \\ 0.09x&=324 \\ x&=\frac{324}{0.09} \\ x&=3{,}600 \end{aligned}$$

The correct answer is c.

2.5 Formulas and Percents

1. Using the perimeter formula:

$$P = 2l + 2w$$
$$300 = 2l + 2(50)$$
$$300 = 2l + 100$$
$$200 = 2l$$
$$l = 100$$

The length is 100 feet.

3. Substituting $x = 3$:

$$2(3) + 3y = 6$$
$$6 + 3y = 6$$
$$6 + (-6) + 3y = 6 + (-6)$$
$$3y = 0$$
$$y = 0$$

5. Substituting $x = 0$:

$$2(0) + 3y = 6$$
$$0 + 3y = 6$$
$$3y = 6$$
$$y = 2$$

7. Substituting $y = 2$:

$$2x - 5(2) = 20$$
$$2x - 10 = 20$$
$$2x - 10 + 10 = 20 + 10$$
$$2x = 30$$
$$x = 15$$

9. Substituting $y = 0$:

$$2x - 5(0) = 20$$
$$2x - 0 = 20$$
$$2x = 20$$
$$x = 10$$

11. Substituting $x = -2$: $y = (-2+1)^2 - 3 = (-1)^2 - 3 = 1 - 3 = -2$

13. Substituting $x = 1$: $y = (1+1)^2 - 3 = (2)^2 - 3 = 4 - 3 = 1$

15. **a.** Substituting $x = 10$: $y = \frac{20}{10} = 2$

b. Substituting $x = 5$: $y = \frac{20}{5} = 4$

17. **a.** Substituting $y = 15$ and $x = 3$:

$$15 = K(3)$$
$$K = \frac{15}{3} = 5$$

b. Substituting $y = 72$ and $x = 4$:

$$72 = K(4)$$
$$K = \frac{72}{4} = 18$$

19. Solving for l:

$$lw = A$$
$$\frac{lw}{w} = \frac{A}{w}$$
$$l = \frac{A}{w}$$

21. Solving for h:

$$lwh = V$$
$$\frac{lwh}{lw} = \frac{V}{lw}$$
$$h = \frac{V}{lw}$$

23. Solving for a:

$$a + b + c = P$$
$$a + b + c - b - c = P - b - c$$
$$a = P - b - c$$

25. Solving for x:

$$x - 3y = -1$$
$$x - 3y + 3y = -1 + 3y$$
$$x = 3y - 1$$

27. Solving for y:

$$-3x+y=6$$
$$-3x+3x+y=6+3x$$
$$y=3x+6$$

29. Solving for y:

$$2x+3y=6$$
$$-2x+2x+3y=-2x+6$$
$$3y=-2x+6$$
$$\frac{1}{3}(3y)=\frac{1}{3}(-2x+6)$$
$$y=-\frac{2}{3}x+2$$

31. Solving for y:

$$y-3=-2(x+4)$$
$$y-3=-2x-8$$
$$y=-2x-5$$

33. Solving for y:

$$y-3=-\frac{2}{3}(x+3)$$
$$y-3=-\frac{2}{3}x-2$$
$$y=-\frac{2}{3}x+1$$

35. Solving for w:

$$2l+2w=P$$
$$2l-2l+2w=P-2l$$
$$2w=P-2l$$
$$\frac{2w}{2}=\frac{P-2l}{2}$$
$$w=\frac{P-2l}{2}$$

37. Solving for v:

$$vt+16t^2=h$$
$$vt+16t^2-16t^2=h-16t^2$$
$$vt=h-16t^2$$
$$\frac{vt}{t}=\frac{h-16t^2}{t}$$
$$v=\frac{h-16t^2}{t}$$

39. Solving for h:

$$\pi r^2+2\pi rh=A$$
$$\pi r^2-\pi r^2+2\pi rh=A-\pi r^2$$
$$2\pi rh=A-\pi r^2$$
$$\frac{2\pi rh}{2\pi r}=\frac{A-\pi r^2}{2\pi r}$$
$$h=\frac{A-\pi r^2}{2\pi r}$$

41. **a.** Solving for y:

$$\frac{y-1}{x}=\frac{3}{5}$$
$$5y-5=3x$$
$$5y=3x+5$$
$$y=\frac{3}{5}x+1$$

b. Solving for y:

$$\frac{y-2}{x}=\frac{1}{2}$$
$$2y-4=x$$
$$2y=x+4$$
$$y=\frac{1}{2}x+2$$

c. Solving for y:

$$\frac{y-3}{x}=4$$
$$y-3=4x$$
$$y=4x+3$$

43. Solving for y:

$$\frac{x}{7}-\frac{y}{3}=1$$
$$-\frac{x}{7}+\frac{x}{7}-\frac{y}{3}=-\frac{x}{7}+1$$
$$-\frac{y}{3}=-\frac{x}{7}+1$$
$$-3\left(-\frac{y}{3}\right)=-3\left(-\frac{x}{7}+1\right)$$
$$y=\frac{3}{7}x-3$$

45. Solving for y:

$$-\frac{1}{4}x+\frac{1}{8}y=1$$
$$-\frac{1}{4}x+\frac{1}{4}x+\frac{1}{8}y=1+\frac{1}{4}x$$
$$\frac{1}{8}y=\frac{1}{4}x+1$$
$$8\left(\frac{1}{8}y\right)=8\left(\frac{1}{4}x+1\right)$$
$$y=2x+8$$

47. Translating into an equation and solving:

$$x=0.25\bullet 40$$
$$x=10$$

The number 10 is 25% of 40.

49. Translating into an equation and solving:

$$x=0.12\bullet 2{,}000$$
$$x=240$$

The number 240 is 12% of 2,000.

51. Translating into an equation and solving:

$$x\bullet 28=7$$
$$28x=7$$
$$\frac{1}{28}(28x)=\frac{1}{28}(7)$$
$$x=0.25=25\%$$

The number 7 is 25% of 28.

53. Translating into an equation and solving:

$$x\bullet 40=14$$
$$40x=14$$
$$\frac{1}{40}(40x)=\frac{1}{40}(14)$$
$$x=0.35=35\%$$

The number 14 is 35% of 40.

55. Translating into an equation and solving:

$$0.50\bullet x=32$$
$$\frac{0.50x}{0.50}=\frac{32}{0.50}$$
$$x=64$$

The number 32 is 50% of 64.

57. Translating into an equation and solving:

$$0.12\bullet x=240$$
$$\frac{0.12x}{0.12}=\frac{240}{0.12}$$
$$x=2{,}000$$

The number 240 is 12% of 2,000.

59. Evaluating when A = 10,000: $T=-0.0035(10{,}000)+70=35$. The temperature is 35°F at an altitude of 10,000 feet.

61. Solving when T = 56:

$$56=-0.0035A+70$$
$$-0.0035A=-14$$
$$A=\frac{-14}{-0.0035}$$
$$A=4{,}000$$

63. Solving for A:

$$T=-0.0035A+70$$
$$-0.0035A=T-70$$
$$A=\frac{T-70}{-0.0035}$$
$$A=\frac{70-T}{0.0035}$$

65. Substituting $F=212$: $C=\frac{5}{9}(212-32)=\frac{5}{9}(180)=100°\text{C}$. This value agrees with the information in Table 1.

67. Substituting $F=68$: $C=\frac{5}{9}(68-32)=\frac{5}{9}(36)=20°\text{C}$. This value agrees with the information in Table 1.

69. Solving for C:

$$\begin{aligned}\frac{9}{5}C+32&=F\\ \frac{9}{5}C+32-32&=F-32\\ \frac{9}{5}C&=F-32\\ \frac{5}{9}\left(\frac{9}{5}C\right)&=\frac{5}{9}(F-32)\\ C&=\frac{5}{9}(F-32)\end{aligned}$$

71. Budd's estimate would be: $F=2(30)+30=60+30=90°\text{F}$

The actual conversion would be: $F=\frac{9}{5}(30)+32=54+32=86°\text{F}$

Budd's estimate is 4°F too high.

73. Solving for C: $C=2\left(\frac{22}{7}\right)(7)=44$ meters

75. Solving for r:

$$\begin{aligned}2\pi r&=C\\ 2\cdot 3.14r&=9.42\\ 6.28r&=9.42\\ \frac{6.28r}{6.28}&=\frac{9.42}{6.28}\\ r&=\frac{3}{2}=1.5\text{ inches}\end{aligned}$$

77. Solving for V: $V=3.14(3)^2(2)=56.52\text{ cm}^3$

79. Solving for h:

$$\begin{aligned}\pi r^2h&=V\\ \frac{22}{7}\left(\frac{7}{22}\right)^2h&=42\\ \frac{7}{22}h&=42\\ \frac{22}{7}\left(\frac{7}{22}h\right)&=\frac{22}{7}(42)\\ h&=132\text{ feet}\end{aligned}$$

81. We need to find what percent of 150 is 90:

$$\begin{aligned}x\cdot 150&=90\\ \frac{1}{150}(150x)&=\frac{1}{150}(90)\\ x&=0.60=60\%\end{aligned}$$

So 60% of the calories in one serving of vanilla ice cream are fat calories.

83. We need to find what percent of 98 is 26:

$$\begin{aligned}x\cdot 98&=26\\ \frac{1}{98}(98x)&=\frac{1}{98}(26)\\ x&\approx 0.265=26.5\%\end{aligned}$$

So 26.5% of one serving of frozen yogurt are carbohydrates.

85. The sum of 4 and 1.

87. The difference of 6 and 2.

89. The difference of a number and 15.

91. Four times the difference of a number and 3.

93. An equivalent expression is: $2(6+3)$

95. An equivalent expression is: $2(5)+3$

97. An equivalent expression is: $x+5$

99. An equivalent expression is: $5(x+7)$

101. Solving for y:

$$\begin{aligned} 2x-3y&=6 \\ -3y&=-2x+6 \\ y&=\frac{-2x+6}{-3} \\ y&=\frac{2}{3}x-2 \end{aligned}$$

The correct answer is a.

2.6 Applications

1. Translating into symbols: $x+5=14$

3. Translating into symbols: $\frac{x}{3}=x+2$

5. Let x represent the number. Translating into symbols: $2(x-9)+5=11$

7. Let x represent the number. Translating into symbols: $\frac{1}{2}(x+5)=3(x-5)$

9. Let x represent the number. The equation is:

$$\begin{aligned} x+5&=13 \\ x&=8 \end{aligned}$$

The number is 8.

11. Let x represent the number. The equation is:

$$\begin{aligned} 2x+4&=14 \\ 2x&=10 \\ x&=5 \end{aligned}$$

The number is 5.

13. Let x represent the number. The equation is:

$$\begin{aligned} 5(x+7)&=30 \\ 5x+35&=30 \\ 5x&=-5 \\ x&=-1 \end{aligned}$$

The number is –1.

15. Let x and $x+2$ represent the two numbers. The equation is:

$$\begin{aligned} x+x+2&=8 \\ 2x+2&=8 \\ 2x&=6 \\ x&=3 \\ x+2&=5 \end{aligned}$$

The two numbers are 3 and 5.

17. Let x and $3x-4$ represent the two numbers. The equation is:

$$\begin{aligned} (x+3x-4)+5&=25 \\ 4x+1&=25 \\ 4x&=24 \\ x&=6 \\ 3x-4=3(6)-4&=14 \end{aligned}$$

The two numbers are 6 and 14.

19. Completing the table:

	Four Years Ago	Now
Shelly	$x+3-4=x-1$	$x+3$
Michele	$x-4$	x

The equation is:

$$\begin{aligned} x-1+x-4&=67 \\ 2x-5&=67 \\ 2x&=72 \\ x&=36 \\ x+3&=39 \end{aligned}$$

Shelly is 39 and Michele is 36.

21. Completing the table:

	Three Years Ago	Now
Cody	$2x-3$	$2x$
Evan	$x-3$	x

The equation is:

$$\begin{aligned} 2x-3+x-3&=27 \\ 3x-6&=27 \\ 3x&=33 \\ x&=11 \\ 2x&=22 \end{aligned}$$

Evan is 11 and Cody is 22.

23. Completing the table:

	Five Years Ago	Now
Fred	$x+4-5=x-1$	$x+4$
Barney	$x-5$	x

The equation is:

$$\begin{aligned} x-1+x-5&=48 \\ 2x-6&=48 \\ 2x&=54 \\ x&=27 \\ x+4&=31 \end{aligned}$$

Barney is 27 and Fred is 31.

25. Completing the table:

	Now	Three Years from Now
Jack	$2x$	$2x+3$
Lacy	x	$x+3$

The equation is:

$$\begin{aligned} 2x+3+x+3&=54 \\ 3x+6&=54 \\ 3x&=48 \\ x&=16 \\ 2x&=32 \end{aligned}$$

Lacy is 16 and Jack is 32.

27. Completing the table:

	Now	Two Years from Now
Pat	$x+20$	$x+20+2=x+22$
Patrick	x	$x+2$

The equation is:

$$\begin{aligned} x+22&=2(x+2) \\ x+22&=2x+4 \\ 22&=x+4 \\ x&=18 \\ x+20&=38 \end{aligned}$$

Patrick is 18 and Pat is 38.

29. Using the formula $P=4s$, the equation is:

$$\begin{aligned} 4s&=36 \\ s&=9 \end{aligned}$$

The length of each side is 9 inches.

31. Using the formula $P=4s$, the equation is:

$$\begin{aligned} 4s&=60 \\ s&=15 \end{aligned}$$

The length of each side is 15 feet.

33. Let x, $3x$, and $x + 7$ represent the sides of the triangle. The equation is:

$$\begin{aligned} x+3x+x+7&=62 \\ 5x+7&=62 \\ 5x&=55 \\ x&=11 \\ x+7&=18 \\ 3x&=33 \end{aligned}$$

The sides are 11 feet, 18 feet, and 33 feet.

35. Let x, $2x$, and $2x - 12$ represent the sides of the triangle. The equation is:

$$\begin{aligned} x+2x+2x-12&=53 \\ 5x-12&=53 \\ 5x&=65 \\ x&=13 \\ 2x&=26 \\ 2x-12&=14 \end{aligned}$$

The sides are 13 feet, 14 feet, and 26 feet.

37. Let w represent the width and $w + 5$ represent the length. The equation is:

$$\begin{aligned} 2w+2(w+5)&=34 \\ 2w+2w+10&=34 \\ 4w+10&=34 \\ 4w&=24 \\ w&=6 \\ w+5&=11 \end{aligned}$$

The length is 11 inches and the width is 6 inches.

39. Let w represent the width and $2w + 7$ represent the length. The equation is:

$$\begin{aligned} 2w+2(2w+7)&=68 \\ 2w+4w+14&=68 \\ 6w+14&=68 \\ 6w&=54 \\ w&=9 \\ 2w+7&=2(9)+7=25 \end{aligned}$$

The length is 25 meters and the width is 9 meters.

41. Let w represent the width and $3w + 6$ represent the length. The equation is:

$$\begin{aligned} 2w+2(3w+6)&=36 \\ 2w+6w+12&=36 \\ 8w+12&=36 \\ 8w&=24 \\ w&=3 \\ 3w+6&=3(3)+6=15 \end{aligned}$$

The length is 15 feet and the width is 3 feet.

43. Completing the table:

	Dimes	Quarters
Number	x	$x+5$
Value (cents)	$10(x)$	$25(x+5)$

The equation is:

$$\begin{aligned} 10(x)+25(x+5)&=440 \\ 10x+25x+125&=440 \\ 35x+125&=440 \\ 35x&=315 \\ x&=9 \\ x+5&=14 \end{aligned}$$

Marissa has 9 dimes and 14 quarters.

45. Completing the table:

	Nickels	Quarters
Number	$x+15$	x
Value (cents)	$5(x+15)$	$25(x)$

The equation is:

$$\begin{aligned} 5(x+15)+25(x)&=435 \\ 5x+75+25x&=435 \\ 30x+75&=435 \\ 30x&=360 \\ x&=12 \\ x+15&=27 \end{aligned}$$

Tanner has 12 quarters and 27 nickels.

47. Completing the table:

	Nickels	Dimes
Number	x	$x+9$
Value (cents)	$5(x)$	$10(x+9)$

The equation is:

$$\begin{aligned} 5(x)+10(x+9)&=210 \\ 5x+10x+90&=210 \\ 15x+90&=210 \\ 15x&=120 \\ x&=8 \\ x+9&=17 \end{aligned}$$

Sue has 8 nickels and 17 dimes.

49. Completing the table:

	Nickels	Dimes	Quarters
Number	x	$x+3$	$x+5$
Value (cents)	$5(x)$	$10(x+3)$	$25(x+5)$

The equation is:

$$\begin{aligned} 5(x)+10(x+3)+25(x+5)&=435 \\ 5x+10x+30+25x+125&=435 \\ 40x+155&=435 \\ 40x&=280 \\ x&=7 \\ x+3&=10 \\ x+5&=12 \end{aligned}$$

Katie has 7 nickels, 10 dimes, and 12 quarters.

51. Completing the table:

	Nickels	Dimes	Quarters
Number	x	$x+6$	$2x$
Value (cents)	$5(x)$	$10(x+6)$	$25(2x)$

The equation is:

$$\begin{aligned} 5(x)+10(x+6)+25(2x)&=255 \\ 5x+10x+60+50x&=255 \\ 65x+60&=255 \\ 65x&=195 \\ x&=3 \\ x+6&=9 \\ 2x&=6 \end{aligned}$$

Corey has 3 nickels, 9 dimes, and 6 quarters.

53. Simplifying: $x+2x+2x=5x$

55. Simplifying: $x+0.075x=1.075x$

57. Simplifying: $0.09(x+2,000)=0.09x+180$

59. Solving the equation:

$$\begin{aligned} 0.05x+0.06(x-1,500) &= 570 \\ 0.05x+0.06x-90 &= 570 \\ 0.11x &= 660 \\ x &= 6,000 \end{aligned}$$

61. Solving the equation:

$$\begin{aligned} x+2x+3x &= 180 \\ 6x &= 180 \\ x &= 30 \end{aligned}$$

63. Let x and $x + 3$ represent the two numbers. The equation is: $x+(x+3)=20$. The correct answer is c.

65. Let x represent the width and $3x - 2$ represent the length. The equation is: $2(3x-2)+2x=36$. The correct answer is d.

2.7 More Applications

1. Let x and $x + 1$ represent the two numbers. The equation is:

$$\begin{aligned} x+x+1 &= 11 \\ 2x+1 &= 11 \\ 2x &= 10 \\ x &= 5 \\ x+1 &= 6 \end{aligned}$$

The numbers are 5 and 6.

3. Let x and $x + 1$ represent the two numbers. The equation is:

$$\begin{aligned} x+x+1 &= -9 \\ 2x+1 &= -9 \\ 2x &= -10 \\ x &= -5 \\ x+1 &= -4 \end{aligned}$$

The numbers are –5 and –4.

5. Let x and $x + 2$ represent the two numbers. The equation is:

$$\begin{aligned} x+x+2 &= 28 \\ 2x+2 &= 28 \\ 2x &= 26 \\ x &= 13 \\ x+2 &= 15 \end{aligned}$$

The numbers are 13 and 15.

7. Let x and $x + 2$ represent the two numbers. The equation is:

$$\begin{aligned} x+x+2 &= 106 \\ 2x+2 &= 106 \\ 2x &= 104 \\ x &= 52 \\ x+2 &= 54 \end{aligned}$$

The numbers are 52 and 54.

9. Let x and $x + 2$ represent the two numbers. The equation is:

$$\begin{aligned} x+x+2 &= -30 \\ 2x+2 &= -30 \\ 2x &= -322 \\ x &= -16 \\ x+2 &= -14 \end{aligned}$$

The numbers are –16 and –14.

11. Let x, $x+2$, and $x+4$ represent the three numbers. The equation is:

$$\begin{aligned} x+x+2+x+4&=57 \\ 3x+6&=57 \\ 3x&=51 \\ x&=17 \\ x+2&=19 \\ x+4&=21 \end{aligned}$$

The numbers are 17, 19, and 21.

13. Let x, $x+2$, and $x+4$ represent the three numbers. The equation is:

$$\begin{aligned} x+x+2+x+4&=132 \\ 3x+6&=132 \\ 3x&=126 \\ x&=42 \\ x+2&=44 \\ x+4&=46 \end{aligned}$$

The numbers are 42, 44, and 46.

15. Completing the table:

	Dollars Invested at 8%	Dollars Invested at 9%
Number of	x	$x+2{,}000$
Interest on	$0.08(x)$	$0.09(x+2{,}000)$

The equation is:

$$\begin{aligned} 0.08(x)+0.09(x+2{,}000)&=860 \\ 0.08x+0.09x+180&=860 \\ 0.17x+180&=860 \\ 0.17x&=680 \\ x&=4{,}000 \\ x+2{,}000&=6{,}000 \end{aligned}$$

You have \$4,000 invested at 8% and \$6,000 invested at 9%.

17. Completing the table:

	Dollars Invested at 10%	Dollars Invested at 12%
Number of	x	$x+500$
Interest on	$0.10(x)$	$0.12(x+500)$

The equation is:

$$\begin{aligned} 0.10(x)+0.12(x+500)&=214 \\ 0.10x+0.12x+60&=214 \\ 0.22x+60&=214 \\ 0.22x&=154 \\ x&=700 \\ x+500&=1{,}200 \end{aligned}$$

Tyler has \$700 invested at 10% and \$1,200 invested at 12%.

19. Completing the table:

	Dollars Invested at 8%	Dollars Invested at 9%	Dollars Invested at 10%
Number of	x	$2x$	$3x$
Interest on	$0.08(x)$	$0.09(2x)$	$0.10(3x)$

The equation is:

$$\begin{aligned} 0.08(x)+0.09(2x)+0.10(3x) &= 280 \\ 0.08x+0.18x+0.30x &= 280 \\ 0.56x &= 280 \\ x &= 500 \\ 2x &= 1,000 \\ 3x &= 1,500 \end{aligned}$$

She has \$500 invested at 8%, \$1,000 invested at 9%, and \$1,500 invested at 10%.

21. Completing the table:

	50% Solution	10% Solution	Mixture
Number of Liters	x	$24-x$	24
Liters of Alcohol	$0.50(x)$	$0.10(24-x)$	$0.30(24)$

The equation is:

$$\begin{aligned} 0.50(x)+0.10(24-x) &= 0.30(24) \\ 0.50x+2.4-0.10x &= 7.2 \\ 0.40x+2.4 &= 7.2 \\ 0.40x &= 4.8 \\ x &= 12 \\ 24-x &= 12 \end{aligned}$$

The mixture contains 12 liters of 50% alcohol solution and 12 liters of 10% alcohol solution.

23. Completing the table:

	10% Solution	5% Solution	Mixture
Number of Gallons	x	$25-x$	25
Gallons of Disinfectant	$0.10(x)$	$0.05(25-x)$	$0.08(25)$

The equation is:

$$\begin{aligned} 0.10(x)+0.05(25-x) &= 0.08(25) \\ 0.10x+1.25-0.05x &= 2 \\ 0.05x+1.25 &= 2 \\ 0.05x &= 0.75 \\ x &= 15 \\ 25-x &= 10 \end{aligned}$$

The mixture contains 15 gallons of 10% disinfectant solution and 10 gallons of 5% disinfectant solution.

25. Completing the table:

	\$9.50 per pound	\$12.00 per pound	Mixture
Number of Pounds	x	$40-x$	40
Cost of Coffee	$9.50(x)$	$12.00(40-x)$	$10.00(40)$

The equation is:

$$\begin{aligned} 9.50(x)+12(40-x)&=10(40) \\ 9.5x+480-12x&=400 \\ -2.5x+480&=400 \\ -2.5x&=-80 \\ x&=32 \\ 40-x&=8 \end{aligned}$$

The mixture contains 32 pounds of \$9.50 coffee beans and 8 pounds of \$12.00 coffee beans.

27. Let x represent the measure of the two equal angles, so $x+x=2x$ represents the measure of the third angle. Since the sum of the three angles is 180°, the equation is:

$$\begin{aligned} x+x+2x&=180° \\ 4x&=180° \\ x&=45° \\ 2x&=90° \end{aligned}$$

The measures of the three angles are 45°, 45°, and 90°.

29. Let x represent the measure of the largest angle. Then $\frac{1}{5}x$ represents the measure of the smallest angle, and $2\left(\frac{1}{5}x\right)=\frac{2}{5}x$ represents the measure of the other angle. Since the sum of the three angles is 180°, the equation is:

$$\begin{aligned} x+\frac{1}{5}x+\frac{2}{5}x&=180° \\ \frac{5}{5}x+\frac{1}{5}x+\frac{2}{5}x&=180° \\ \frac{8}{5}x&=180° \\ x&=112.5° \\ \frac{1}{5}x&=22.5° \\ \frac{2}{5}x&=45° \end{aligned}$$

The measures of the three angles are 22.5°, 45°, and 112.5°.

31. Let x represent the measure of the smallest angle, so $x + 20$ represents the measure of the second angle and $2x$ represents the measure of the third angle. Since the sum of the three angles is 180°, the equation is:

$$\begin{aligned} x+x+20+2x&=180° \\ 4x+20&=180° \\ 4x&=160° \\ x&=40° \\ x+20&=60° \\ 2x&=80° \end{aligned}$$

The measures of the three angles are 40°, 60°, and 80°.

33. Completing the table:

	Adult	Child
Number	x	$x+6$
Income	$6(x)$	$4(x+6)$

The equation is:

$$\begin{aligned} 6(x)+4(x+6) &= 184 \\ 6x+4x+24 &= 184 \\ 10x+24 &= 184 \\ 10x &= 160 \\ x &= 16 \\ x+6 &= 22 \end{aligned}$$

Miguel sold 16 adult and 22 children's tickets.

35. Let x represent the total minutes for the call. Then \$0.41 is charged for the first minute, and \$0.32 is charged for the additional $x-1$ minutes. The equation is:

$$\begin{aligned} 0.41(1)+0.32(x-1) &= 5.21 \\ 0.41+0.32x-0.32 &= 5.21 \\ 0.32x+0.09 &= 5.21 \\ 0.32x &= 5.12 \\ x &= 16 \end{aligned}$$

The call was 16 minutes long.

37. Let x represent the hours JoAnn worked that week. Then \$12/hour is paid for the first 35 hours and \$18/hour is paid for the additional $x-35$ hours. The equation is:

$$\begin{aligned} 12(35)+18(x-35) &= 492 \\ 420+18x-630 &= 492 \\ 18x-210 &= 492 \\ 18x &= 702 \\ x &= 39 \end{aligned}$$

JoAnn worked 39 hours that week.

39. Let x and $x+2$ represent the two office numbers. The equation is:

$$\begin{aligned} x+x+2 &= 14{,}660 \\ 2x+2 &= 14{,}660 \\ 2x &= 14{,}658 \\ x &= 7329 \\ x+2 &= 7331 \end{aligned}$$

They are in offices 7329 and 7331.

41. Let x represent Kendra's age and $x+2$ represent Marissa's age. The equation is:

$$\begin{aligned} x+2+2x &= 26 \\ 3x+2 &= 26 \\ 3x &= 24 \\ x &= 8 \\ x+2 &= 10 \end{aligned}$$

Kendra is 8 years old and Marissa is 10 years old.

43. For Jeff, the total time traveled is $\dfrac{425 \text{ miles}}{55 \text{ miles/hour}} \approx 7.72 \text{ hours} \approx 463 \text{ minutes}$. Since he left at 11:00 AM, he will arrive at 6:43 PM. For Carla, the total time traveled is $\dfrac{425 \text{ miles}}{65 \text{ miles/hour}} \approx 6.54 \text{ hours} \approx 392 \text{ minutes}$. Since she left at 1:00 PM, she will arrive at 7:32 PM. Thus Jeff will arrive in Lake Tahoe first.

45. Since $\frac{1}{5}$ mile $= 0.2$ mile, the taxi charge is \$1.25 for the first $\frac{1}{5}$ mile and \$0.25 per fifth mile for the remaining $7.5 - 0.2 = 7.3$ miles. Since $7.3 \text{ miles} = \frac{7.3}{0.2} = 36.5$ fifths, the total charge is: $\$1.25 + \$0.25(36.5) \approx \$10.38$

47. Let w represent the width and $w + 2$ represent the length. The equation is:

$$\begin{aligned} 2w + 2(w+2) &= 44 \\ 2w + 2w + 4 &= 44 \\ 4w + 4 &= 44 \\ 4w &= 40 \\ w &= 10 \\ w + 2 &= 12 \end{aligned}$$

The length is 12 meters and the width is 10 meters.

49. Let x, $x + 1$, and $x + 2$ represent the measures of the three angles. Since the sum of the three angles is 180°, the equation is:

$$\begin{aligned} x + x + 1 + x + 2 &= 180° \\ 3x + 3 &= 180° \\ 3x &= 177° \\ x &= 59° \\ x + 1 &= 60° \\ x + 2 &= 61° \end{aligned}$$

The measures of the three angles are 59°, 60°, and 61°.

51. If all 36 people are Elk's Lodge members (which would be the least amount), the cost of the lessons would be $\$3(36) = \$108$. Since half of the money is paid to Ike and Nancy, the least amount they could make is $\frac{1}{2}(\$108) = \$54$.

53. Yes. The total receipts were \$160, which is possible if there were 10 Elk's members and 26 nonmembers. Computing the total receipts: $10(\$3) + 26(\$5) = \$30 + \$130 = \$160$

55.

a. Solving the equation:

$$\begin{aligned} x - 3 &= 6 \\ x &= 9 \end{aligned}$$

b. Solving the equation:

$$\begin{aligned} x + 3 &= 6 \\ x &= 3 \end{aligned}$$

c. Solving the equation:

$$\begin{aligned} -x - 3 &= 6 \\ -x &= 9 \\ x &= -9 \end{aligned}$$

d. Solving the equation:

$$\begin{aligned} -x + 3 &= 6 \\ -x &= 3 \\ x &= -3 \end{aligned}$$

57.

a. Solving the equation:

$$\begin{aligned} \frac{x}{4} &= -2 \\ x &= -2(4) = -8 \end{aligned}$$

b. Solving the equation:

$$\begin{aligned} -\frac{x}{4} &= -2 \\ x &= -2(-4) = 8 \end{aligned}$$

c. Solving the equation:

$$\begin{aligned} \frac{x}{4} &= 2 \\ x &= 2(4) = 8 \end{aligned}$$

d. Solving the equation:

$$\begin{aligned} -\frac{x}{4} &= 2 \\ x &= 2(-4) = -8 \end{aligned}$$

59. Solving the equation:

$$2.5x - 3.48 = 4.9x + 2.07$$
$$-2.4x - 3.48 = 2.07$$
$$-2.4x = 5.55$$
$$x = -2.3125$$

61. Solving the equation:

$$3(x-4) = -2$$
$$3x - 12 = -2$$
$$3x = 10$$
$$x = \frac{10}{3}$$

63. Let x and $x + 2$ represent the two numbers. The equation is: $x+(x+2)=272$. The correct answer is a.

65. Completing the table:

	25% Solution	60% Solution	Mixture
Number of Gallons	x	$40-x$	40
Gallons of Alcohol	$0.25(x)$	$0.60(40-x)$	$0.35(40)$

The equation is: $0.25x + 0.60(40-x) = 0.35(40)$. The correct answer is a.

2.8 Linear Inequalities

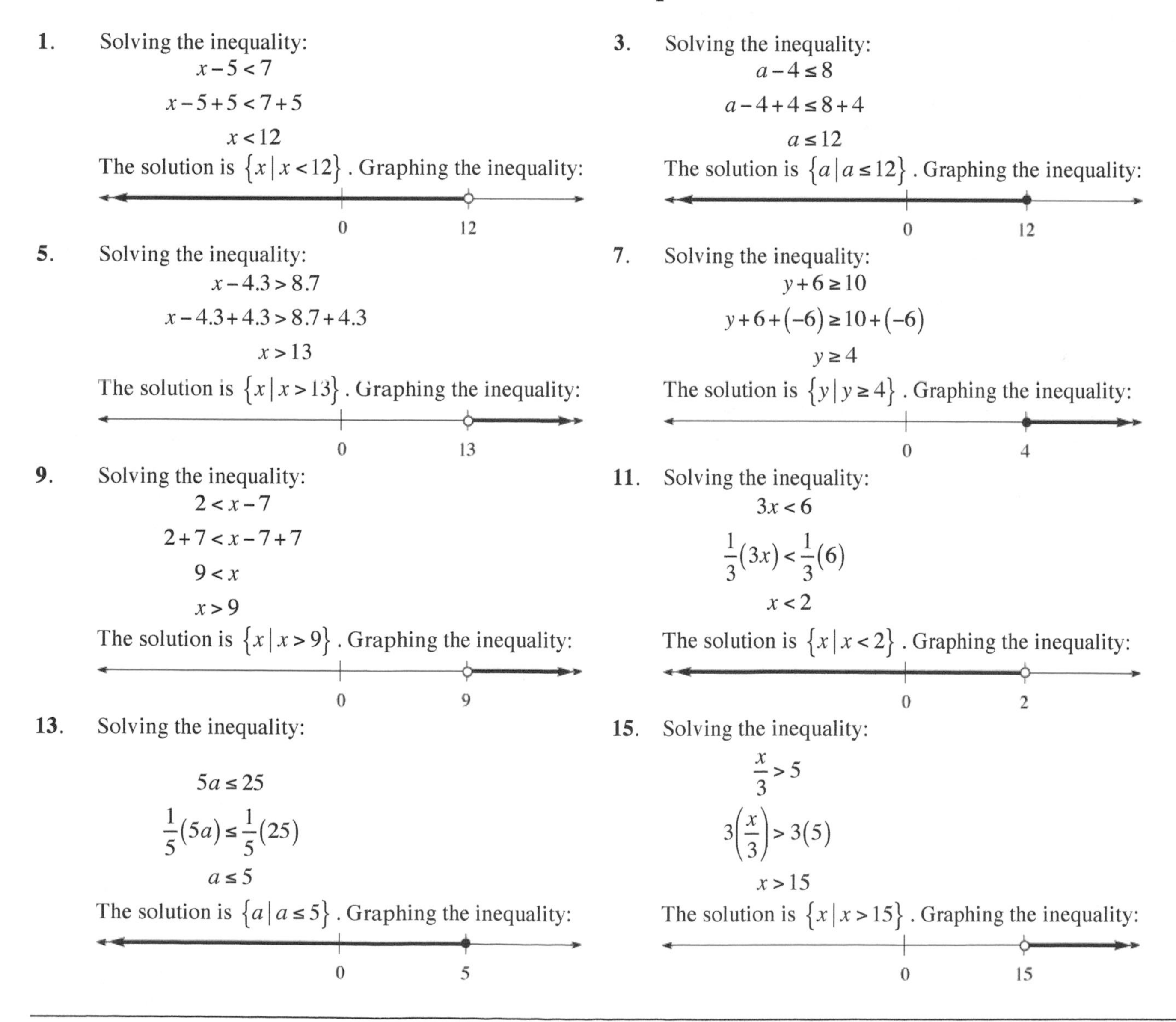

1. Solving the inequality:

$$x - 5 < 7$$
$$x - 5 + 5 < 7 + 5$$
$$x < 12$$

The solution is $\{x \mid x < 12\}$. Graphing the inequality:

3. Solving the inequality:

$$a - 4 \le 8$$
$$a - 4 + 4 \le 8 + 4$$
$$a \le 12$$

The solution is $\{a \mid a \le 12\}$. Graphing the inequality:

5. Solving the inequality:

$$x - 4.3 > 8.7$$
$$x - 4.3 + 4.3 > 8.7 + 4.3$$
$$x > 13$$

The solution is $\{x \mid x > 13\}$. Graphing the inequality:

7. Solving the inequality:

$$y + 6 \ge 10$$
$$y + 6 + (-6) \ge 10 + (-6)$$
$$y \ge 4$$

The solution is $\{y \mid y \ge 4\}$. Graphing the inequality:

9. Solving the inequality:

$$2 < x - 7$$
$$2 + 7 < x - 7 + 7$$
$$9 < x$$
$$x > 9$$

The solution is $\{x \mid x > 9\}$. Graphing the inequality:

11. Solving the inequality:

$$3x < 6$$
$$\frac{1}{3}(3x) < \frac{1}{3}(6)$$
$$x < 2$$

The solution is $\{x \mid x < 2\}$. Graphing the inequality:

13. Solving the inequality:

$$5a \le 25$$
$$\frac{1}{5}(5a) \le \frac{1}{5}(25)$$
$$a \le 5$$

The solution is $\{a \mid a \le 5\}$. Graphing the inequality:

15. Solving the inequality:

$$\frac{x}{3} > 5$$
$$3\left(\frac{x}{3}\right) > 3(5)$$
$$x > 15$$

The solution is $\{x \mid x > 15\}$. Graphing the inequality:

17. Solving the inequality:

$$-2x > 6$$
$$-\frac{1}{2}(-2x) < -\frac{1}{2}(6)$$
$$x < -3$$

The solution is $\{x \mid x < -3\}$. Graphing the inequality:

-3 0

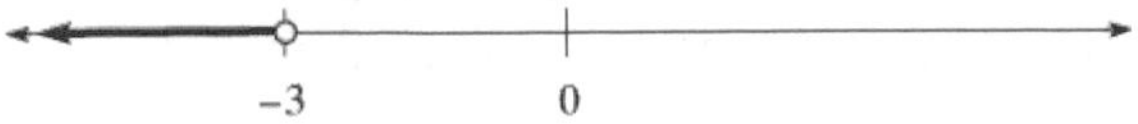

19. Solving the inequality:

$$-3x \geq -18$$
$$-\frac{1}{3}(-3x) \leq -\frac{1}{3}(-18)$$
$$x \leq 6$$

The solution is $\{x \mid x \leq 6\}$. Graphing the inequality:

0 6

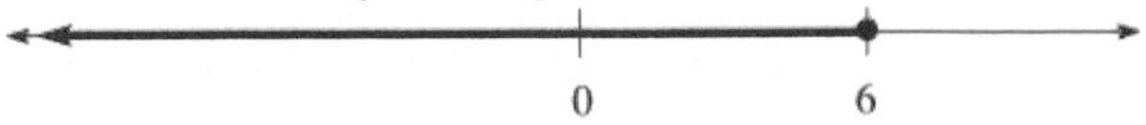

21. Solving the inequality:

$$-\frac{x}{5} \leq 10$$
$$-5\left(-\frac{x}{5}\right) \geq -5(10)$$
$$x \geq -50$$

The solution is $\{x \mid x \geq -50\}$. Graphing the inequality:

-50 0

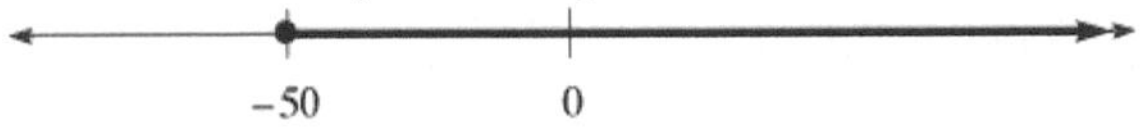

23. Solving the inequality:

$$-\frac{2}{3}y > 4$$
$$-\frac{3}{2}\left(-\frac{2}{3}y\right) < -\frac{3}{2}(4)$$
$$y < -6$$

The solution is $\{y \mid y < -6\}$. Graphing the inequality:

-6 0

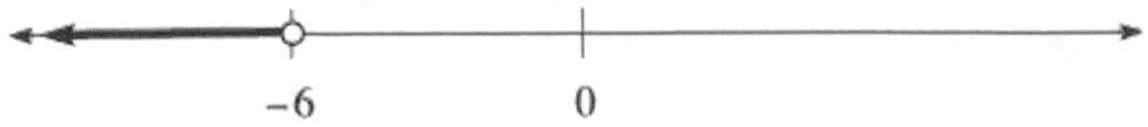

25. Solving the inequality:

$$2x - 3 < 9$$
$$2x - 3 + 3 < 9 + 3$$
$$2x < 12$$
$$\frac{1}{2}(2x) < \frac{1}{2}(12)$$
$$x < 6$$

The solution is $\{x \mid x < 6\}$. Graphing the inequality:

0 6

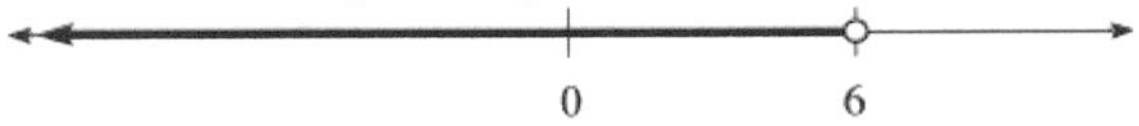

27. Solving the inequality:

$$-\frac{1}{5}y - \frac{1}{3} \leq \frac{2}{3}$$
$$-\frac{1}{5}y - \frac{1}{3} + \frac{1}{3} \leq \frac{2}{3} + \frac{1}{3}$$
$$-\frac{1}{5}y \leq 1$$
$$-5\left(-\frac{1}{5}y\right) \geq -5(1)$$
$$y \geq -5$$

The solution is $\{y \mid y \geq -5\}$. Graphing the inequality:

-5 0

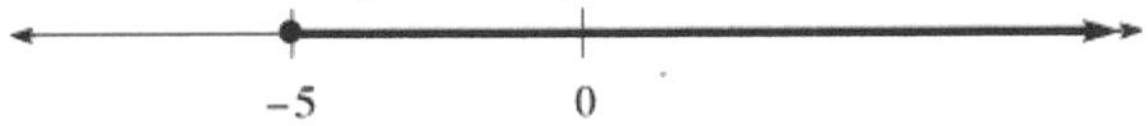

29. Solving the inequality:

$$-7.2x + 1.8 > -19.8$$
$$-7.2x + 1.8 - 1.8 > -19.8 - 1.8$$
$$-7.2x > -21.6$$
$$\frac{-7.2x}{-7.2} < \frac{-21.6}{-7.2}$$
$$x < 3$$

The solution is $\{x \mid x < 3\}$. Graphing the inequality:

0 3

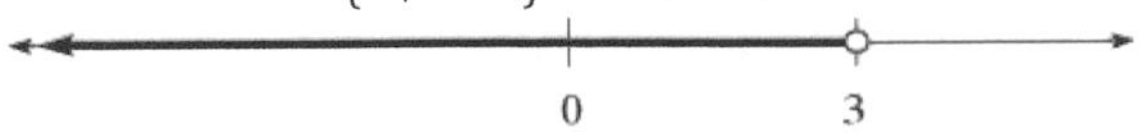

31. Solving the inequality:

$$\frac{2}{3}x - 5 \leq 7$$
$$\frac{2}{3}x - 5 + 5 \leq 7 + 5$$
$$\frac{2}{3}x \leq 12$$
$$\frac{3}{2}\left(\frac{2}{3}x\right) \leq \frac{3}{2}(12)$$
$$x \leq 18$$

The solution is $\{x \mid x \leq 18\}$. Graphing the inequality:

0 18

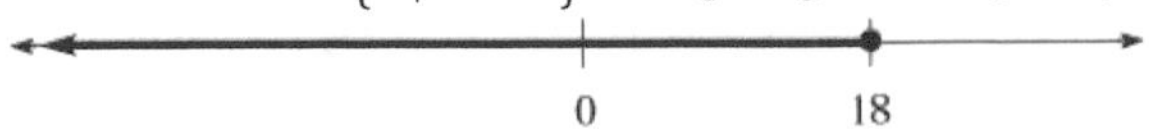

33. Solving the inequality:

$$-\frac{2}{5}a-3>5$$
$$-\frac{2}{5}a-3+3>5+3$$
$$-\frac{2}{5}a>8$$
$$-\frac{5}{2}\left(-\frac{2}{5}a\right)<-\frac{5}{2}(8)$$
$$a<-20$$

The solution is $\{a \mid a<-20\}$. Graphing the inequality:

-20 0

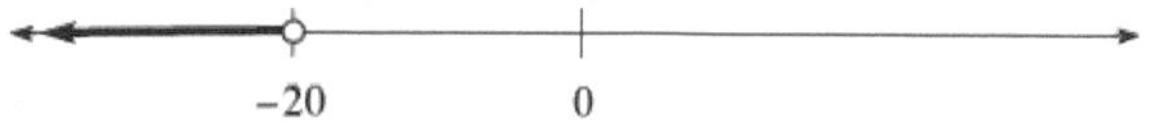

35. Solving the inequality:

$$5-\frac{3}{5}y>-10$$
$$-5+5-\frac{3}{5}y>-5+(-10)$$
$$-\frac{3}{5}y>-15$$
$$-\frac{5}{3}\left(-\frac{3}{5}y\right)<-\frac{5}{3}(-15)$$
$$y<25$$

The solution is $\{y \mid y<25\}$. Graphing the inequality:

0 25

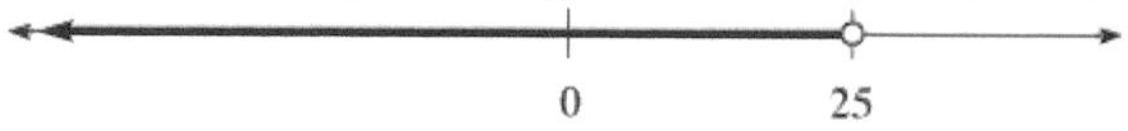

37. Solving the inequality:

$$0.3(a+1)\le 1.2$$
$$0.3a+0.3\le 1.2$$
$$0.3a+0.3+(-0.3)\le 1.2+(-0.3)$$
$$0.3a\le 0.9$$
$$\frac{0.3a}{0.3}\le\frac{0.9}{0.3}$$
$$a\le 3$$

The solution is $\{a \mid a\le 3\}$. Graphing the inequality:

0 3

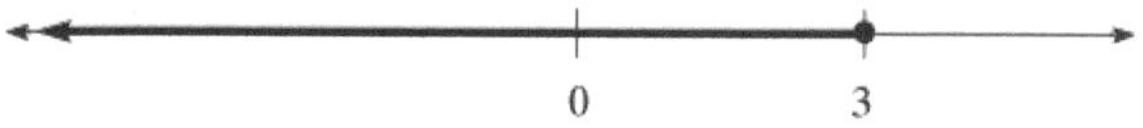

39. Solving the inequality:

$$2(5-2x)\le -20$$
$$10-4x\le -20$$
$$-10+10-4x\le -10+(-20)$$
$$-4x\le -30$$
$$-\frac{1}{4}(-4x)\ge -\frac{1}{4}(-30)$$
$$x\ge\frac{15}{2}$$

The solution is $\left\{x \mid x\ge\frac{15}{2}\right\}$. Graphing the inequality:

0 15/2

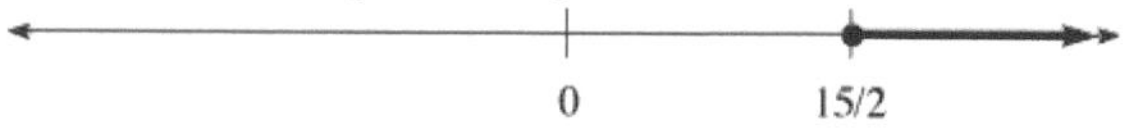

41. Solving the inequality:

$$3x-5>8x$$
$$-3x+3x-5>-3x+8x$$
$$-5>5x$$
$$\frac{1}{5}(-5)>\frac{1}{5}(5x)$$
$$-1>x$$
$$x<-1$$

The solution is $\{x \mid x<-1\}$. Graphing the inequality:

-1 0

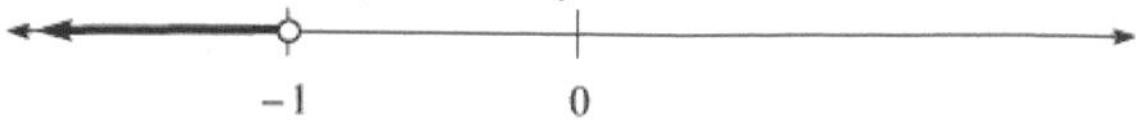

43. First multiply by 6 to clear the inequality of fractions:

$$\frac{1}{3}y-\frac{1}{2}\le\frac{5}{6}y+\frac{1}{2}$$
$$6\left(\frac{1}{3}y-\frac{1}{2}\right)\le 6\left(\frac{5}{6}y+\frac{1}{2}\right)$$
$$2y-3\le 5y+3$$
$$-5y+2y-3\le -5y+5y+3$$
$$-3y-3\le 3$$
$$-3y-3+3\le 3+3$$
$$-3y\le 6$$
$$-\frac{1}{3}(-3y)\ge -\frac{1}{3}(6)$$
$$y\ge -2$$

The solution is $\{y \mid y\ge -2\}$. Graphing the inequality:

-2 0

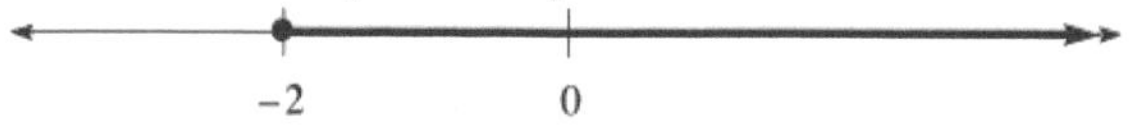

45. Solving the inequality:

$$-2.8x+8.4<-14x-2.8$$
$$-2.8x+14x+8.4<-14x+14x-2.8$$
$$11.2x+8.4<-2.8$$
$$11.2x+8.4-8.4<-2.8-8.4$$
$$11.2x<-11.2$$
$$\frac{11.2x}{11.2}<\frac{-11.2}{11.2}$$
$$x<-1$$

The solution is $\{x \mid x<-1\}$. Graphing the inequality:

-1 0

47. Solving the inequality:

$$3(m-2)-4\ge 7m+14$$
$$3m-6-4\ge 7m+14$$
$$3m-10\ge 7m+14$$
$$-7m+3m-10\ge -7m+7m+14$$
$$-4m-10\ge 14$$
$$-4m-10+10\ge 14+10$$
$$-4m\ge 24$$
$$-\frac{1}{4}(-4m)\le -\frac{1}{4}(24)$$
$$m\le -6$$

The solution is $\{m \mid m\le -6\}$. Graphing the inequality:

-6 0

49. Solving the inequality:

$$3-4(x-2)\le -5x+6$$
$$3-4x+8\le -5x+6$$
$$-4x+11\le -5x+6$$
$$-4x+5x+11\le -5x+5x+6$$
$$x+11\le 6$$
$$x+11+(-11)\le 6+(-11)$$
$$x\le -5$$

The solution is $\{x \mid x\le -5\}$. Graphing the inequality:

-5 0

51. Solving for y:

$$3x+2y<6$$
$$2y<-3x+6$$
$$y<-\frac{3}{2}x+3$$

53. Solving for y:

$$2x-5y>10$$
$$-5y>-2x+10$$
$$y<\frac{2}{5}x-2$$

55. Solving for y:

$$-3x+7y\le 21$$
$$7y\le 3x+21$$
$$y\le\frac{3}{7}x+3$$

57. Solving for y:

$$2x-4y\ge -4$$
$$-4y\ge -2x-4$$
$$y\le\frac{1}{2}x+1$$

59. **a.** Evaluating when $x = 0$:

$$-5x+3=-5(0)+3=0+3=3$$

b. Solving the equation:

$$-5x+3=-7$$
$$-5x+3-3=-7-3$$
$$-5x=-10$$
$$-\frac{1}{5}(-5x)=-\frac{1}{5}(-10)$$
$$x=2$$

c. Substituting $x = 0$:

$$-5(0)+3<-7$$
$$3<-7 \text{ (false)}$$

No, $x = 0$ is not a solution to the inequality.

d. Solving the inequality:

$$-5x+3<-7$$
$$-5x+3-3<-7-3$$
$$-5x<-10$$
$$-\frac{1}{5}(-5x)>-\frac{1}{5}(-10)$$
$$x>2$$

The solution is $\{x \mid x>2\}$.

61. The inequality is $x<3$.

63. The inequality is $x \le 3$.

65. Let x and $x + 1$ represent the integers. Solving the inequality:

$$x+x+1 \ge 583$$
$$2x+1 \ge 583$$
$$2x \ge 582$$
$$x \ge 291$$

The two numbers are at least 291.

67. Let x represent the number. Solving the inequality:

$$2x+6<10$$
$$2x<4$$
$$x<2$$

The solution is $\{x \mid x<2\}$.

69. Let x represent the number. Solving the inequality:

$$4x>x-8$$
$$3x>-8$$
$$x>-\frac{8}{3}$$

The solution is $\left\{x \mid x>-\frac{8}{3}\right\}$.

71. Let w represent the width, so $3w$ represents the length. Using the formula for perimeter:

$$2(w)+2(3w) \ge 48$$
$$2w+6w \ge 48$$
$$8w \ge 48$$
$$w \ge 6$$

The width is at least 6 meters.

73. Let x, $x + 2$, and $x + 4$ represent the sides of the triangle. The inequality is:

$$x+(x+2)+(x+4)>24$$
$$3x+6>24$$
$$3x>18$$
$$x>6$$

The shortest side is an even number greater than 6 inches (greater than or equal to 8 inches).

75. Finding the number: $0.25(32)=8$

77. Finding the number: $0.20(120)=24$

79. Finding the percent:

$$p \cdot 36=9$$
$$p=\frac{9}{36}=0.25=25\%$$

81. Finding the percent:

$$p \cdot 50=5$$
$$p=\frac{5}{50}=0.10=10\%$$

83. Finding the number:

$$0.20n = 16$$
$$n = \frac{16}{0.20} = 80$$

85. Finding the number:

$$0.02n = 8$$
$$n = \frac{8}{0.02} = 400$$

87. Simplifying the expression: $-|-5| = -(5) = -5$

89. Simplifying the expression: $-3-4(-2) = -3+8 = 5$

91. Simplifying the expression: $5|3-8|-6|2-5| = 5|-5|-6|-3| = 5(5)-6(3) = 25-18 = 7$

93. Simplifying the expression: $5-2[-3(5-7)-8] = 5-2[-3(-2)-8] = 5-2(6-8) = 5-2(-2) = 5+4 = 9$

95. The expression is: $-3-(-9) = -3+9 = 6$

97. Applying the distributive property: $\frac{1}{2}(4x-6) = \frac{1}{2} \cdot 4x - \frac{1}{2} \cdot 6 = 2x-3$

99. The integers are: $-3, 0, 2$

101. Solving the inequality:

$$3(9-5x) < 72$$
$$27-15x < 72$$
$$-15x < 45$$
$$-\frac{1}{15}(-15x) > -\frac{1}{15}(45)$$
$$x > -3$$

The solution is $\{x \mid x > -3\}$. The correct answer is b.

Chapter 2 Test

1. Simplifying: $5y-3-6y+4 = 5y-6y-3+4 = -y+1$

2. Simplifying: $3x-4+x+3 = 3x+x-4+3 = 4x-1$

3. Simplifying: $4-2(y-3)-6 = 4-2y+6-6 = -2y+4$

4. Simplifying: $3(3x-4)-2(4x+5) = 9x-12-8x-10 = 9x-8x-12-10 = x-22$

5. Evaluating when $x = -3$: $3x+12+2x = 5x+12 = 5(-3)+12 = -15+12 = -3$

6. Evaluating when $x = -2$ and $y = -4$: $x^2-3xy+y^2 = (-2)^2-3(-2)(-4)+(-4)^2 = 4-24+16 = -4$

7. **a.** Completing the table:

n	1	2	3	4
$(n+2)^2$	9	16	25	36

b. Completing the table:

n	1	2	3	4
n^2+2	3	6	11	18

8. Solving the equation:

$$3x-2 = 7$$
$$3x-2+2 = 7+2$$
$$3x = 9$$
$$x = 3$$

9. Solving the equation:

$$4y+15 = y$$
$$4y-4y+15 = y-4y$$
$$15 = -3y$$
$$y = -5$$

10. First clear the equation of fractions by multiplying by 12:

$$\frac{1}{4}x-\frac{1}{12}=\frac{1}{3}x+\frac{1}{6}$$
$$12\left(\frac{1}{4}x-\frac{1}{12}\right)=12\left(\frac{1}{3}x+\frac{1}{6}\right)$$
$$3x-1=4x+2$$
$$3x-4x-1=4x-4x+2$$
$$-x-1=2$$
$$-x-1+1=2+1$$
$$-x=3$$
$$x=-3$$

11. Solving the equation:

$$-3(3-2x)-7=8$$
$$-9+6x-7=8$$
$$6x-16=8$$
$$6x-16+16=8+16$$
$$6x=24$$
$$x=4$$

12. Solving the equation:

$$3x-9=-6$$
$$3x-9+9=-6+9$$
$$3x=3$$
$$x=1$$

13. Solving the equation:

$$0.05+0.07(100-x)=3.2$$
$$0.05+7-0.07x=3.2$$
$$-0.07x+7.05=3.2$$
$$-0.07x+7.05-7.05=3.2-7.05$$
$$-0.07x=-3.85$$
$$x=\frac{-3.85}{-0.07}=55$$

14. Solving the equation:

$$4(t-3)+2(t+4)=2t-16$$
$$4t-12+2t+8=2t-16$$
$$6t-4=2t-16$$
$$6t-2t-4=2t-2t-16$$
$$4t-4=-16$$
$$4t-4+4=-16+4$$
$$4t=-12$$
$$t=-3$$

15. Solving the equation:

$$4x-2(3x-1)=2x-8$$
$$4x-6x+2=2x-8$$
$$-2x+2=2x-8$$
$$-2x-2x+2=2x-2x-8$$
$$-4x+2=-8$$
$$-4x+2-2=-8-2$$
$$-4x=-10$$
$$x=\frac{10}{4}=\frac{5}{2}$$

16. Writing the equation: $x=0.40\cdot 56$

17. Writing the equation: $0.24\cdot x=720$

18. Substituting $x=4$:

$$3(4)-4y=16$$
$$12-4y=16$$
$$-4y=4$$
$$y=-1$$

19. Substituting $y=2$:

$$3x-4(2)=16$$
$$3x-8=16$$
$$3x=24$$
$$x=8$$

20. Solving for y:

$$2x+6y=12$$
$$6y=-2x+12$$
$$y=-\frac{1}{3}x+2$$

21. Solving for v:

$$v^2+2ad=x^2$$
$$2ad=x^2-v^2$$
$$a=\frac{x^2-v^2}{2d}$$

22. Completing the table:

	Five Years Ago	Now
Paul	$2x-5$	$2x$
Becca	$x-5$	x

The equation is:

$$2x-5+x-5=44$$
$$3x-10=44$$
$$3x=54$$
$$x=18$$
$$2x=36$$

Paul is 36 years old and Becca is 18 years old.

23. Let w represent the width and $3w-5$ represent the length. Using the perimeter formula:

$$2(w)+2(3w-5)=150$$
$$2w+6w-10=150$$
$$8w-10=150$$
$$8w=160$$
$$w=20$$
$$3w-5=55$$

The width is 20 cm and the length is 55 cm.

24. Completing the table:

	Dimes	Nickels
Number	$x+8$	x
Value (cents)	$10(x+8)$	$5(x)$

The equation is:

$$10(x+8)+5(x)=170$$
$$10x+80+5x=170$$
$$15x+80=170$$
$$15x=90$$
$$x=6$$
$$x+8=14$$

He has 6 nickels and 14 dimes in his collection.

25. Completing the table:

	Dollars Invested at 6%	Dollars Invested at 12%
Number of	x	$x+500$
Interest on	$0.06(x)$	$0.12(x+500)$

The equation is:

$$0.06(x)+0.12(x+500)=186$$
$$0.06x+0.12x+60=186$$
$$0.18x+60=186$$
$$0.18x=126$$
$$x=700$$
$$x+500=1{,}200$$

She invested \$700 at 6% and \$1,200 at 12%.

26. Solving the inequality:

$$\frac{1}{2}x - 2 > 3$$
$$\frac{1}{2}x > 5$$
$$x > 10$$

The solution is $\{x \mid x > 10\}$. Graphing the inequality:

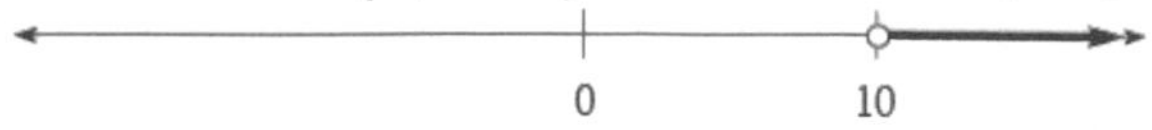

27. Solving the inequality:

$$-6y \le 24$$
$$-\frac{1}{6}(-6y) \ge -\frac{1}{6}(24)$$
$$y \ge -4$$

The solution is $\{y \mid y \ge -4\}$. Graphing the inequality:

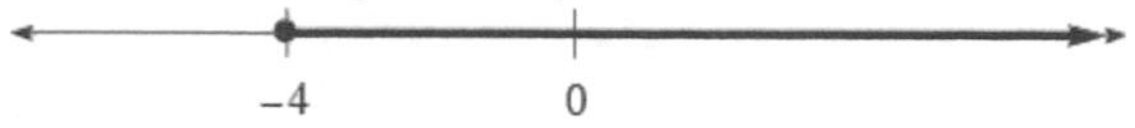

28. Solving the inequality:

$$0.3 - 0.2x < 1.1$$
$$-0.2x < 0.8$$
$$\frac{-0.2x}{-0.2} > \frac{0.8}{-0.2}$$
$$x > -4$$

The solution is $\{x \mid x > -4\}$. Graphing the inequality:

29. Solving the inequality:

$$3 - 2(n-1) \ge 9$$
$$3 - 2n + 2 \ge 9$$
$$-2n + 5 \ge 9$$
$$-2n \ge 4$$
$$-\frac{1}{2}(-2n) \le -\frac{1}{2}(4)$$
$$n \le -2$$

The solution is $\{n \mid n \le -2\}$. Graphing the inequality:

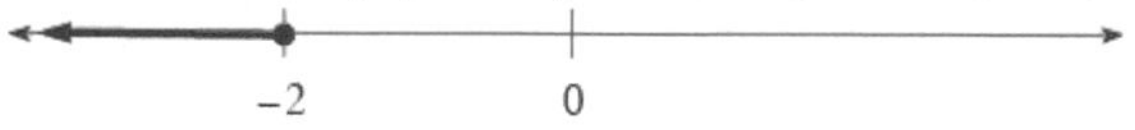

Chapter 3
Linear Equations and Inequalities in Two Variables

3.1 Paired Data and Graphing Ordered Pairs

1. The point lies in quadrant I:

3. The point lies in quadrant II:

5. The point lies in quadrant I:

7. The point lies in quadrant I:

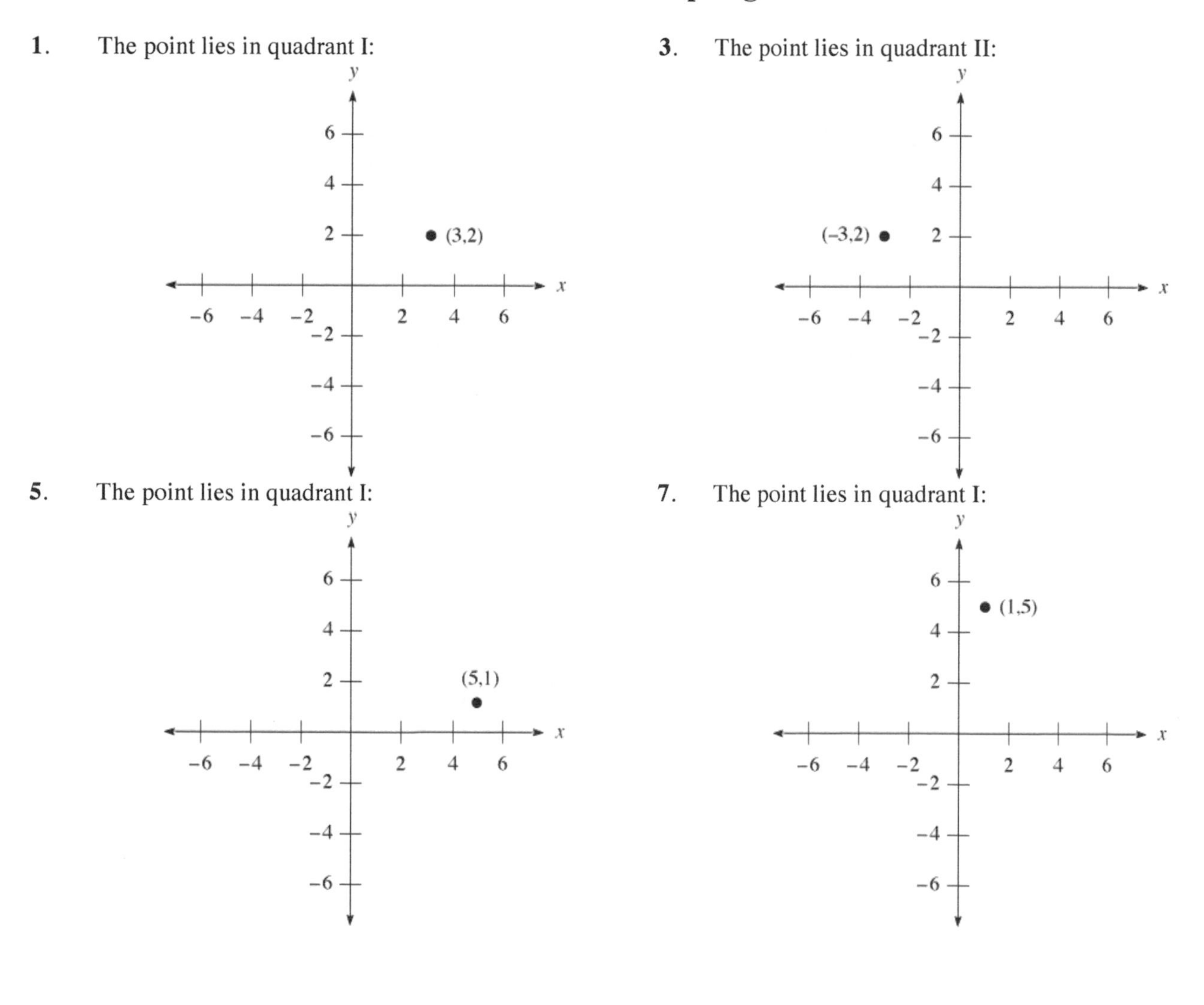

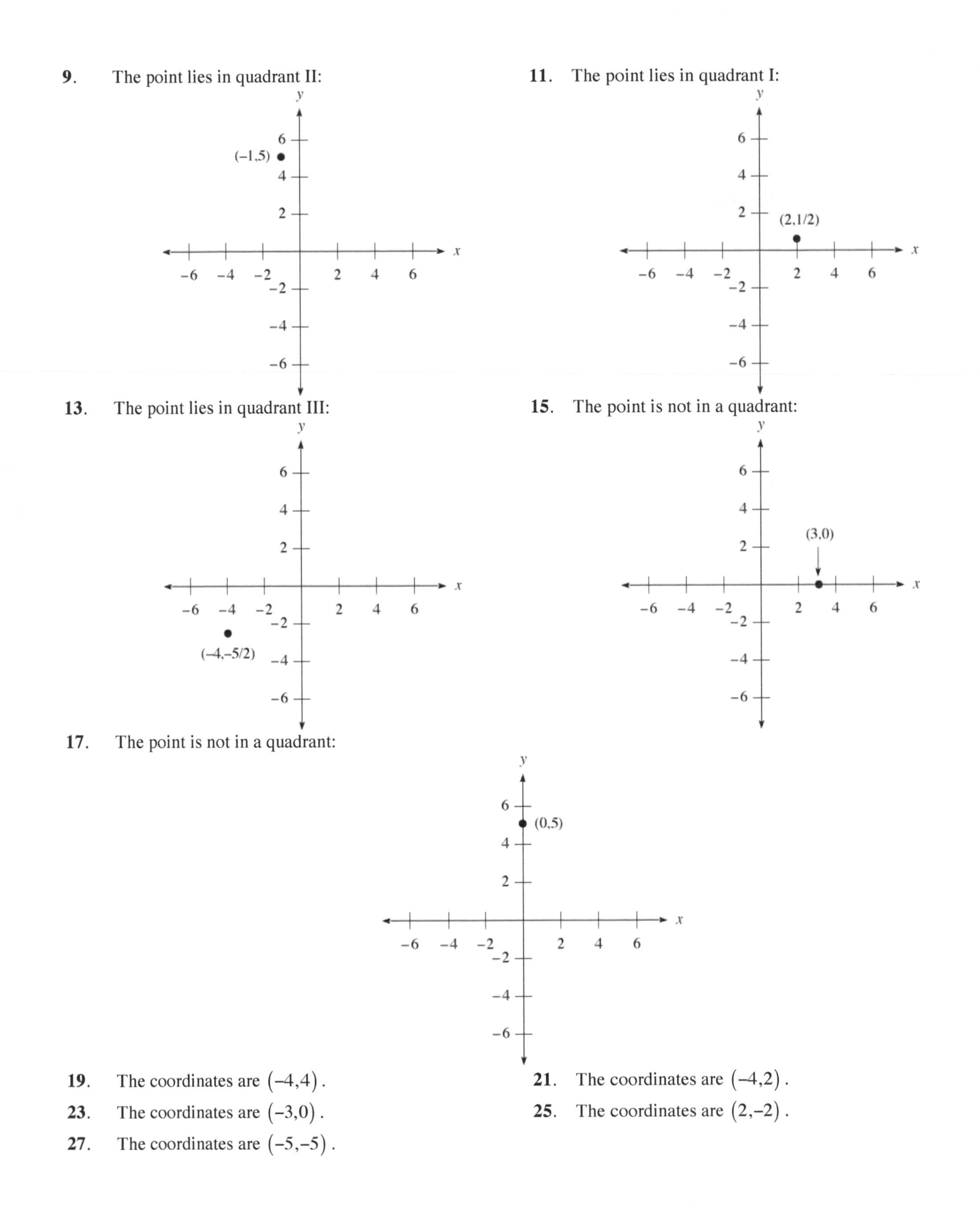

9. The point lies in quadrant II:

11. The point lies in quadrant I:

13. The point lies in quadrant III:

15. The point is not in a quadrant:

17. The point is not in a quadrant:

19. The coordinates are $(-4,4)$.

21. The coordinates are $(-4,2)$.

23. The coordinates are $(-3,0)$.

25. The coordinates are $(2,-2)$.

27. The coordinates are $(-5,-5)$.

29. Graphing the line:

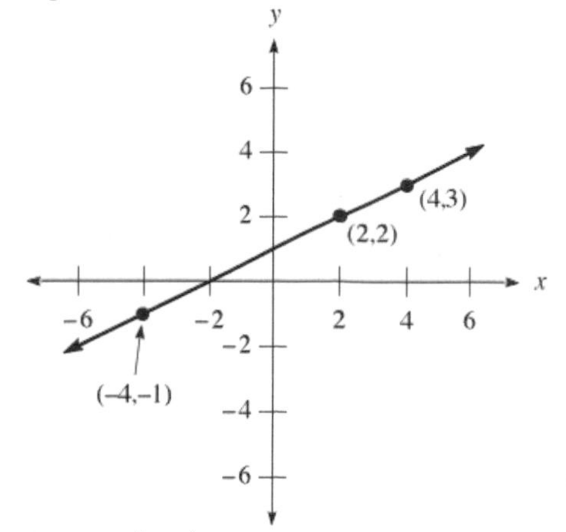

Yes, the point $(2,2)$ lies on the line.

31. Graphing the line:

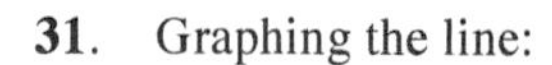

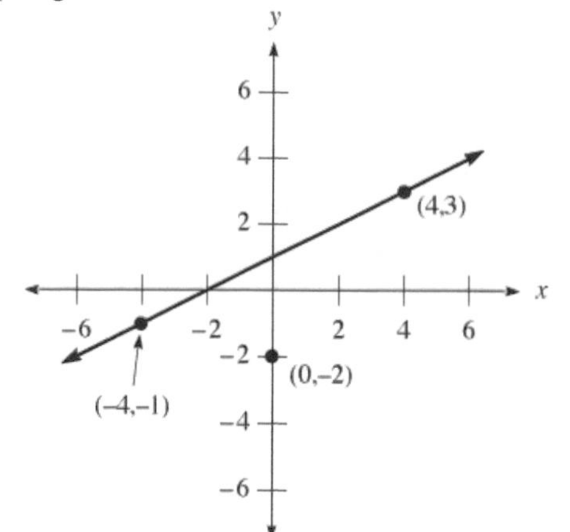

No, the point $(0,-2)$ does not lie on the line.

33. Graphing the line:

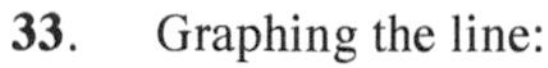

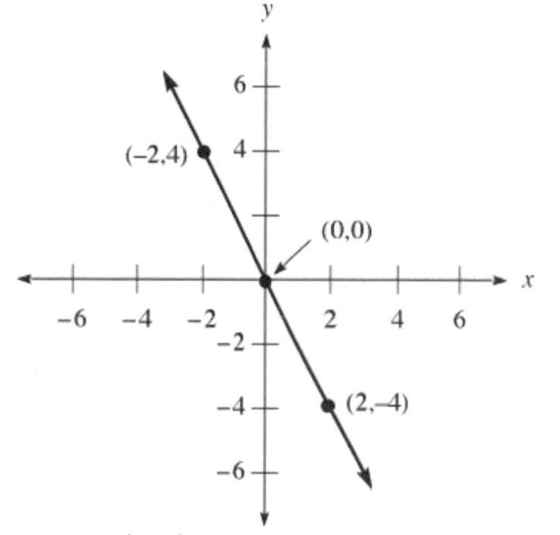

Yes, the point $(0,0)$ lies on the line.

35. Graphing the line:

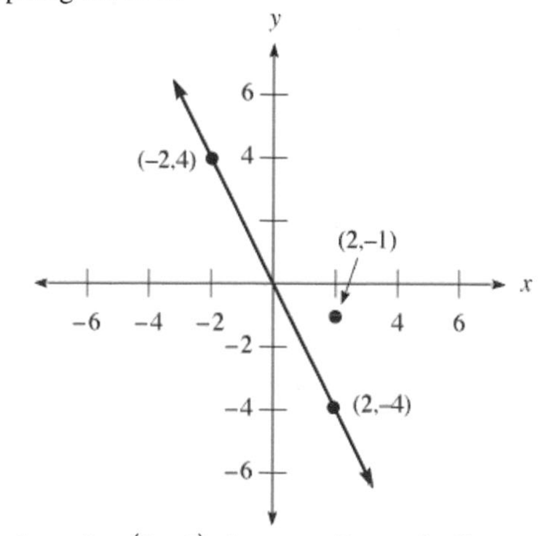

No, the point $(2,-1)$ does not lie on the line.

37. Graphing the line:

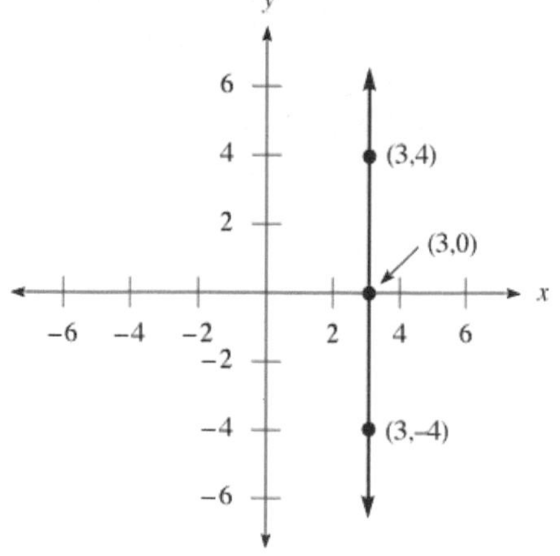

Yes, the point $(3,0)$ lies on the line.

39. No, the x-coordinate of every point on this line is 3.

41. Graphing the line:

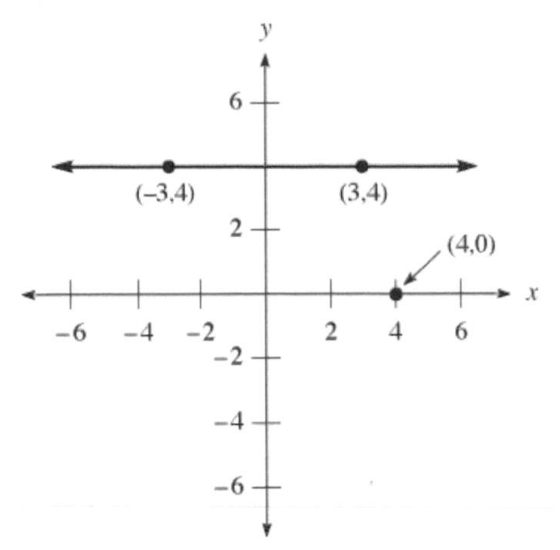

No, the point $(4,0)$ does not lie on the line.

43. No, the y-coordinate of every point on this line is 4.

45. Constructing a scatter diagram:

47. Sketching a line graph:

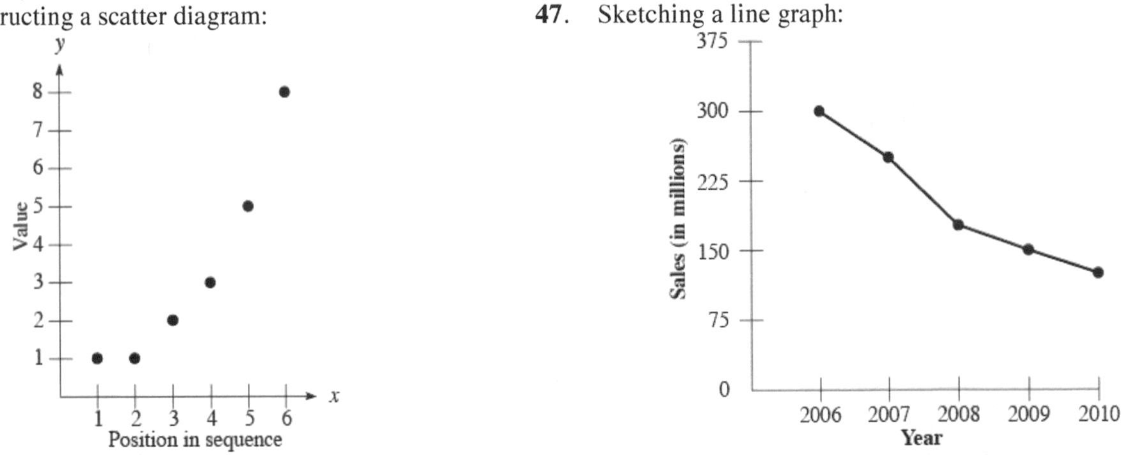

49. **a.** Three ordered pairs on the graph are (5, 40), (10, 80), and (20, 160).

b. She will earn \$320 for working 40 hours.

c. If her check is \$240, she worked 30 hours that week.

d. No. She should be paid \$280 for working 35 hours, not \$260.

51. Five ordered pairs on the graph are (1995, 44.8), (2000, 65.4), (2005, 104), (2010, 112.7), and (2015, 137.9).

53. Point A is (7 – 5, 2) = (2, 2), point B is (2, 2 + 3) = (2, 5), and point C is (2 + 5, 5) = (7, 5).

55. **a.** Substituting $y = 4$:

$$2x+3(4)=6$$
$$2x+12=6$$
$$2x=-6$$
$$x=-3$$

b. Substituting $y = -2$:

$$2x+3(-2)=6$$
$$2x-6=6$$
$$2x=12$$
$$x=6$$

c. Substituting $x = 3$:

$$2(3)+3y=6$$
$$6+3y=6$$
$$3y=0$$
$$y=0$$

d. Substituting $x = 9$:

$$2(9)+3y=6$$
$$18+3y=6$$
$$3y=-12$$
$$y=-4$$

57. **a.** Substituting $y = 7$:

$$2x-1=7$$
$$2x=8$$
$$x=4$$

b. Substituting $y = 3$:

$$2x-1=3$$
$$2x=4$$
$$x=2$$

c. Substituting $x = 0$: $y=2(0)-1=0-1=-1$

d. Substituting $x = 5$: $y=2(5)-1=10-1=9$

59. Solving for y:

$$5x+y=4$$
$$y=-5x+4$$

61. Solving for y:

$$3x-2y=6$$
$$-2y=-3x+6$$
$$y=\frac{3}{2}x-3$$

63. The ordered pair (3, 11) appears in the scatter diagram. The correct answer is c.

65. The correct graph is d.

3.2 Graphing Linear Equations in Two Variables

1. Substituting each ordered pair into the equation:

$$(2,3):\ 2(2))-5(3)=4-15=-11\neq 10$$
$$(0,-2):\ 2(0)-5(-2)=0+10=10$$
$$\left(\frac{5}{2},1\right):\ 2\left(\frac{5}{2}\right)-5(1)=5-5=0\neq 10$$

Only the ordered pair $(0,-2)$ is a solution.

3. Substituting each ordered pair into the equation:

$$(1,5):\ 7(1)-2=7-2=5$$
$$(0,-2):\ 7(0)-2=0-2=-2$$
$$(-2,-16):\ 7(-2)-2=-14-2=-16$$

All the ordered pairs $(1,5)$, $(0,-2)$ and $(-2,-16)$ are solutions.

5. Substituting each ordered pair into the equation:

$$(1,1):\ 1+1=2\neq 0$$
$$(2,-2):\ 2+(-2)=0$$
$$(3,3):\ 3+3=6\neq 0$$

Only the ordered pair $(2,-2)$ is a solution.

7. Since $x = 3$, the ordered pair $(5,3)$ cannot be a solution. The ordered pairs $(3,0)$ and $(3,-3)$ are solutions.

9. Substituting $x = 0$, $y = 0$, and $y = -6$:

$$2(0)+y=6 \qquad 2x+0=6 \qquad 2x+(-6)=6$$
$$0+y=6 \qquad 2x=6 \qquad 2x=12$$
$$y=6 \qquad x=3 \qquad x=6$$

The ordered pairs are $(0,6)$, $(3,0)$, and $(6,-6)$.

11. Substituting $x = 0$, $y = 0$, and $x = -4$:

$$3(0)+4y=12 \qquad 3x+4(0)=12 \qquad 3(-4)+4y=12$$
$$0+4y=12 \qquad 3x+0=12 \qquad -12+4y=12$$
$$4y=12 \qquad 3x=12 \qquad 4y=24$$
$$y=3 \qquad x=4 \qquad y=6$$

The ordered pairs are $(0,3)$, $(4,0)$, and $(-4,6)$.

13. Substituting $x = 1$, $y = 0$, and $x = 5$:

$y = 4(1) - 3$	$0 = 4x - 3$	$y = 4(5) - 3$
$y = 4 - 3$	$3 = 4x$	$y = 20 - 3$
$y = 1$	$x = \frac{3}{4}$	$y = 17$

The ordered pairs are $(1,1)$, $\left(\frac{3}{4},0\right)$, and $(5,17)$.

15. Substituting $x = 2$, $y = 6$, and $x = 0$:

$y = 7(2) - 1$	$6 = 7x - 1$	$y = 7(0) - 1$
$y = 14 - 1$	$7 = 7x$	$y = 0 - 1$
$y = 13$	$x = 1$	$y = -1$

The ordered pairs are $(2,13)$, $(1,6)$, and $(0,-1)$.

17. Substituting $y = 4$, $y = -3$, and $y = 0$ results (in each case) in $x = -5$. The ordered pairs are $(-5,4)$, $(-5,-3)$, and $(-5,0)$.

19. Completing the table:

x	y
1	3
–3	–9
4	12
6	18

21. Completing the table:

x	y
2	3
3	2
5	0
9	–4

23. Completing the table:

x	y
2	0
3	2
1	–2
–3	–10

25. Completing the table:

x	y
0	–1
–1	–7
–3	–19
$\frac{3}{2}$	8

27. The ordered pairs are $(0,4)$, $(2,2)$, and $(4,0)$:

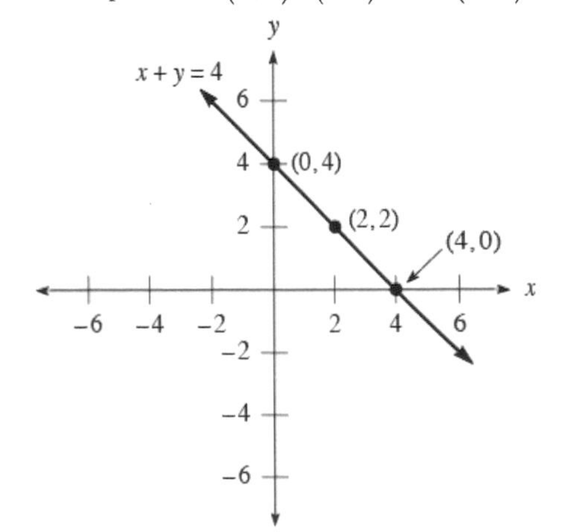

29. The ordered pairs are $(0,0)$, $(-2,-4)$, and $(2,4)$:

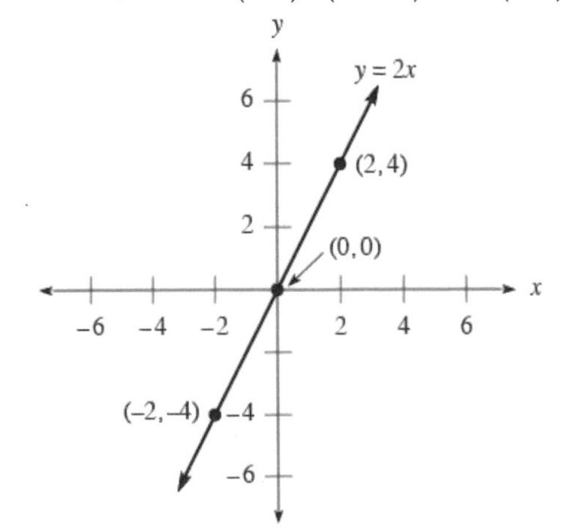

31. The ordered pairs are $(-3,-1)$, $(0,0)$, and $(3,1)$:

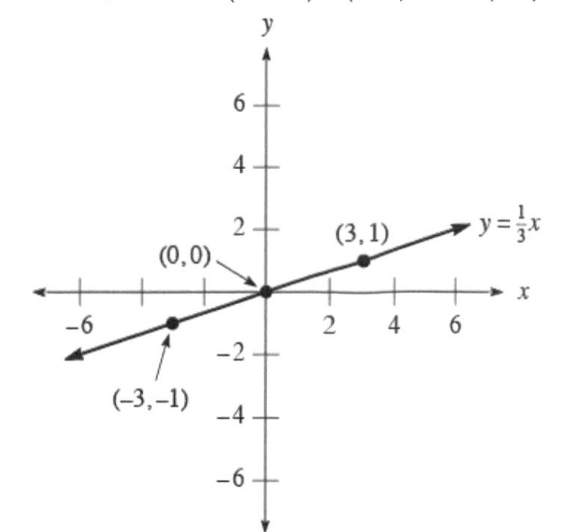

33. The ordered pairs are $(0,1)$, $(-1,-1)$, and $(1,3)$:

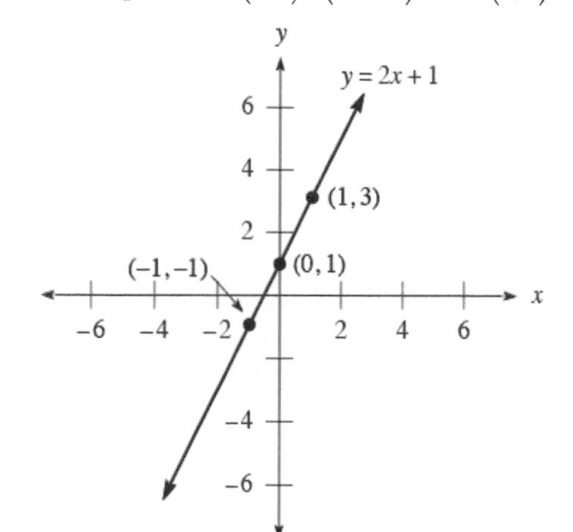

35. The ordered pairs are $(0,4)$, $(-1,4)$, and $(2,4)$:

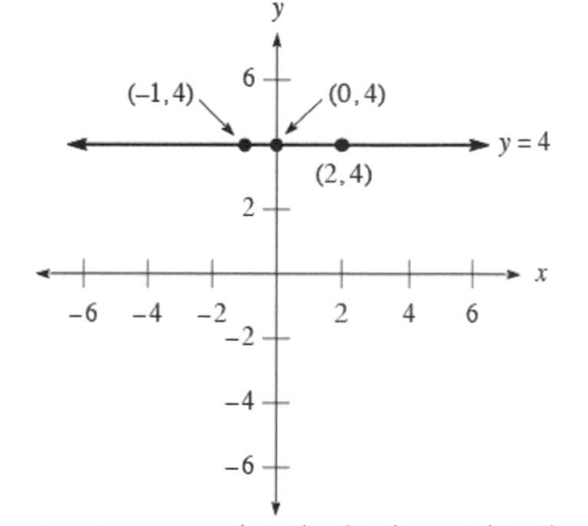

37. The ordered pairs are $(-2,2)$, $(0,3)$, and $(2,4)$:

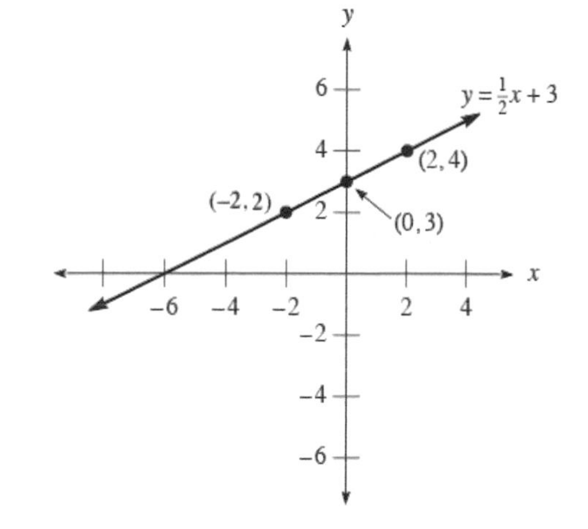

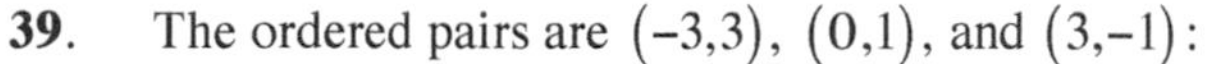
39. The ordered pairs are $(-3,3)$, $(0,1)$, and $(3,-1)$:

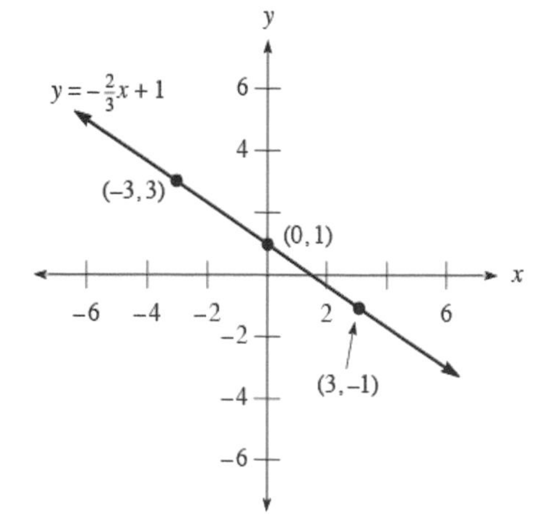

41. Solving for y:

$$2x+y=3$$
$$y=-2x+3$$

The ordered pairs are $(-1,5)$, $(0,3)$, and $(1,1)$:

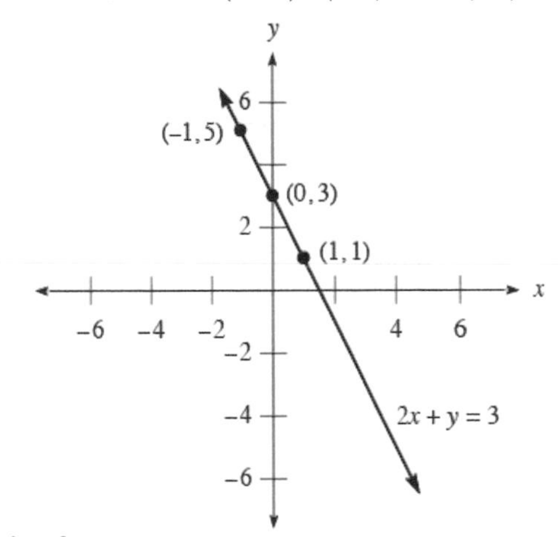

43. Solving for y:

$$3x+2y=6$$
$$2y=-3x+6$$
$$y=-\frac{3}{2}x+3$$

The ordered pairs are $(0,3)$, $(2,0)$, and $(4,-3)$:

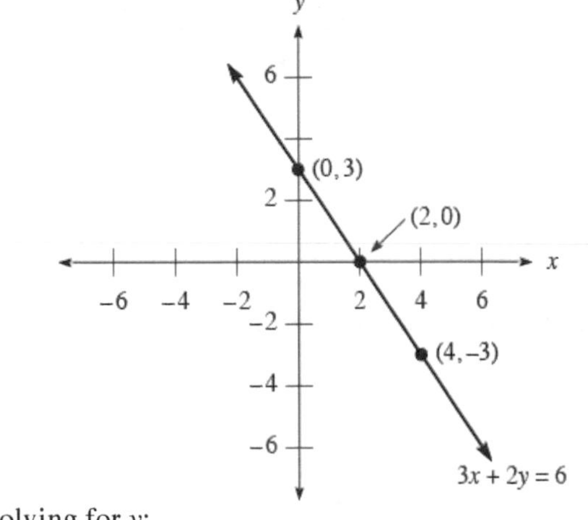

45. Solving for y:

$$-x+2y=6$$
$$2y=x+6$$
$$y=\frac{1}{2}x+3$$

The ordered pairs are $(-2,2)$, $(0,3)$, and $(2,4)$:

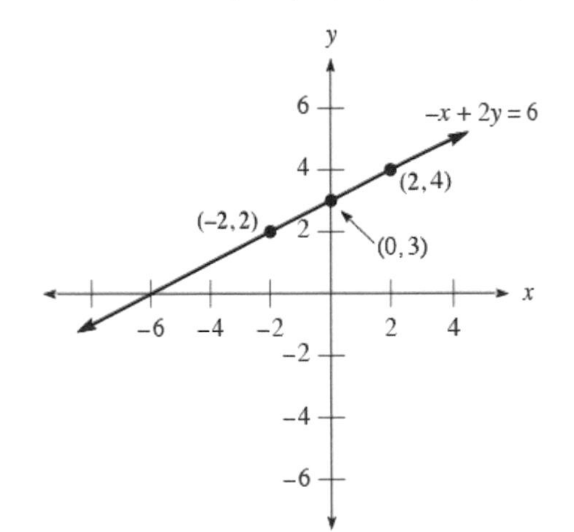

47. Solving for y:

$$4y=2$$
$$y=\frac{1}{2}$$

The ordered pairs are $\left(-4,\frac{1}{2}\right)$, $\left(0,\frac{1}{2}\right)$, and $\left(4,\frac{1}{2}\right)$:

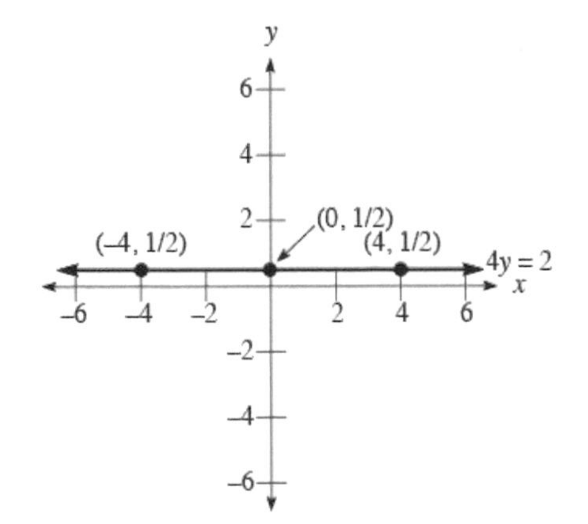

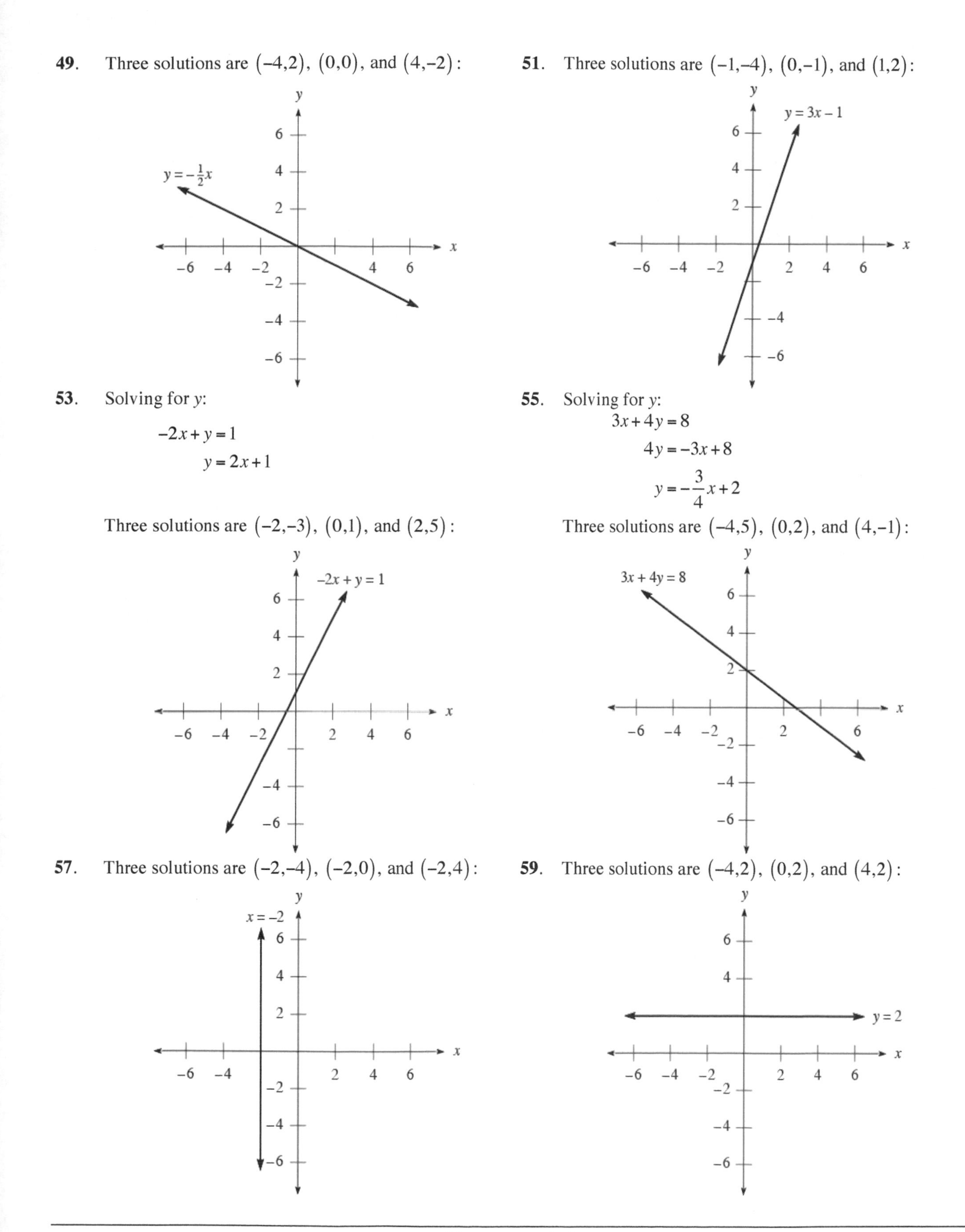

49. Three solutions are $(-4,2)$, $(0,0)$, and $(4,-2)$:

51. Three solutions are $(-1,-4)$, $(0,-1)$, and $(1,2)$:

53. Solving for *y*:

$$-2x+y=1$$
$$y=2x+1$$

Three solutions are $(-2,-3)$, $(0,1)$, and $(2,5)$:

55. Solving for *y*:

$$3x+4y=8$$
$$4y=-3x+8$$
$$y=-\frac{3}{4}x+2$$

Three solutions are $(-4,5)$, $(0,2)$, and $(4,-1)$:

57. Three solutions are $(-2,-4)$, $(-2,0)$, and $(-2,4)$:

59. Three solutions are $(-4,2)$, $(0,2)$, and $(4,2)$:

61. Graphing the equation:

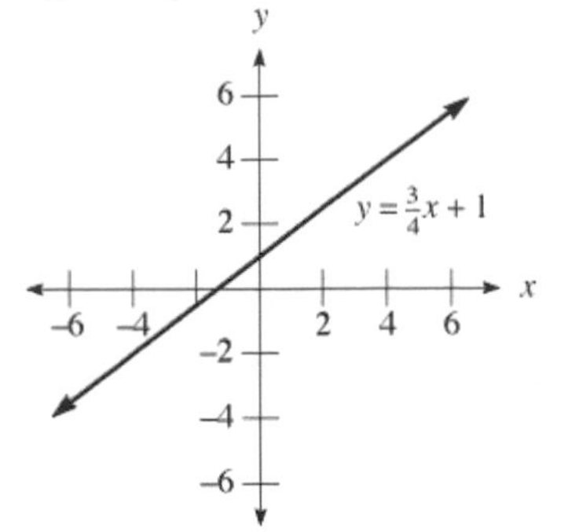

63. Graphing the equation:

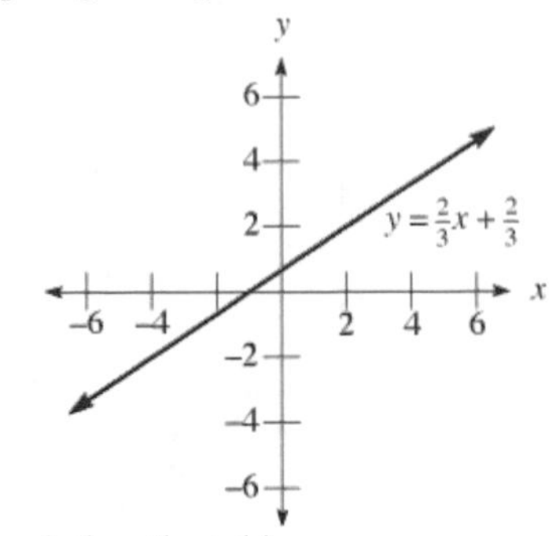

65. Completing the table:

Equation	H, V, and/or O
$x = 3$	V
$y = 3$	H
$y = 3x$	O
$y = 0$	O,H

67. Completing the table:

Equation	H, V, and/or O
$x = -\frac{3}{5}$	V
$y = -\frac{3}{5}$	H
$y = -\frac{3}{5}x$	O
$x = 0$	O,V

69. Completing the table:

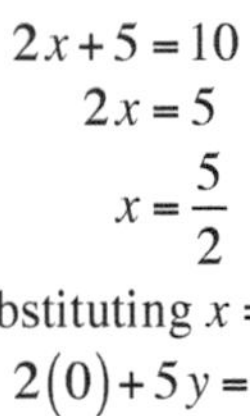

x	y
−4	−3
−2	−2
0	−1
2	0
6	2

71. **a.** Solving the equation:

$$2x + 5 = 10$$
$$2x = 5$$
$$x = \frac{5}{2}$$

b. Substituting $y = 0$:

$$2x + 5(0) = 10$$
$$2x = 10$$
$$x = 5$$

c. Substituting $x = 0$:

$$2(0) + 5y = 10$$
$$5y = 10$$
$$y = 2$$

d. Graphing the line:

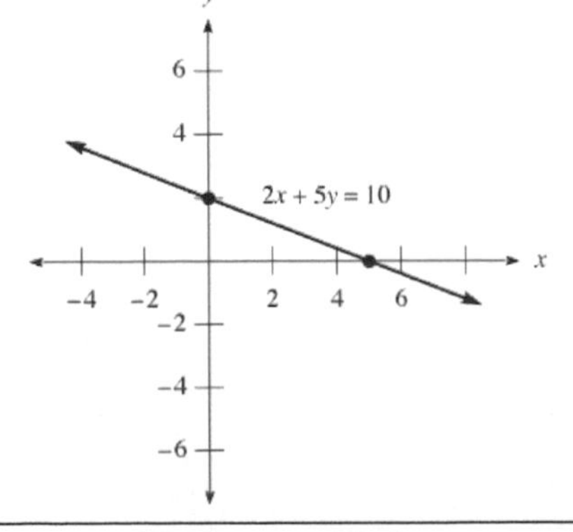

e. Solving for y:

$$2x+5y=10$$
$$5y=-2x+10$$
$$y=-\frac{2}{5}x+2$$

73. Substituting $w = 3$:

$$2l+2(3)=30$$
$$2l+6=30$$
$$2l=24$$
$$l=12$$

The length is 12 inches.

75. **a.** This is correct, since; $y=12(5)=\$60$

b. This is not correct, since: $y=12(9)=\$108$. Her check should be for $108.

c. This is not correct, since: $y=12(7)=\$84$. Her check should be for $84.

d. This is correct, since: $y=12(14)=\$168$

77. **a.** Substituting $t = 5$: $V=-45{,}000(5)+600{,}000=-225{,}000+600{,}000=\$375{,}000$

b. Solving when $V = 330{,}000$:

$$-45{,}000t+600{,}000=330{,}000$$
$$-45{,}000t=-270{,}000$$
$$t=6$$

The crane will be worth $330,000 at the end of 6 years.

c. Substituting $t = 9$: $V=-45{,}000(9)+600{,}000=-405{,}000+600{,}000=\$195{,}000$

No, the crane will be worth $195,000 after 9 years.

d. The crane cost $600,000 (the value when $t = 0$).

79. **a.** Substituting $y = 0$:

$$3x+2(0)=6$$
$$3x+0=6$$
$$3x=6$$
$$x=2$$

b. Substituting $x = 0$:

$$3(0)+2y=6$$
$$0+2y=6$$
$$2y=6$$
$$y=3$$

81. **a.** Substituting $y = 0$:

$$-x+2(0)=4$$
$$-x+0=4$$
$$-x=4$$
$$x=-4$$

b. Substituting $x = 0$:

$$-(0)+2y=4$$
$$0+2y=4$$
$$2y=4$$
$$y=2$$

83. **a.** Substituting $y = 0$:

$$0=-\frac{1}{3}x+2$$
$$-\frac{1}{3}x=-2$$
$$x=6$$

b. Substituting $x = 0$:

$$y=-\frac{1}{3}(0)+2$$
$$y=2$$

85. Substituting each ordered pair into the equation:

$$(0,2):\ 3(0)-4(2)=0-8=-8\neq 8$$
$$(4,-1):\ 3(4)-4(-1)=12+4=16\neq 8$$
$$(-2,1):\ 3(-2)-4(1)=-6-4=-10\neq 8$$
$$(-4,-5):\ 3(-4)-4(-5)=-12+20=8$$

The ordered pair $(-4,-5)$ is a solution. The correct answer is d.

87. This is the graph of $2x+y=-4$, so the correct answer is c.

3.3 More on Graphing: Intercepts

1. To find the x-intercept, let $y = 0$:

$$2x+0=4$$
$$2x=4$$
$$x=2$$

To find the y-intercept, let $x = 0$:

$$2(0)+y=4$$
$$0+y=4$$
$$y=4$$

Graphing the line:

3. To find the x-intercept, let $y = 0$:

$$-x+0=3$$
$$-x=3$$
$$x=-3$$

To find the y-intercept, let $x = 0$:

$$-0+y=3$$
$$y=3$$

Graphing the line:

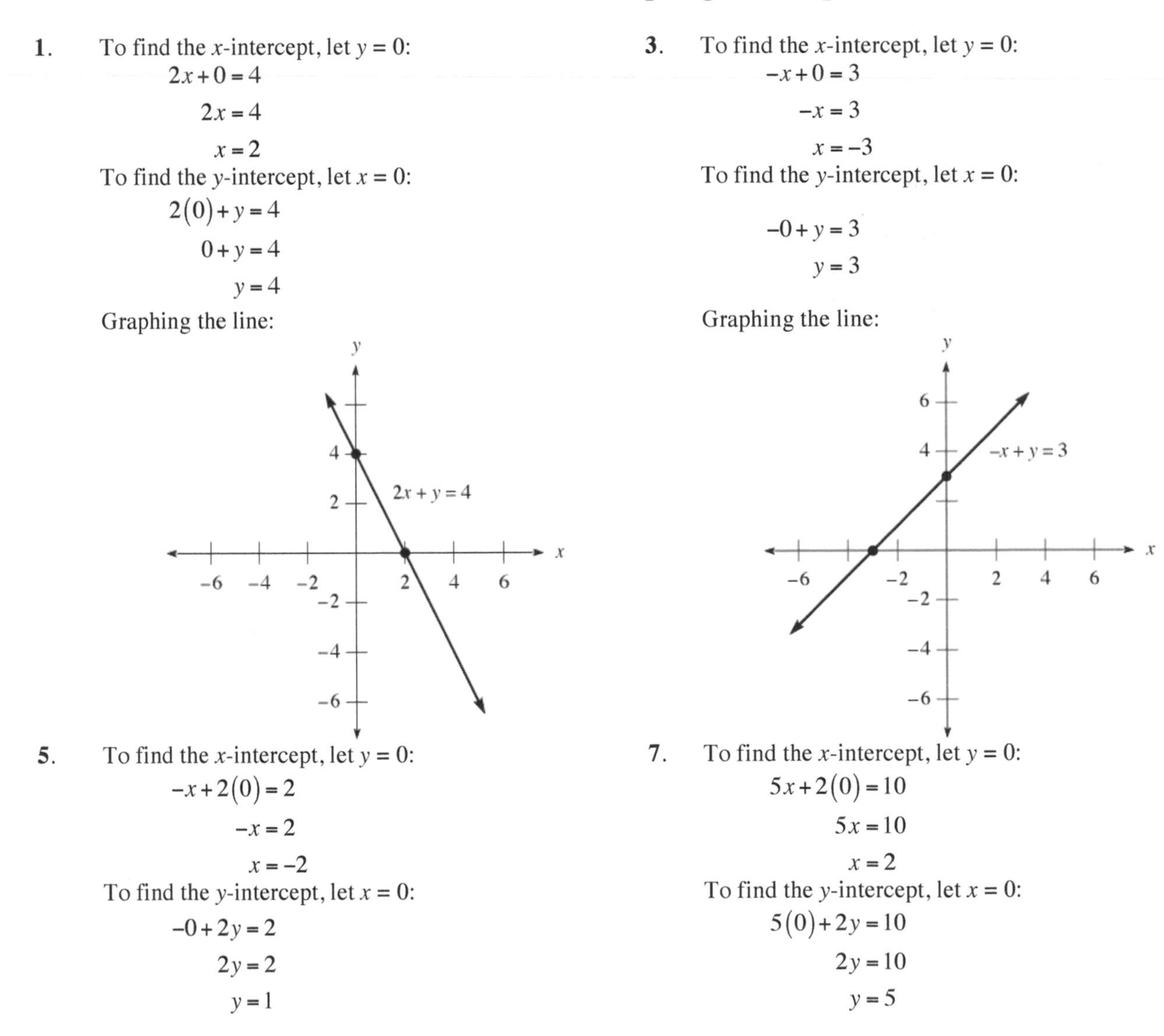

5. To find the x-intercept, let $y = 0$:

$$-x+2(0)=2$$
$$-x=2$$
$$x=-2$$

To find the y-intercept, let $x = 0$:

$$-0+2y=2$$
$$2y=2$$
$$y=1$$

7. To find the x-intercept, let $y = 0$:

$$5x+2(0)=10$$
$$5x=10$$
$$x=2$$

To find the y-intercept, let $x = 0$:

$$5(0)+2y=10$$
$$2y=10$$
$$y=5$$

Graphing the line:

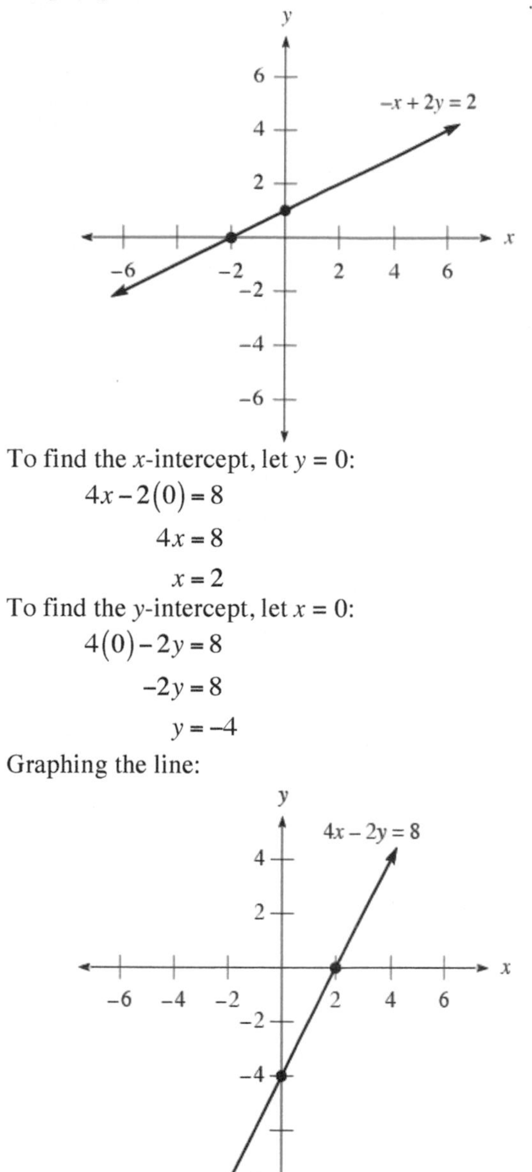

9. To find the x-intercept, let $y = 0$:

$$4x-2(0)=8$$
$$4x=8$$
$$x=2$$

To find the y-intercept, let $x = 0$:

$$4(0)-2y=8$$
$$-2y=8$$
$$y=-4$$

Graphing the line:

Graphing the line:

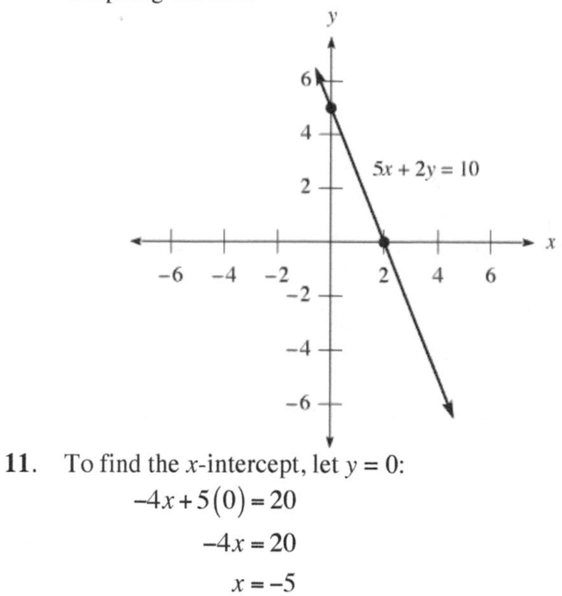

11. To find the x-intercept, let $y = 0$:

$$-4x+5(0)=20$$
$$-4x=20$$
$$x=-5$$

To find the y-intercept, let $x = 0$:

$$-4(0)+5y=20$$
$$5y=20$$
$$y=4$$

Graphing the line:

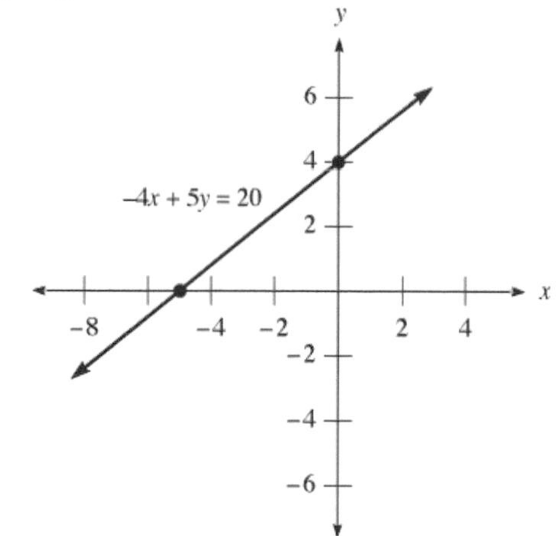

13. To find the x-intercept, let $y = 0$:

$$2x-6=0$$
$$2x=6$$
$$x=3$$

To find the y-intercept, let $x = 0$:

$$y=2(0)-6=-6$$

Graphing the line:

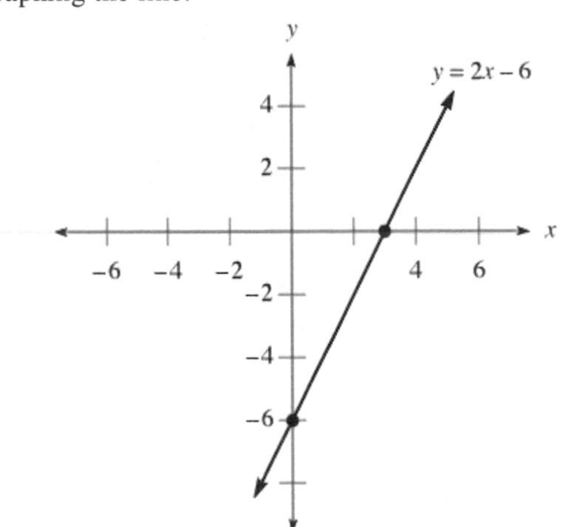

15. To find the x-intercept, let $y = 0$:

$$2x+2=0$$
$$2x=-2$$
$$x=-1$$

To find the y-intercept, let $x = 0$:

$$y=2(0)+2=2$$

Graphing the line:

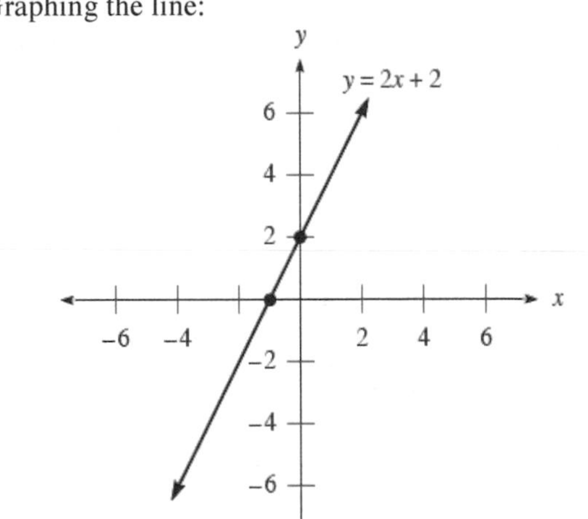

17. To find the x-intercept, let $y = 0$:

$$2x-1=0$$
$$2x=1$$
$$x=\frac{1}{2}$$

To find the y-intercept, let $x = 0$:

$$y=2(0)-1=-1$$

Graphing the line:

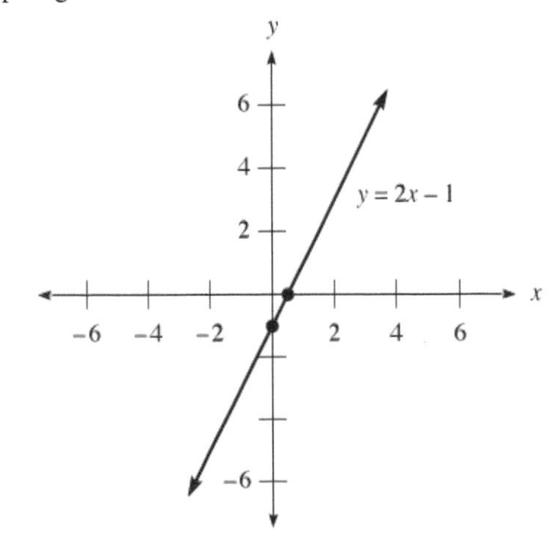

19. To find the x-intercept, let $y = 0$:

$$\frac{1}{2}x+3=0$$
$$\frac{1}{2}x=-3$$
$$x=-6$$

To find the y-intercept, let $x = 0$:

$$y=\frac{1}{2}(0)+3=3$$

Graphing the line:

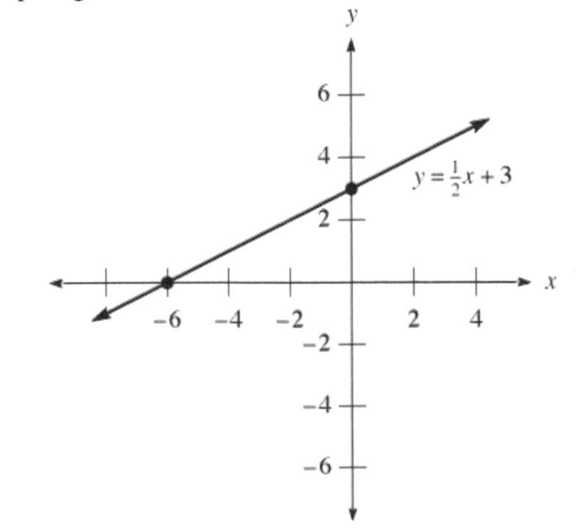

21. To find the x-intercept, let $y = 0$:

$$-\frac{1}{3}x - 2 = 0$$
$$-\frac{1}{3}x = 2$$
$$x = -6$$

To find the y-intercept, let $x = 0$: $y = -\frac{1}{3}(0) - 2 = -2$

Graphing the line:

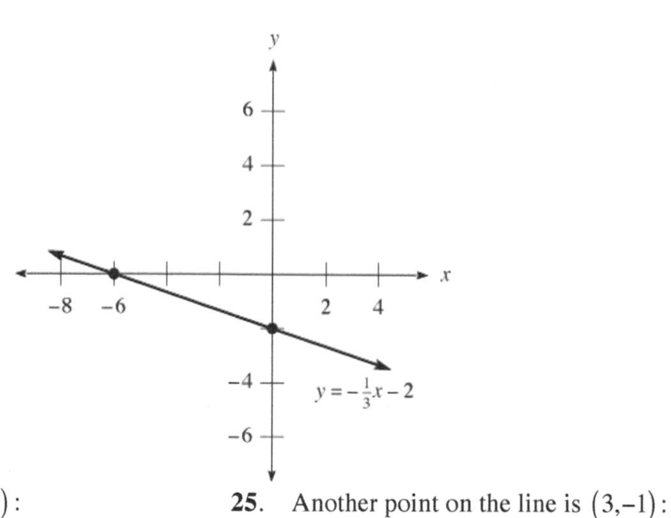

23. Another point on the line is $(2,-4)$:

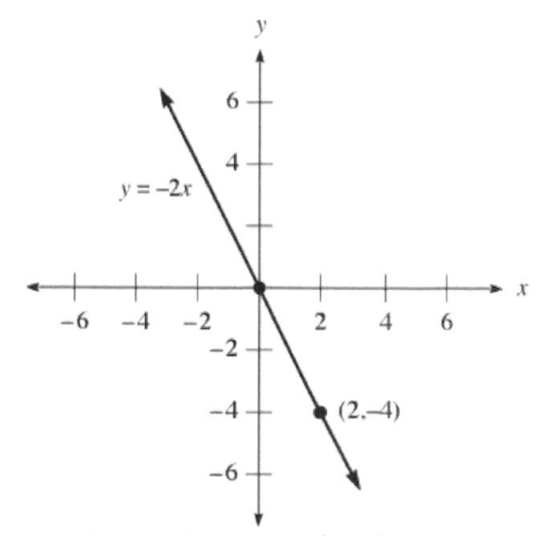

25. Another point on the line is $(3,-1)$:

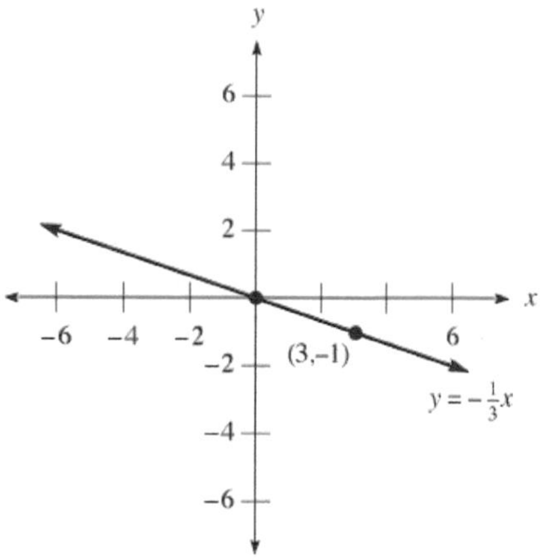

27. Another point on the line is $(3,2)$:

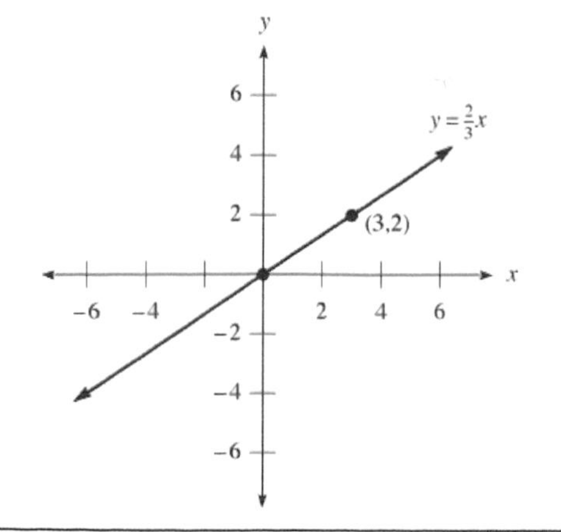

29. Completing the table:

Equation	x – intercept	y – intercept
$3x+4y=12$	4	3
$3x+4y=4$	$\frac{4}{3}$	1
$3x+4y=3$	1	$\frac{3}{4}$
$3x+4y=2$	$\frac{2}{3}$	$\frac{1}{2}$

31. Completing the table:

Equation	x – intercept	y – intercept
$x-3y=2$	2	$-\frac{2}{3}$
$y=\frac{1}{3}x-\frac{2}{3}$	2	$-\frac{2}{3}$
$x-3y=0$	0	0
$y=\frac{1}{3}x$	0	0

33. Substituting $x = 0$: $y=2(0)+5=0+5=5$

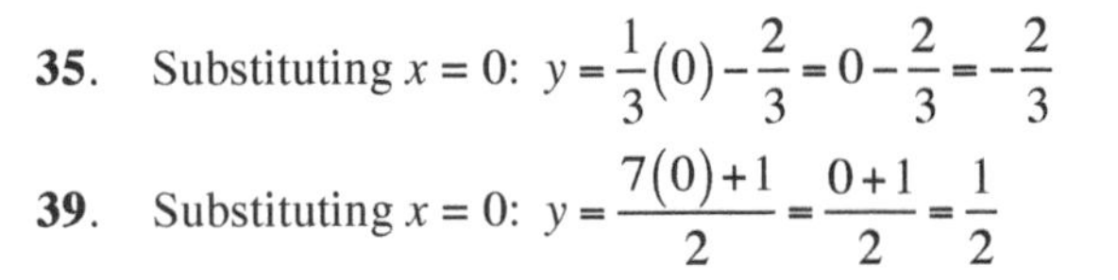

35. Substituting $x = 0$: $y=\frac{1}{3}(0)-\frac{2}{3}=0-\frac{2}{3}=-\frac{2}{3}$

37. Substituting $x = 0$: $y=4-0=4$

39. Substituting $x = 0$: $y=\frac{7(0)+1}{2}=\frac{0+1}{2}=\frac{1}{2}$

41. **a.** Solving the equation:

$$\begin{aligned}2x-3&=-3\\2x&=0\\x&=0\end{aligned}$$

b. Substituting $y = 0$:

$$\begin{aligned}2x-3(0)&=-3\\2x-0&=-3\\2x&=-3\\x&=-\frac{3}{2}\end{aligned}$$

c. Substituting $x = 0$:

$$\begin{aligned}2(0)-3y&=-3\\0-3y&=-3\\-3y&=-3\\y&=1\end{aligned}$$

d. Graphing the equation:

e. Solving for y:

$$2x-3y=-3$$
$$-3y=-2x-3$$
$$y=\frac{2}{3}x+1$$

43. The x-intercept is 3 and the y-intercept is 5.

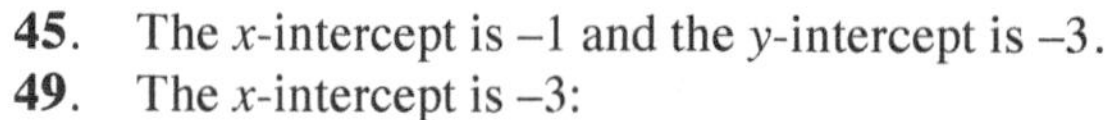

45. The x-intercept is -1 and the y-intercept is -3.

47. The y-intercept is -4:

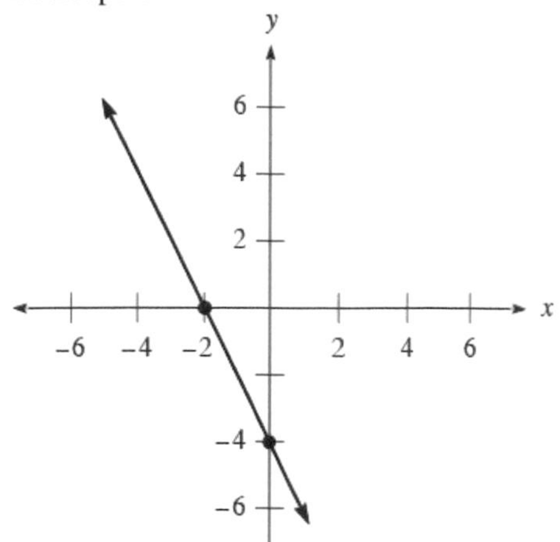

49. The x-intercept is -3:

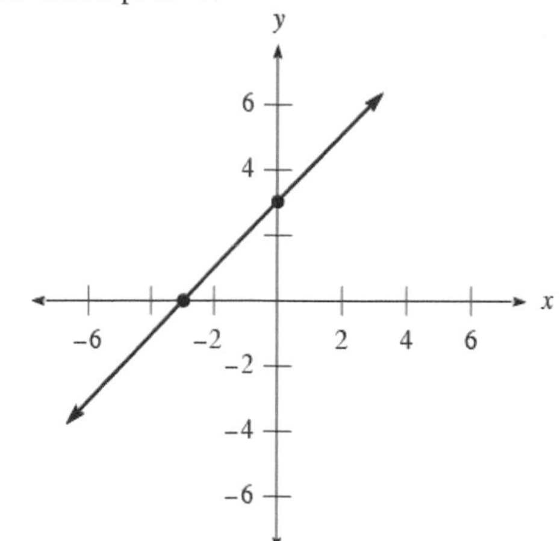

51. The x- and y-intercepts are both 3:

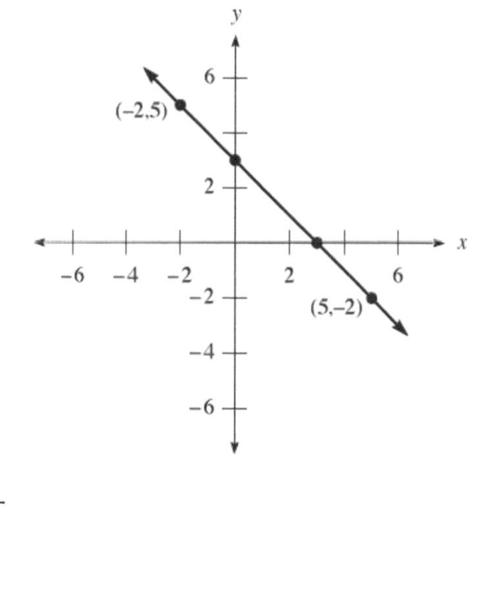

53. Completing the table:

x	y
−2	1
0	−1
−1	0
1	−2

55. The x-intercept is 3:

57. The y-intercept is 4:

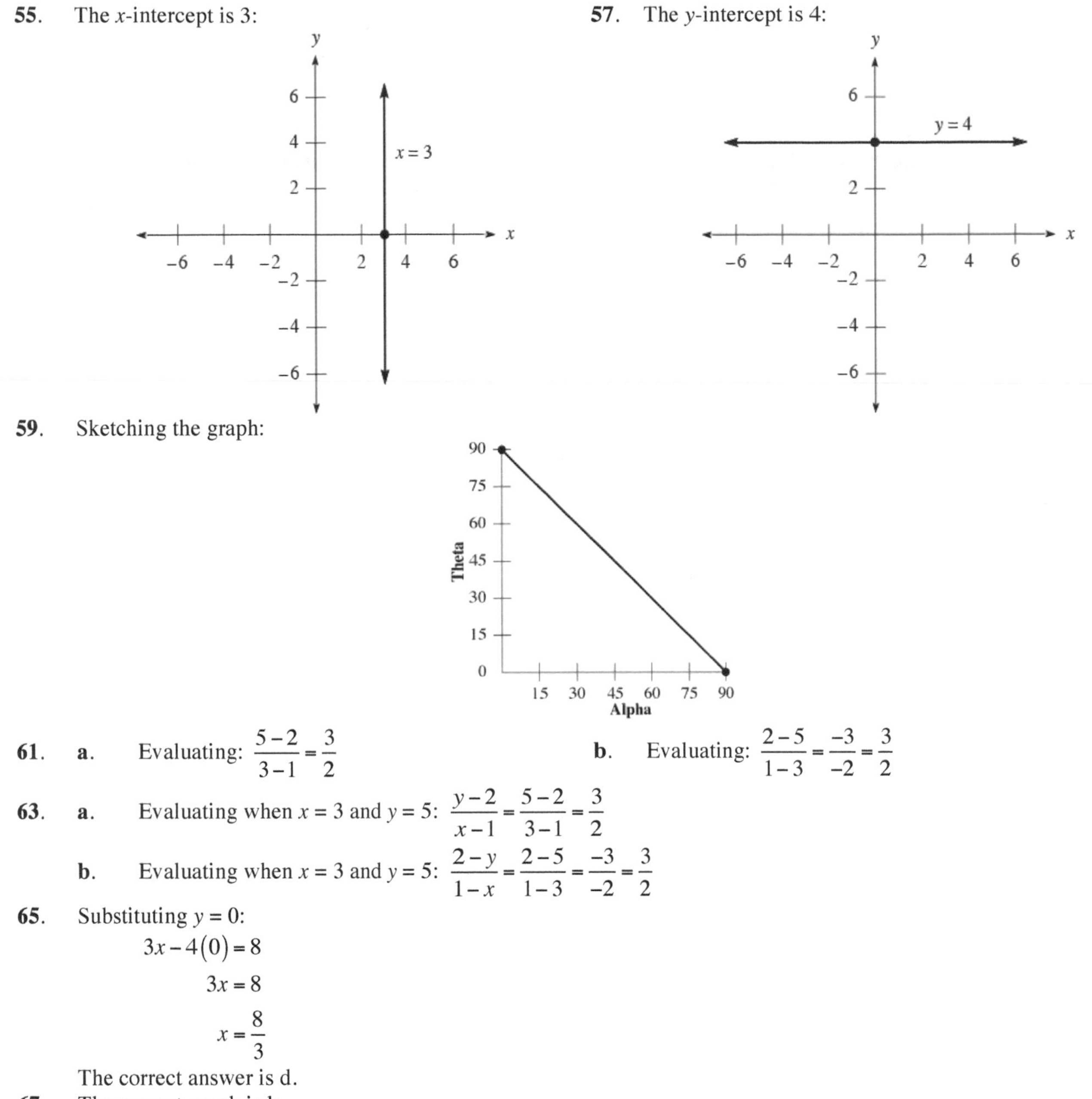

59. Sketching the graph:

61. **a.** Evaluating: $\frac{5-2}{3-1}=\frac{3}{2}$

b. Evaluating: $\frac{2-5}{1-3}=\frac{-3}{-2}=\frac{3}{2}$

63. **a.** Evaluating when $x = 3$ and $y = 5$: $\frac{y-2}{x-1}=\frac{5-2}{3-1}=\frac{3}{2}$

b. Evaluating when $x = 3$ and $y = 5$: $\frac{2-y}{1-x}=\frac{2-5}{1-3}=\frac{-3}{-2}=\frac{3}{2}$

65. Substituting $y = 0$:

$$3x-4(0)=8$$
$$3x=8$$
$$x=\frac{8}{3}$$

The correct answer is d.

67. The correct graph is b.

3.4 The Slope of a Line

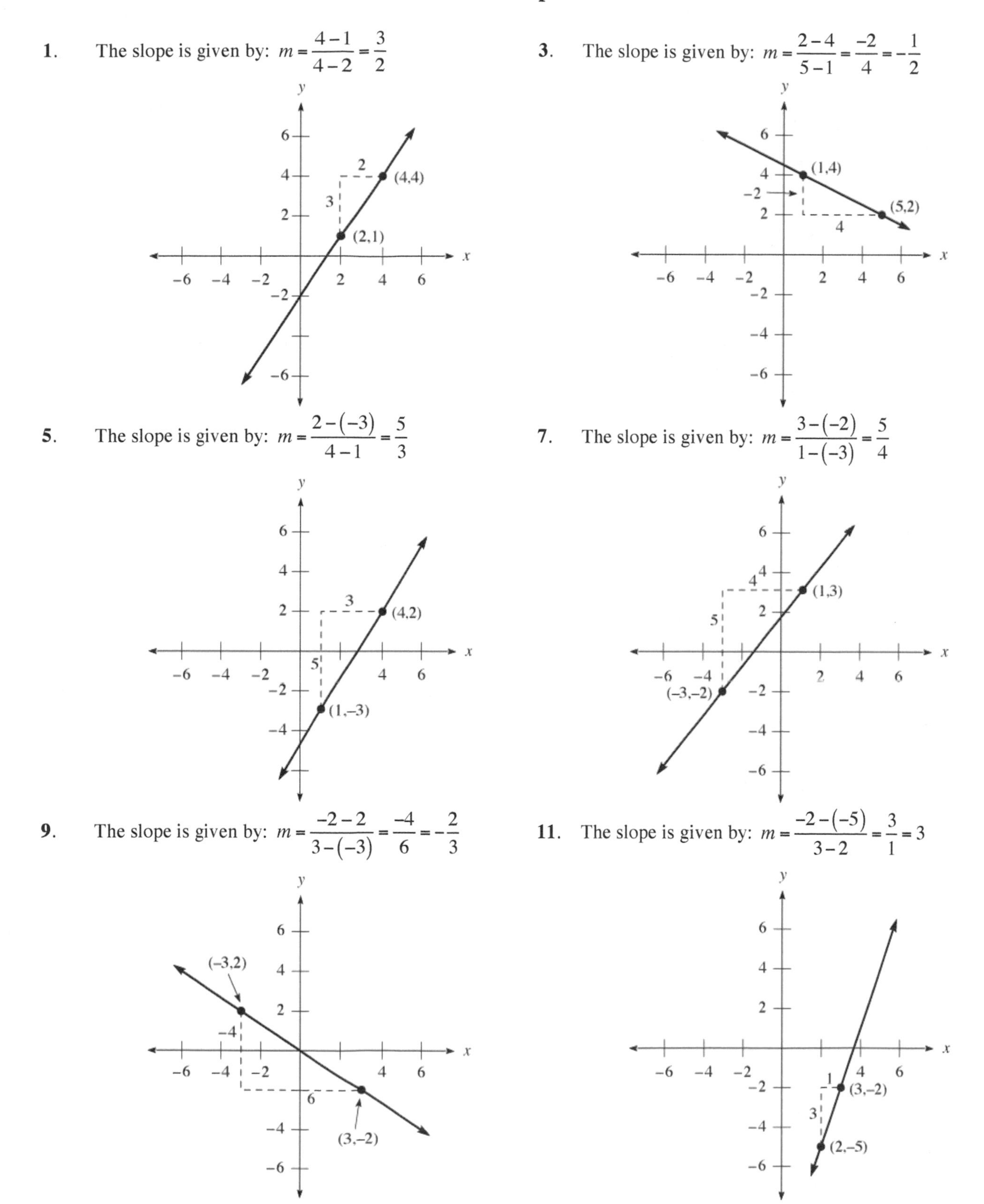

1. The slope is given by: $m=\dfrac{4-1}{4-2}=\dfrac{3}{2}$

3. The slope is given by: $m=\dfrac{2-4}{5-1}=\dfrac{-2}{4}=-\dfrac{1}{2}$

5. The slope is given by: $m=\dfrac{2-(-3)}{4-1}=\dfrac{5}{3}$

7. The slope is given by: $m=\dfrac{3-(-2)}{1-(-3)}=\dfrac{5}{4}$

9. The slope is given by: $m=\dfrac{-2-2}{3-(-3)}=\dfrac{-4}{6}=-\dfrac{2}{3}$

11. The slope is given by: $m=\dfrac{-2-(-5)}{3-2}=\dfrac{3}{1}=3$

13. The slope is given by: $m = \frac{1-(-5)}{4-4} = \frac{6}{0}$, which is undefined

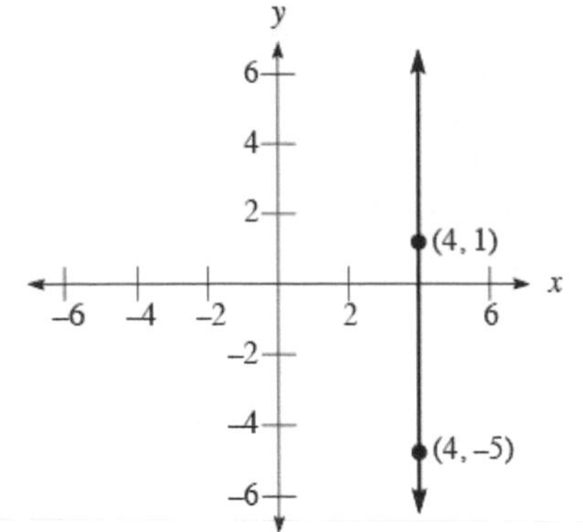

15. The slope is given by: $m = \frac{-2-(-2)}{-1-(-4)} = \frac{0}{3} = 0$

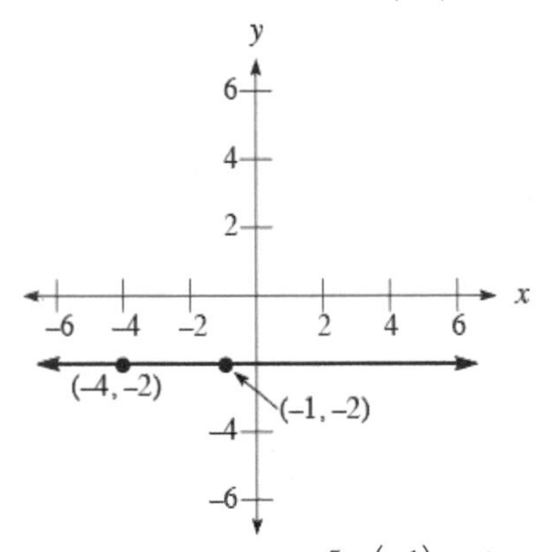

17. The slope is given by: $m = \frac{-1-0}{-3-0} = \frac{-1}{-3} = \frac{1}{3}$

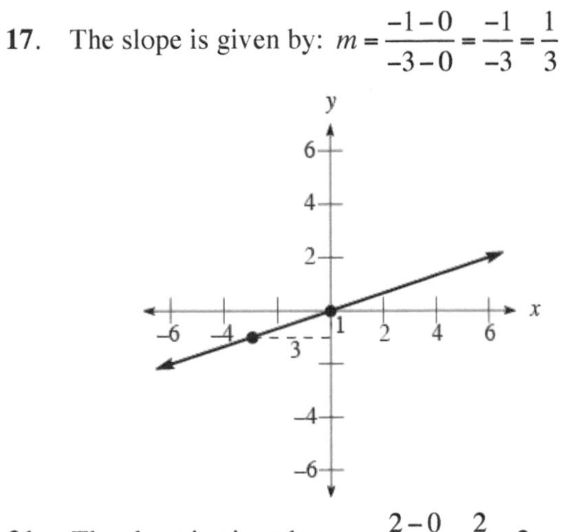

19. The slope is given by: $m = \frac{5-(-1)}{1-(-1)} = \frac{6}{2} = 3$

21. The slope is given by: $m = \frac{2-0}{2-1} = \frac{2}{1} = 2$

23. Graphing the line:

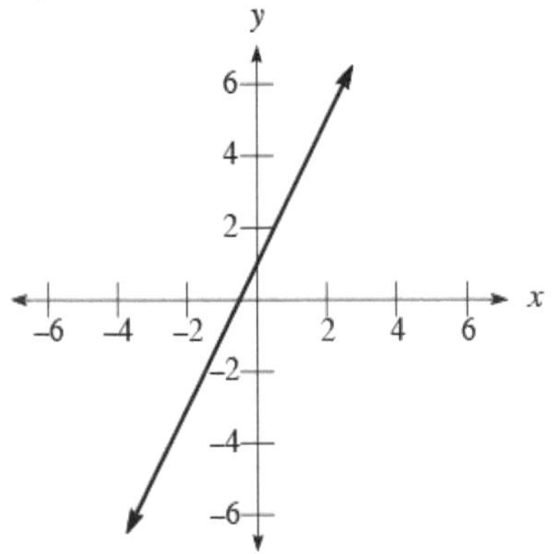

25. Graphing the line:

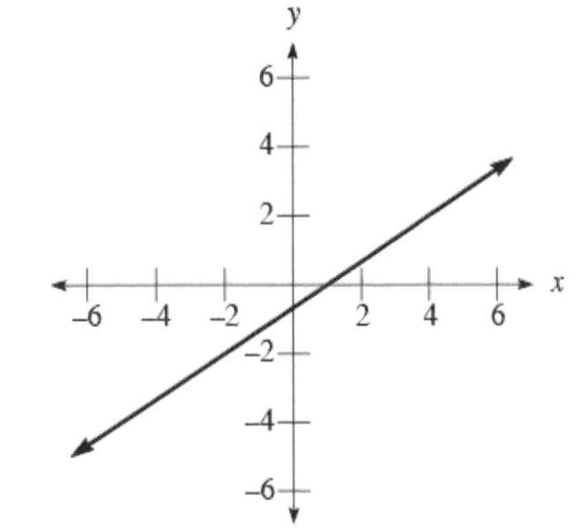

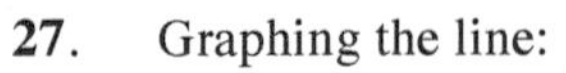

27. Graphing the line:

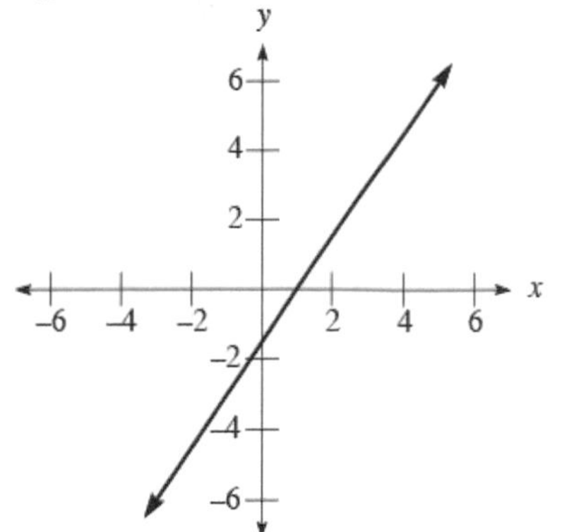

29. Graphing the line:

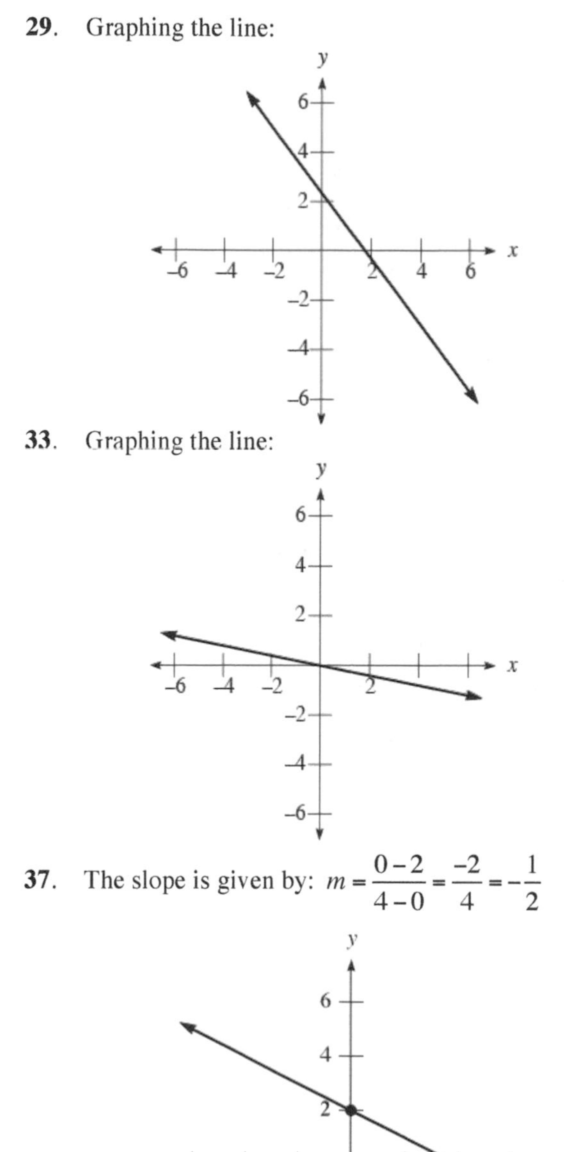

31. Graphing the line:

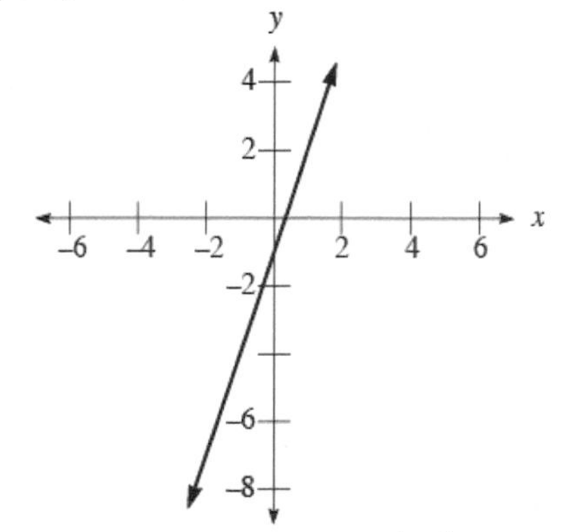

33. Graphing the line:

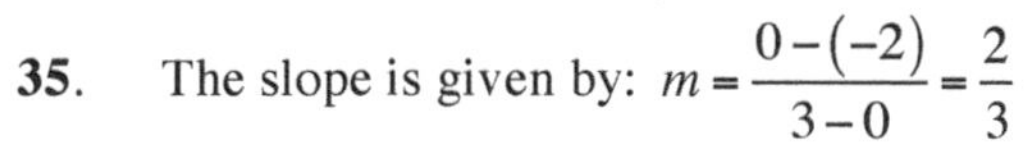

35. The slope is given by: $m=\frac{0-(-2)}{3-0}=\frac{2}{3}$

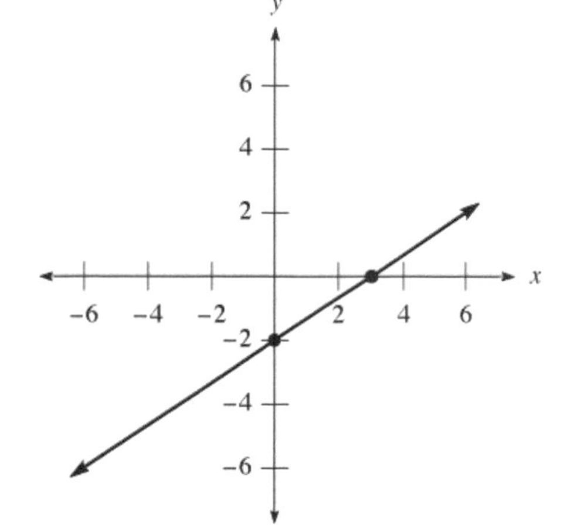

37. The slope is given by: $m=\frac{0-2}{4-0}=\frac{-2}{4}=-\frac{1}{2}$

39. The slope is 2 and the y-intercept is -3:

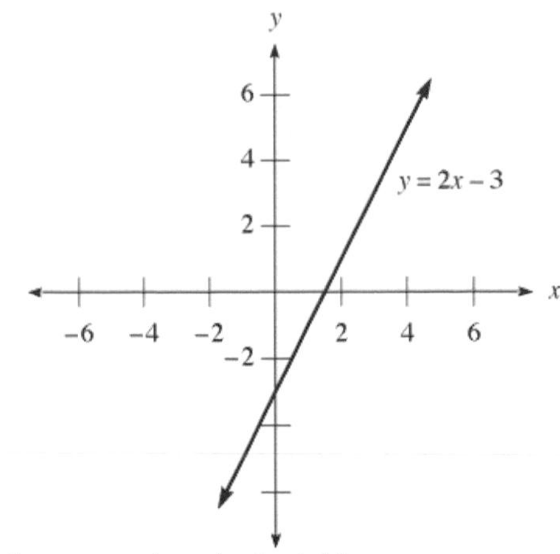

41. The slope is $\frac{1}{2}$ and the y-intercept is 1:

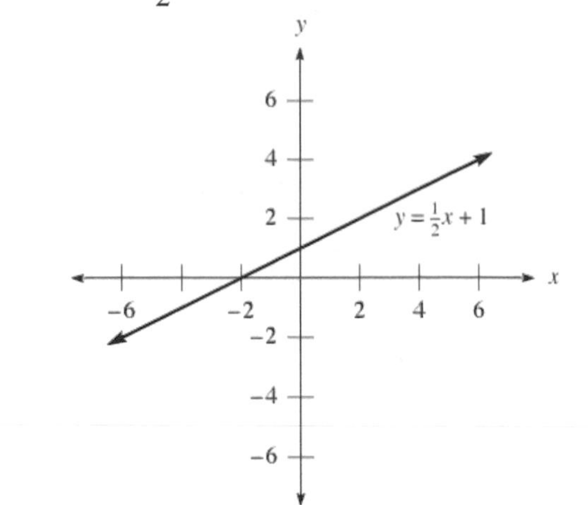

43. The slopes are given in the table:

Equation	Slope
$x = 3$	undefined
$y = 3$	0
$y = 3x$	3

45. The slopes are given in the table:

Equation	Slope
$y = -\frac{2}{3}$	0
$x = -\frac{2}{3}$	undefined
$y = -\frac{2}{3}x$	$-\frac{2}{3}$

47. **a.** The parallel line slope is $\frac{2}{3}$. **b.** The perpendicular line slope is $-\frac{3}{2}$.

49. **a.** The parallel line slope is $-\frac{1}{4}$. **b.** The perpendicular line slope is 4.

51. **a.** The parallel line slope is 2. **b.** The perpendicular line slope is $-\frac{1}{2}$.

53. **a.** The parallel line slope is 0. **b.** The perpendicular line slope is undefined.

55. Finding each slope:

$$m_1 = \frac{8-2}{6-(-3)} = \frac{6}{9} = \frac{2}{3} \qquad m_2 = \frac{1-(-1)}{3-0} = \frac{2}{3}$$

Since the slopes are equal, the two lines are parallel.

57. Finding each slope:

$$m_1 = \frac{4-6}{2-(-2)} = \frac{-2}{4} = -\frac{1}{2} \qquad m_2 = \frac{3-(-5)}{1-(-3)} = \frac{8}{4} = 2$$

Since the slopes are negative reciprocals, the two lines are perpendicular.

59. Finding each slope:

$$m_1 = \frac{5-0}{1-(-4)} = \frac{5}{5} = 1 \qquad m_2 = \frac{-6-(-4)}{2-0} = \frac{-2}{2} = -1$$

Since the slopes are negative reciprocals, the two lines are perpendicular.

61. Using the slope formula:

$$\frac{y-2}{6-4}=2$$
$$\frac{y-2}{2}=2$$
$$y-2=4$$
$$y=6$$

63. **a.** Finding the slopes:

A: $\frac{121-88}{1970-1960}=\frac{33}{10}=3.3$

B: $\frac{152-121}{1980-1970}=\frac{31}{10}=3.1$

C: $\frac{205-152}{1990-1980}=\frac{53}{10}=5.3$

D: $\frac{224-205}{2000-1990}=\frac{19}{10}=1.9$

E: $\frac{230-224}{2010-2000}=\frac{6}{10}=0.6$

b. The annual production of garbage in the United States increased at a rate of 3.3 million tons per year from 1960 to 1970.

65. **a.** Finding the slopes:

A: $\frac{250-300}{2007-2006}=-50$

B: $\frac{175-250}{2008-2007}=-75$

C: $\frac{125-150}{2010-2009}=-25$

b. Non-camera phone sales decreased at a rate of 75 million phones per year between 2007 and 2008.

67. Solving for y:

$$-2x+y=4$$
$$y=2x+4$$

69. Solving for y:

$$2x+y=3$$
$$y=-2x+3$$

71. Solving for y:

$$4x-5y=20$$
$$-5y=-4x+20$$
$$y=\frac{4}{5}x-4$$

73. Finding the slope: $m=\frac{-4-2}{1-(-3)}=\frac{-6}{4}=-\frac{3}{2}$. The correct answer is a.

3.5 Slope-Intercept Form

1. The slope is $m=5$ and the y-intercept is $b=-3$.

3. The slope is $m=-\frac{2}{3}$ and the y-intercept is $b=\frac{7}{3}$.

5. The slope is $m=1$ and the y-intercept is $b=9$.

7. The slope is $m=\frac{1}{2}$ and the y-intercept is $b=-\frac{5}{2}$.

9. The slope is $m=-2$ and the y-intercept is $b=\frac{1}{4}$.

11. The slope is $m=3$ and the y-intercept is $b=0$.

13. The slope is $m=0$ and the y-intercept is $b=-10$.

15. Graphing the line:

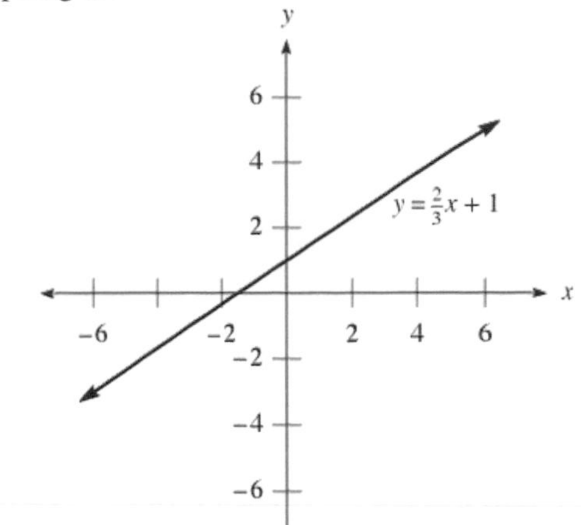

17. Graphing the line:

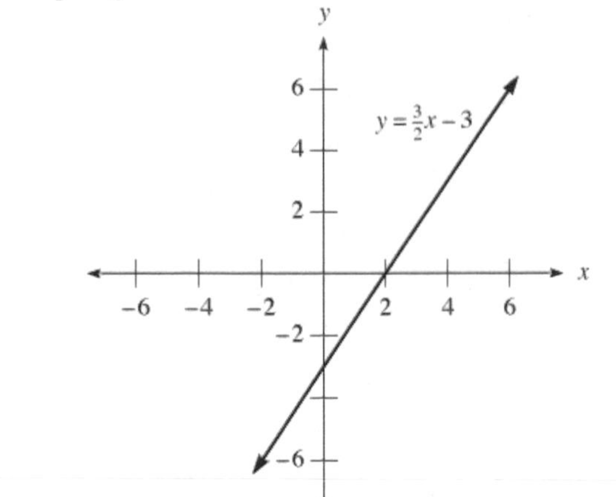

19. Graphing the line:

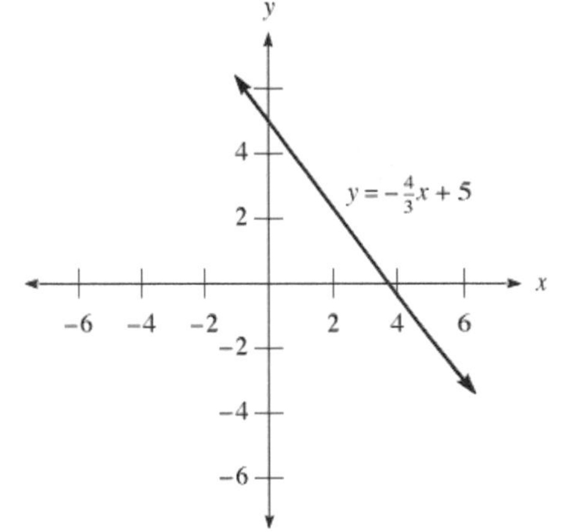

21. Graphing the line:

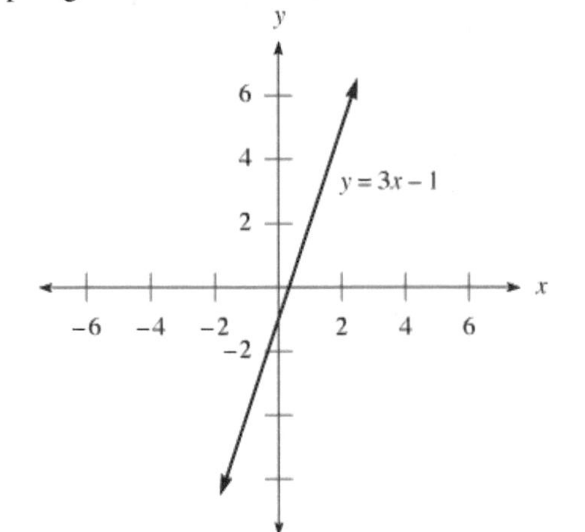

23. Graphing the line:

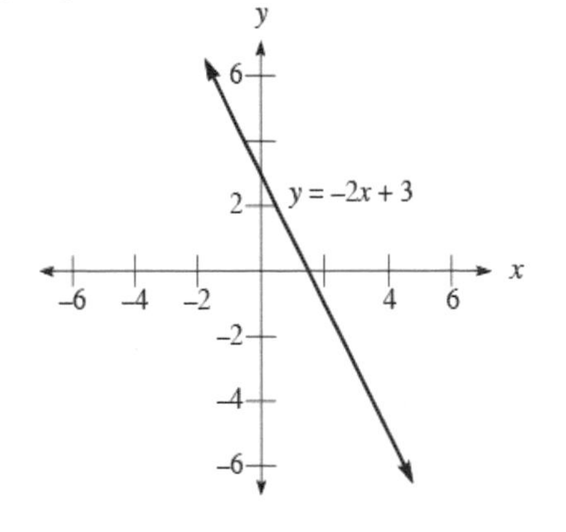

25. Graphing the line:

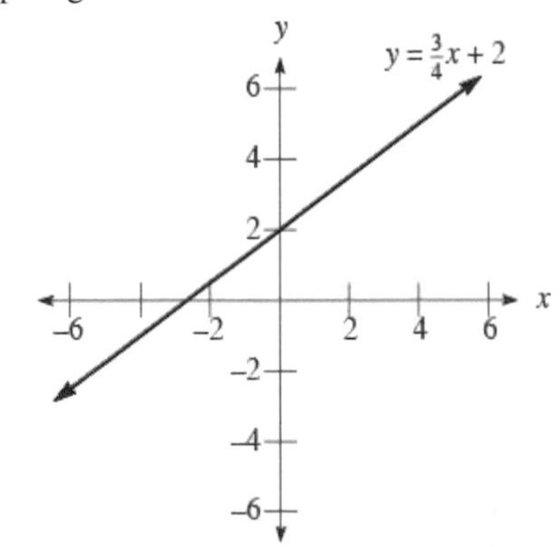

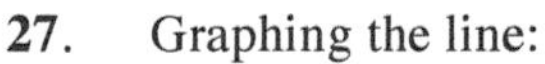

27. Graphing the line:

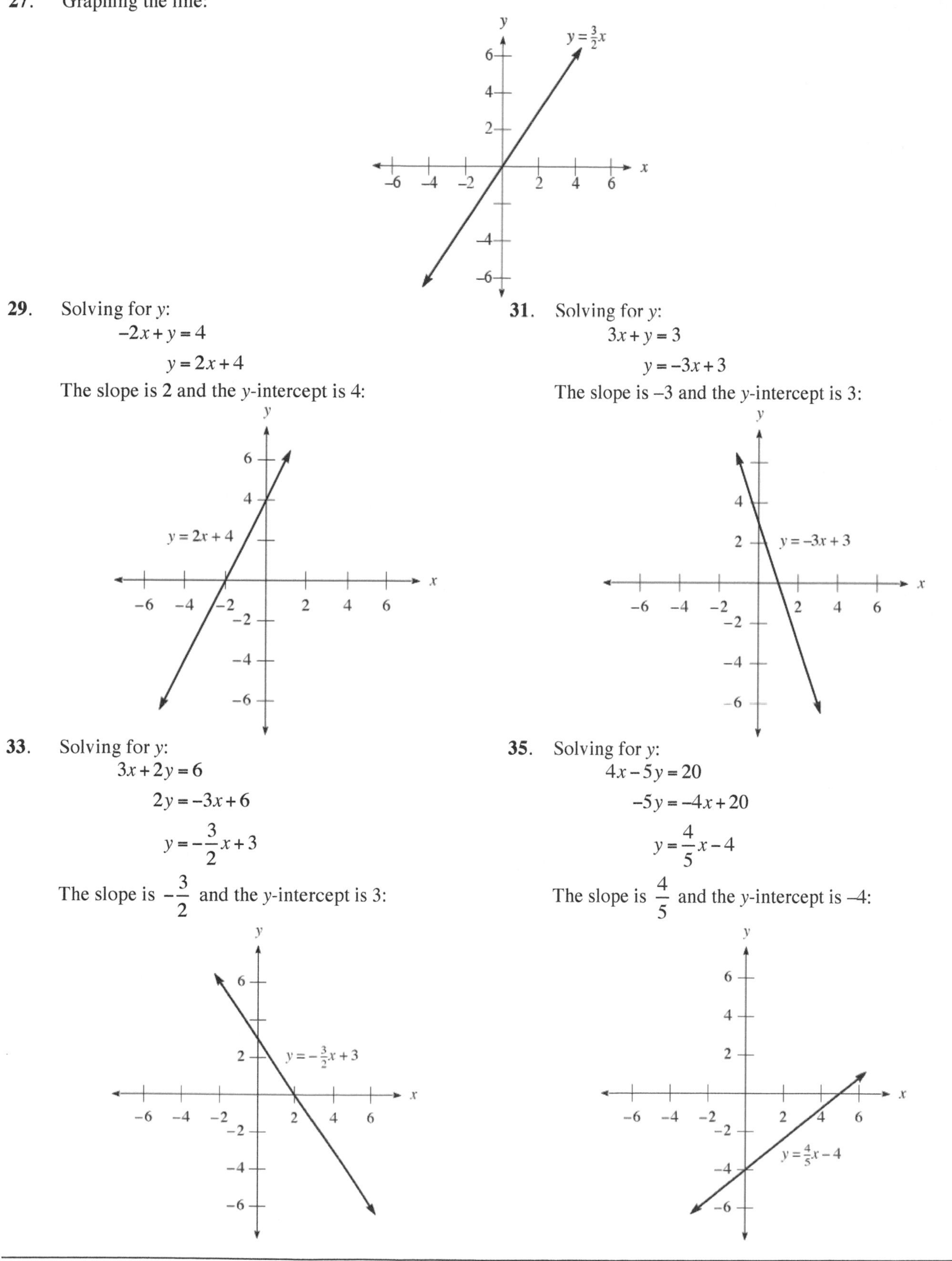

29. Solving for y:

$$-2x + y = 4$$
$$y = 2x + 4$$

The slope is 2 and the y-intercept is 4:

31. Solving for y:

$$3x + y = 3$$
$$y = -3x + 3$$

The slope is –3 and the y-intercept is 3:

33. Solving for y:

$$3x + 2y = 6$$
$$2y = -3x + 6$$
$$y = -\frac{3}{2}x + 3$$

The slope is $-\frac{3}{2}$ and the y-intercept is 3:

35. Solving for y:

$$4x - 5y = 20$$
$$-5y = -4x + 20$$
$$y = \frac{4}{5}x - 4$$

The slope is $\frac{4}{5}$ and the y-intercept is –4:

37. Solving for y:

$$-2x-5y=10$$
$$-5y=2x+10$$
$$y=-\frac{2}{5}x-2$$

The slope is $-\frac{2}{5}$ and the y-intercept is -2:

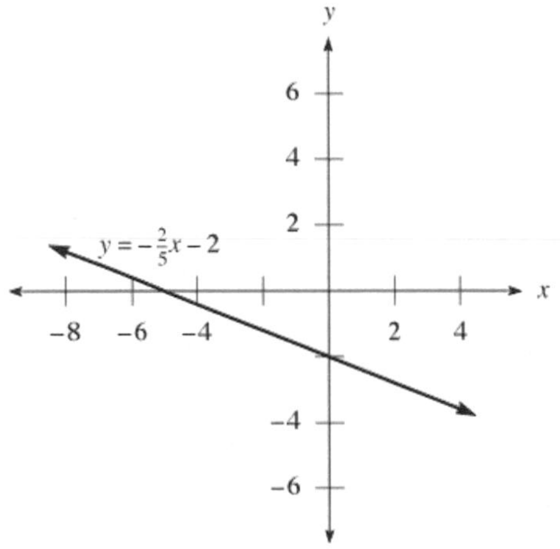

39. The slope-intercept form is $y=\frac{2}{3}x+1$.

41. The slope-intercept form is $y=\frac{3}{2}x-1$.

43. The slope-intercept form is $y=-\frac{2}{3}x+3$.

45. The slope-intercept form is $y=2x-4$.

47. The y-intercept is 3, and the slope is: $m=\frac{3-0}{0-(-1)}=\frac{3}{1}=3$. The slope-intercept form is $y=3x+3$.

49. The y-intercept is -1, and the slope is: $m=\frac{0-(-1)}{4-0}=\frac{0+1}{4}=\frac{1}{4}$. The slope-intercept form is $y=\frac{1}{4}x-1$.

51. Since $m=2$, we can substitute the point $(2,3)$ into the slope-intercept form:

$$y=2x+b$$
$$3=2(2)+b$$
$$3=4+b$$
$$b=-1$$

Writing the equation in standard form:

$$y=2x-1$$
$$-2x+y=-1$$
$$2x-y=1$$

53. Since $m=-\frac{1}{3}$, we can substitute the point $(1,-4)$ into the slope-intercept form:

$$y=-\frac{1}{3}x+b$$
$$-4=-\frac{1}{3}(1)+b$$
$$-4=-\frac{1}{3}+b$$
$$b=-\frac{11}{3}$$

Writing the equation in standard form:

$$y = -\frac{1}{3}x - \frac{11}{3}$$
$$3y = -x - 11$$
$$x + 3y = -11$$

55. Since $m = \frac{3}{5}$, we can substitute the point $(-2,6)$ into the slope-intercept form:

$$y = \frac{3}{5}x + b$$
$$6 = \frac{3}{5}(-2) + b$$
$$6 = -\frac{6}{5} + b$$
$$b = \frac{36}{5}$$

Writing the equation in standard form:

$$y = \frac{3}{5}x + \frac{36}{5}$$
$$5y = 3x + 36$$
$$-3x + 5y = 36$$
$$3x - 5y = -36$$

57. Since $m = -\frac{4}{3}$, we can substitute the point $(4,0)$ into the slope-intercept form:

$$y = -\frac{4}{3}x + b$$
$$0 = -\frac{4}{3}(4) + b$$
$$0 = -\frac{16}{3} + b$$
$$b = \frac{16}{3}$$

Writing the equation in standard form:

$$y = -\frac{4}{3}x + \frac{16}{3}$$
$$3y = -4x + 16$$
$$4x + 3y = 16$$

59. **a.** Since $m = 3$, we can substitute the point $(1,2)$ into the slope-intercept form:

$$y = 3x + b$$
$$2 = 3(1) + b$$
$$2 = 3 + b$$
$$b = -1$$

The slope-intercept form is $y = 3x - 1$.

b. Since $m=-\frac{1}{3}$, we can substitute the point $(1,2)$ into the slope-intercept form:

$$y=-\frac{1}{3}x+b$$
$$2=-\frac{1}{3}(1)+b$$
$$2=-\frac{1}{3}+b$$
$$b=\frac{7}{3}$$

The slope-intercept form is $y=-\frac{1}{3}x+\frac{7}{3}$.

61. First solve for y to find the slope:

$$x+2y=6$$
$$2y=-x+6$$
$$y=-\frac{1}{2}x+3$$

a. Since $m=-\frac{1}{2}$, we can substitute the point $(-4,-1)$ into the slope-intercept form:

$$y=-\frac{1}{2}x+b$$
$$-1=-\frac{1}{2}(-4)+b$$
$$-1=2+b$$
$$b=-3$$

The slope-intercept form is $y=-\frac{1}{2}x-3$.

b. Since $m=2$, we can substitute the point $(-4,-1)$ into the slope-intercept form:

$$y=2x+b$$
$$-1=2(-4)+b$$
$$-1=-8+b$$
$$b=7$$

The slope-intercept form is $y=2x+7$.

63. First solve for y to find the slope:

$$2x+3y=6$$
$$3y=-2x+6$$
$$y=-\frac{2}{3}x+2$$

a. Since $m=-\frac{2}{3}$, we can substitute the point $(3,4)$ into the slope-intercept form:

$$y=-\frac{2}{3}x+b$$
$$4=-\frac{2}{3}(3)+b$$
$$4=-2+b$$
$$b=6$$

The slope-intercept form is $y=-\frac{2}{3}x+6$.

b. Since $m=\frac{3}{2}$, we can substitute the point $(3,4)$ into the slope-intercept form:

$$y=\frac{3}{2}x+b$$
$$4=\frac{3}{2}(3)+b$$
$$4=\frac{9}{2}+b$$
$$b=-\frac{1}{2}$$

The slope-intercept form is $y=\frac{3}{2}x-\frac{1}{2}$.

65. Since the slope is 0 and the line passes through $(3,2)$, the equation is $y=2$.

67. Since the slope is 0 and the line passes through $(-1,5)$, the equation is $y=5$.

69. **a.** Solving the equation:

$$-2x+1=6$$
$$-2x=5$$
$$x=-\frac{5}{2}$$

b. Writing in slope-intercept form:

$$-2x+y=6$$
$$y=2x+6$$

c. Substituting $x = 0$:

$$-2(0)+y=6$$
$$0+y=6$$
$$y=6$$

d. Finding the slope:

$$-2x+y=6$$
$$y=2x+6$$

The slope is 2.

e. Graphing the line:

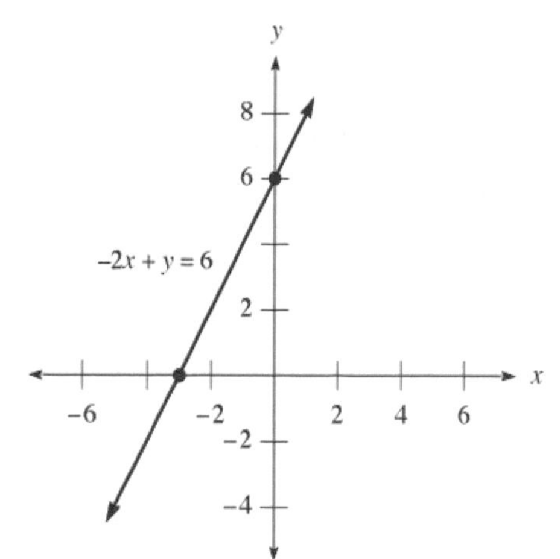

71. **a.** After 5 years, the copier is worth $6,000.

b. The copier is worth $12,000 after 3 years.

c. The slope of the line is –3,000.

d. The copier is decreasing in value by $3,000 per year.

e. The equation is $V=-3{,}000t+21{,}000$.

73. Solving for y:

$$-y-3=-2(x+4)$$
$$-y-3=-2x-8$$
$$-y=-2x-5$$
$$y=2x+5$$

75. Solving for y:

$$-y-3=-\frac{2}{3}(x+3)$$
$$-y-3=-\frac{2}{3}x-2$$
$$-y=-\frac{2}{3}x+1$$
$$y=\frac{2}{3}x-1$$

77. Solving for y:

$$-\frac{y-1}{x}=\frac{3}{2}$$
$$-2y+2=3x$$
$$-2y=3x-2$$
$$y=-\frac{3}{2}x+1$$

79. Finding the slope: $m=\frac{-1-3}{3-(-3)}=\frac{-4}{6}=-\frac{2}{3}$

81. Finding the slope: $m=\frac{1-0}{4-0}=\frac{1}{4}$

83. Finding the slope: $m=\frac{5-1}{2-2}=\frac{4}{0}$, which is undefined

85. The slope is $-\frac{5}{6}$. The correct answer is c.

87. Since $m=3$, we can substitute the point $(-4,2)$ into the slope-intercept form:

$$y=3x+b$$
$$2=3(-4)+b$$
$$2=-12+b$$
$$b=14$$

The slope-intercept form is $y=3x+14$. The correct answer is c.

3.6 Point-Slope Form

1. Using the point-slope formula:

$$y-(-5)=2(x-(-2))$$
$$y+5=2(x+2)$$
$$y+5=2x+4$$
$$-2x+y=-1$$
$$2x-y=1$$

3. Using the point-slope formula:

$$y-1=-\frac{1}{2}(x-(-4))$$
$$y-1=-\frac{1}{2}(x+4)$$
$$y-1=-\frac{1}{2}x-2$$
$$2y-2=-x-4$$
$$x+2y=-2$$

5. Using the point-slope formula:

$$y-(-3)=\frac{3}{2}(x-2)$$
$$y+3=\frac{3}{2}x-3$$
$$2y+6=3x-6$$
$$-3x+2y=-12$$
$$3x-2y=12$$

7. Using the point-slope formula:

$$y-4=-3(x-(-1))$$
$$y-4=-3(x+1)$$
$$y-4=-3x-3$$
$$3x+y=1$$

9. Using the point-slope formula:

$$y-0=-\frac{2}{3}(x-6)$$
$$y=-\frac{2}{3}x+4$$
$$3y=-2x+12$$
$$2x+3y=12$$

11. Using the point-slope formula:

$$y-(-1)=\frac{1}{5}(x-0)$$
$$y+1=\frac{1}{5}x$$
$$5y+5=x$$
$$-x+5y=-5$$
$$x-5y=5$$

13. Using the point-slope formula:

$$y-3=0(x-5)$$
$$y-3=0$$
$$y=3$$

15. Using the point-slope formula:

$$y-0=\frac{3}{2}(x-0)$$
$$y=\frac{3}{2}x$$
$$2y=3x$$
$$-3x+2y=0$$
$$3x-2y=0$$

17. Finding the slope: $m=\frac{-1-(-4)}{1-(-2)}=\frac{-1+4}{1+2}=\frac{3}{3}=1$

Using the point-slope formula:

$$y-(-4)=1(x-(-2))$$
$$y+4=x+2$$
$$y=x-2$$

19. Finding the slope: $m=\frac{1-(-5)}{2-(-1)}=\frac{1+5}{2+1}=\frac{6}{3}=2$

Using the point-slope formula:

$$y-1=2(x-2)$$
$$y-1=2x-4$$
$$y=2x-3$$

21. Finding the slope: $m=\frac{6-(-2)}{3-(-3)}=\frac{6+2}{3+3}=\frac{8}{6}=\frac{4}{3}$

Using the point-slope formula:

$$y-6=\frac{4}{3}(x-3)$$
$$y-6=\frac{4}{3}x-4$$
$$y=\frac{4}{3}x+2$$

23. Finding the slope: $m=\frac{-5-(-1)}{3-(-3)}=\frac{-5+1}{3+3}=\frac{-4}{6}=-\frac{2}{3}$

Using the point-slope formula:

$$y-(-5)=-\frac{2}{3}(x-3)$$
$$y+5=-\frac{2}{3}x+2$$
$$y=-\frac{2}{3}x-3$$

25. Finding the slope: $m=\frac{3-0}{0-2}=-\frac{3}{2}$

Using the point-slope formula:

$$y-3=-\frac{3}{2}(x-0)$$
$$y-3=-\frac{3}{2}x$$
$$y=-\frac{3}{2}x+3$$

27. Finding the slope: $m=\frac{5-0}{-1-(-1)}=\frac{5}{0}$, which is undefined. Since this is a vertical line, its equation is $x=-1$.

29. Finding the slope: $m=\frac{1-1}{5-1}=\frac{0}{4}=0$. Since this is a horizontal line, its equation is $y=1$.

31. The slope is given by: $m=\dfrac{0-2}{3-0}=-\dfrac{2}{3}$. Since $b = 2$, the equation is $y=-\dfrac{2}{3}x+2$.

33. The slope is given by: $m=\dfrac{0-(-5)}{-2-0}=-\dfrac{5}{2}$. Since $b = -5$, the equation is $y=-\dfrac{5}{2}x-5$.

35. Since this is a vertical line, its equation is $x=3$.

37.
a. This is a horizontal line, so its equation is $y=3$.
b. This is a vertical line, so its equation is $x=-2$.

39.
a. This is a horizontal line, so its equation is $y=2$.
b. This is a vertical line, so its equation is $x=6$.

41.
a. Since $m=2$, we can substitute the point $(3,-1)$ into the point-slope form:

$$\begin{aligned} y-(-1)&=2(x-3)\\ y+1&=2x-6\\ -2x+y&=-7\\ 2x-y&=7 \end{aligned}$$

b. Since $m=-\dfrac{1}{2}$, we can substitute the point $(3,-1)$ into the point-slope form:

$$\begin{aligned} y-(-1)&=-\frac{1}{2}(x-3)\\ y+1&=-\frac{1}{2}x+\frac{3}{2}\\ 2y+2&=-x+3\\ x+2y&=1 \end{aligned}$$

43.
a. Since $m=-\dfrac{1}{3}$, we can substitute the point $(-4,-3)$ into the point-slope form:

$$\begin{aligned} y-(-3)&=-\frac{1}{3}(x-(-4))\\ y+3&=-\frac{1}{3}x-\frac{4}{3}\\ 3y+9&=-x-4\\ x+3y&=-13 \end{aligned}$$

b. Since $m=3$, we can substitute the point $(-4,-3)$ into the point-slope form:

$$\begin{aligned} y-(-3)&=3(x-(-4))\\ y+3&=3x+12\\ -3x+y&=9\\ 3x-y&=-9 \end{aligned}$$

45. First solve for y to find the slope:

$$\begin{aligned} 3x+4y&=12\\ 4y&=-3x+12\\ y&=-\frac{3}{4}x+3 \end{aligned}$$

a. Since $m = -\frac{3}{4}$, we can substitute the point $(0,5)$ into the point-slope form:

$$y - 5 = -\frac{3}{4}(x - 0)$$
$$y - 5 = -\frac{3}{4}x$$
$$4y - 20 = -3x$$
$$3x + 4y = 20$$

b. Since $m = \frac{4}{3}$, we can substitute the point $(0,5)$ into the slope-intercept form:

$$y - 5 = \frac{4}{3}(x - 0)$$
$$y - 5 = \frac{4}{3}x$$
$$3y - 15 = 4x$$
$$-4x + 3y = 15$$
$$4x - 3y = -15$$

47. First solve for y to find the slope:

$$2x - 3y = 6$$
$$-3y = -2x + 6$$
$$y = \frac{2}{3}x - 2$$

a. Since $m = \frac{2}{3}$, we can substitute the point $(1,0)$ into the point-slope form:

$$y - 0 = \frac{2}{3}(x - 1)$$
$$y = \frac{2}{3}x - \frac{2}{3}$$
$$3y = 2x - 2$$
$$-2x + 3y = -2$$
$$2x - 3y = 2$$

b. Since $m = -\frac{3}{2}$, we can substitute the point $(1,0)$ into the slope-intercept form:

$$y - 0 = -\frac{3}{2}(x - 1)$$
$$y = -\frac{3}{2}x + \frac{3}{2}$$
$$2y = -3x + 3$$
$$3x + 2y = 3$$

49. **a.** Since this slope is 0, the parallel line is horizontal. Its equation is $y = 2$.

b. Since this slope is 0, the perpendicular line is vertical. Its equation is $x = 4$.

51. **a.** The point $(32,0)$ means the Celsius temperature is 0°C.

b. Finding the slope: $m = \frac{30-0}{86-32} = \frac{30}{54} = \frac{5}{9}$

c. Substituting the point $(32,0)$ into the point-slope form:

$$y - 0 = \frac{5}{9}(x - 32)$$
$$y = \frac{5}{9}x - \frac{160}{9}$$

d. Substituting $x = 212$: $y = \frac{5}{9}(212) - \frac{160}{9} = \frac{1{,}060}{9} - \frac{160}{9} = \frac{900}{9} = 100$

Water boils at 100°C.

53. Graphing the line:

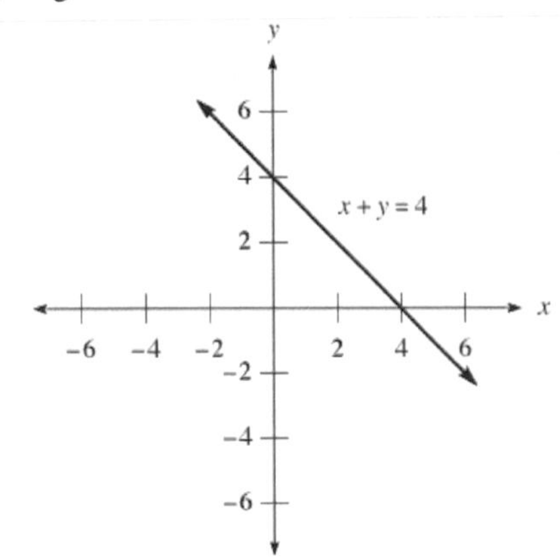

55. Graphing the line:

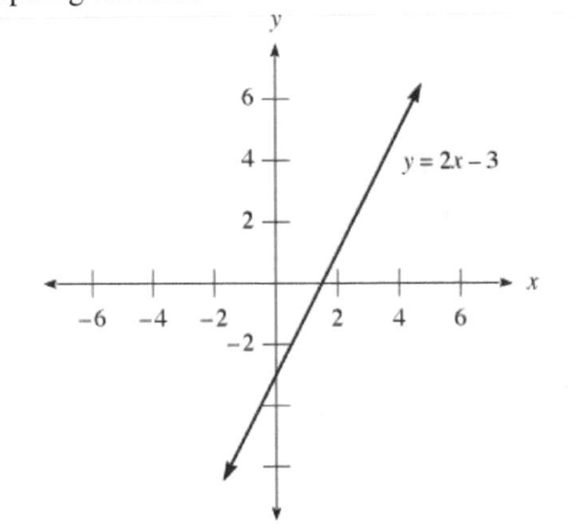

57. Graphing the line:

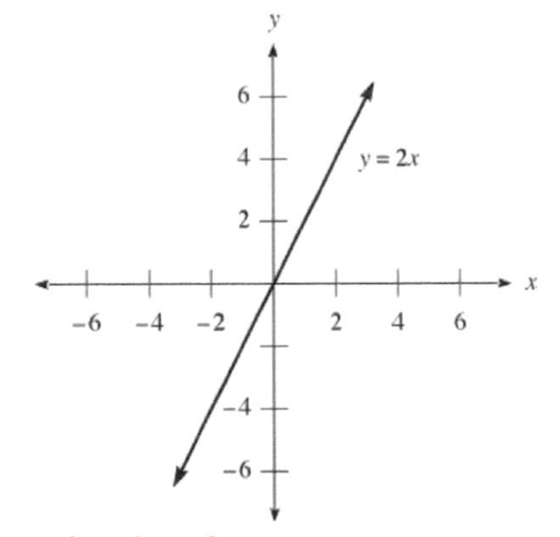

59. Substituting the point $(2,-4)$ into the point-slope form:

$$y - (-4) = -3(x - 2)$$
$$y + 4 = -3x + 6$$
$$y = -3x + 2$$

The correct answer is b.

3.7 Linear Inequalities in Two Variables

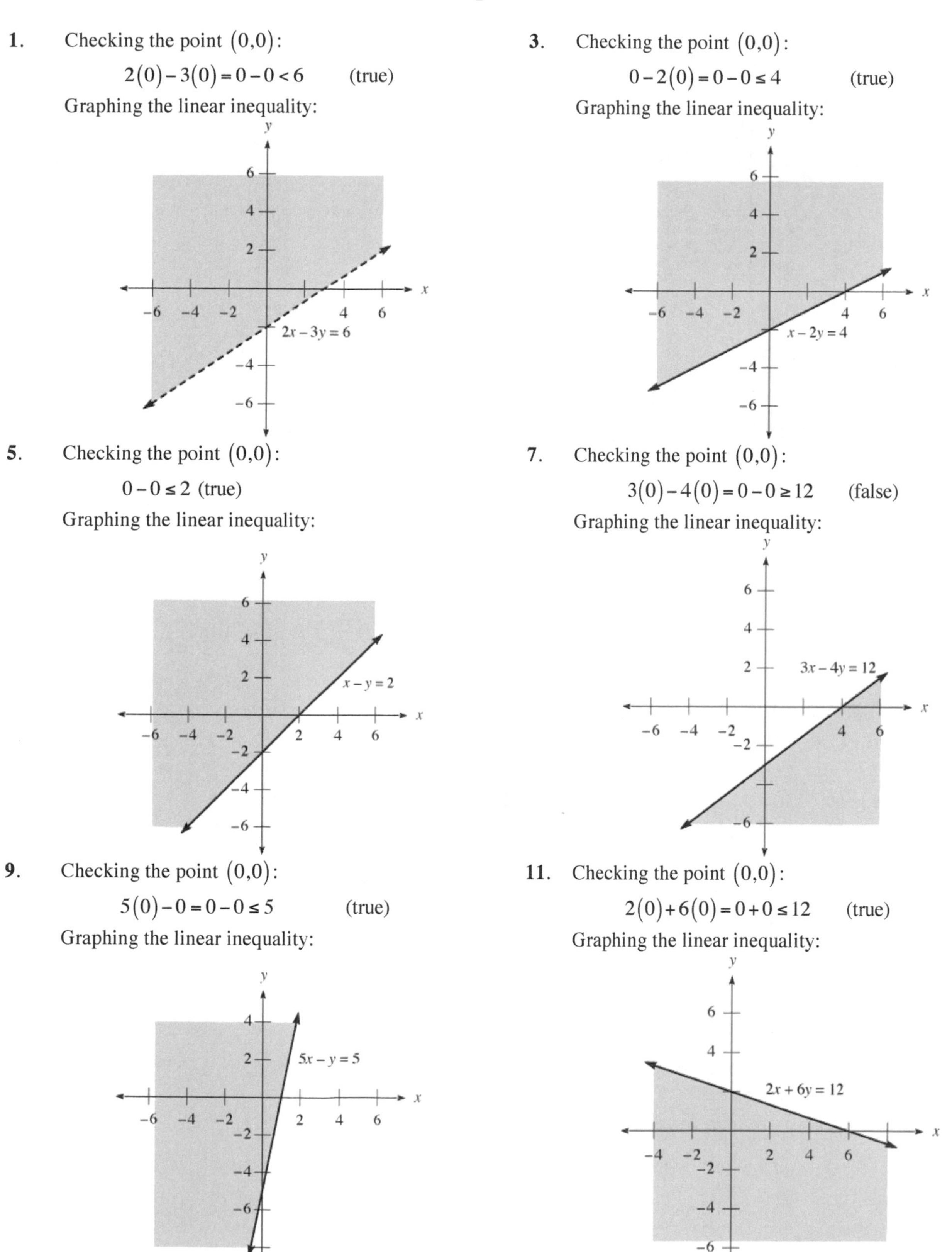

1. Checking the point $(0,0)$:

$$2(0)-3(0)=0-0<6 \qquad \text{(true)}$$

Graphing the linear inequality:

3. Checking the point $(0,0)$:

$$0-2(0)=0-0\leq 4 \qquad \text{(true)}$$

Graphing the linear inequality:

5. Checking the point $(0,0)$:

$$0-0\leq 2 \text{ (true)}$$

Graphing the linear inequality:

7. Checking the point $(0,0)$:

$$3(0)-4(0)=0-0\geq 12 \qquad \text{(false)}$$

Graphing the linear inequality:

9. Checking the point $(0,0)$:

$$5(0)-0=0-0\leq 5 \qquad \text{(true)}$$

Graphing the linear inequality:

11. Checking the point $(0,0)$:

$$2(0)+6(0)=0+0\leq 12 \qquad \text{(true)}$$

Graphing the linear inequality:

13. Graphing the linear inequality:

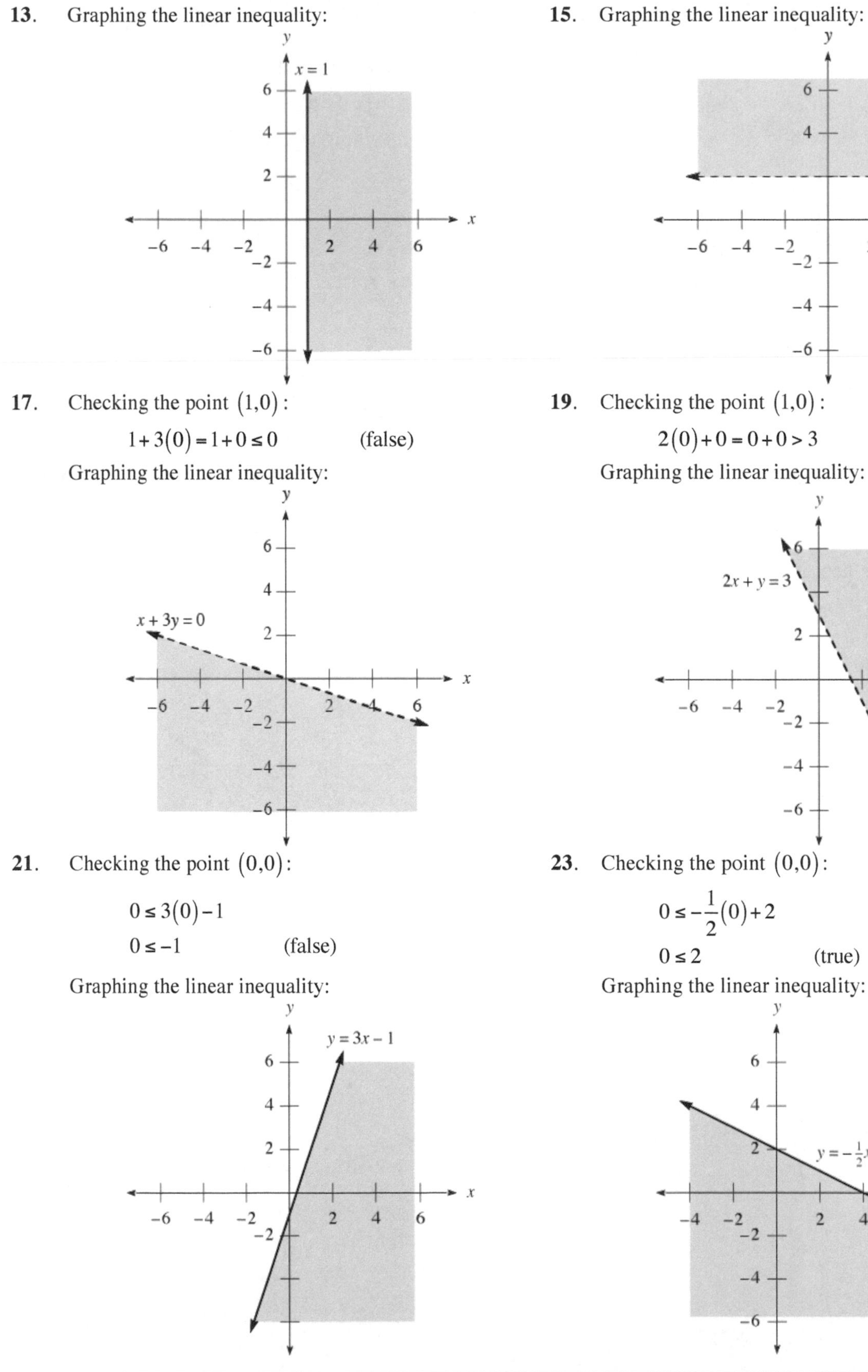

15. Graphing the linear inequality:

17. Checking the point $(1,0)$:

$$1 + 3(0) = 1 + 0 \leq 0 \qquad \text{(false)}$$

Graphing the linear inequality:

19. Checking the point $(1,0)$:

$$2(0) + 0 = 0 + 0 > 3 \qquad \text{(false)}$$

Graphing the linear inequality:

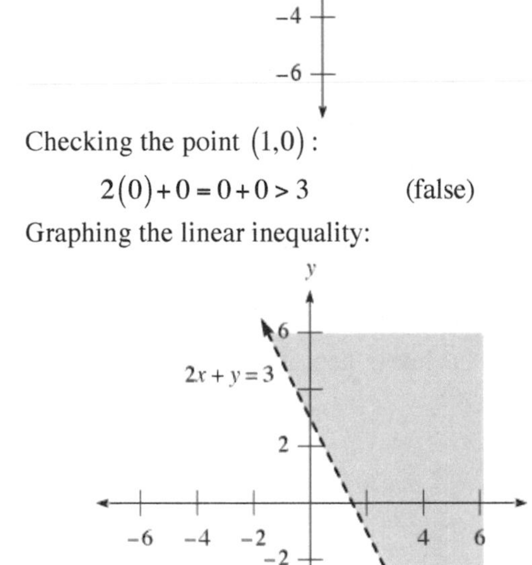

21. Checking the point $(0,0)$:

$$0 \leq 3(0) - 1$$
$$0 \leq -1 \qquad \text{(false)}$$

Graphing the linear inequality:

23. Checking the point $(0,0)$:

$$0 \leq -\frac{1}{2}(0) + 2$$
$$0 \leq 2 \qquad \text{(true)}$$

Graphing the linear inequality:

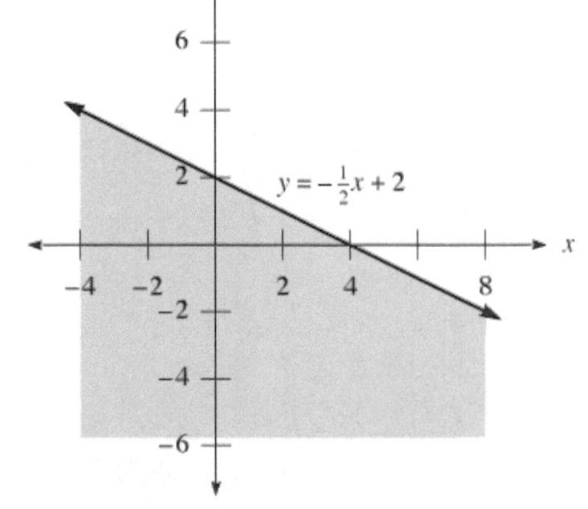

25. **a.** Solving the inequality:

$$4+3y<12$$
$$3y<8$$
$$y<\frac{8}{3}$$

b. Solving the inequality:

$$4-3y<12$$
$$-3y<8$$
$$y>-\frac{8}{3}$$

c. Solving for y:

$$4x+3y=12$$
$$3y=-4x+12$$
$$y=-\frac{4}{3}x+4$$

d. Graphing the inequality:

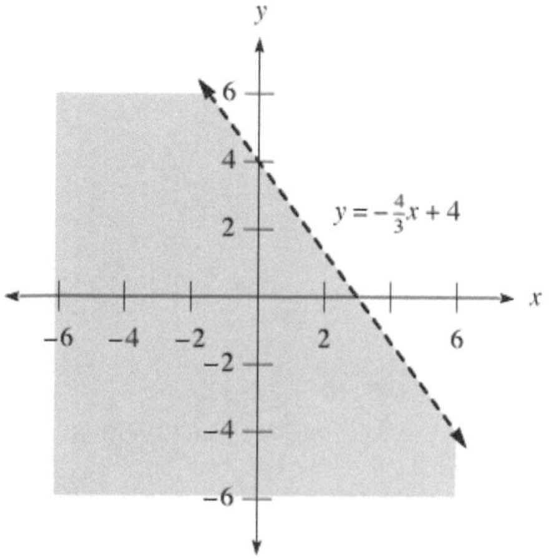

27. **a.** The slope is $\frac{2}{5}$ and the y-intercept is 2, so the equation is $y=\frac{2}{5}x+2$.

b. Since the shading is below the line, the inequality is $y<\frac{2}{5}x+2$.

c. Since the shading is above the line, the inequality is $y>\frac{2}{5}x+2$.

29. Simplifying the expression: $7-3(2x-4)-8=7-6x+12-8=-6x+11$

31. Solving the equation:

$$-\frac{3}{2}x=12$$
$$-\frac{2}{3}\left(-\frac{3}{2}x\right)=-\frac{2}{3}(12)$$
$$x=-8$$

33. Solving the equation:

$$8-2(x+7)=2$$
$$8-2x-14=2$$
$$-2x-6=2$$
$$-2x=8$$
$$x=-4$$

35. Solving for w:

$$2l+2w=P$$
$$2w=P-2l$$
$$w=\frac{P-2l}{2}$$

37. Solving the inequality:

$$3-2x>5$$
$$3-3-2x>5-3$$
$$-2x>2$$
$$-\frac{1}{2}(-2x)<-\frac{1}{2}(2)$$
$$x<-1$$

The solution set is $\{x \mid x<-1\}$. Graphing the inequality:

−1 0

39. Solving for y:

$$3x-2y\le 12$$
$$3x-3x-2y\le -3x+12$$
$$-2y\le -3x+12$$
$$-\frac{1}{2}(-2y)\ge -\frac{1}{2}(-3x+12)$$
$$y\ge \frac{3}{2}x-6$$

41. Let w represent the width and $3w + 5$ represent the length. Using the perimeter formula:

$$2(w)+2(3w+5)=26$$
$$2w+6w+10=26$$
$$8w+10=26$$
$$8w=16$$
$$w=2$$
$$3w+5=3(2)+5=11$$

The width is 2 inches and the length is 11 inches.

43. Since $(0,0)$ does not check the inequality, the inequality must shade below the line $x-2y=4$. The correct answer is c.

Chapter 3 Test

1. Graphing the ordered pair:

2. Graphing the ordered pair:

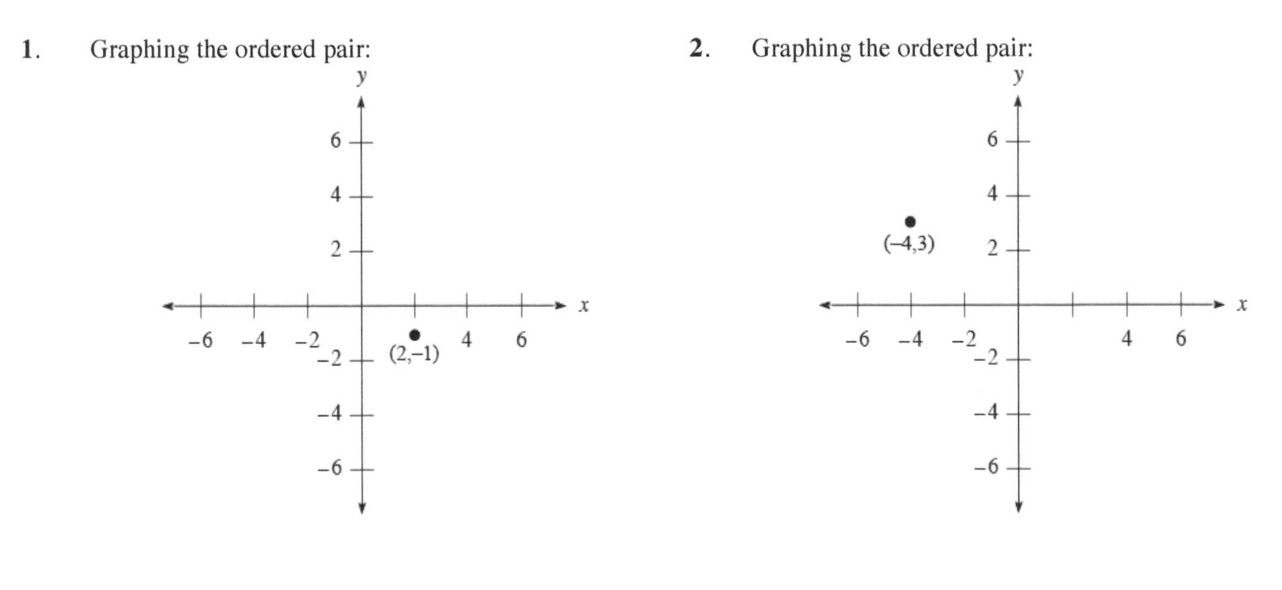

3. Graphing the ordered pair:

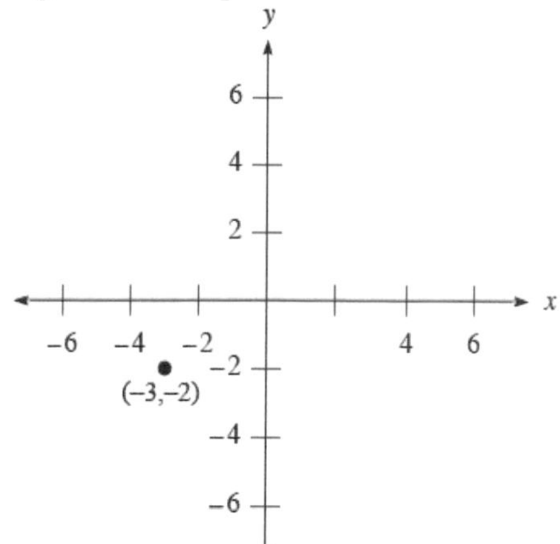

4. Graphing the ordered pair:

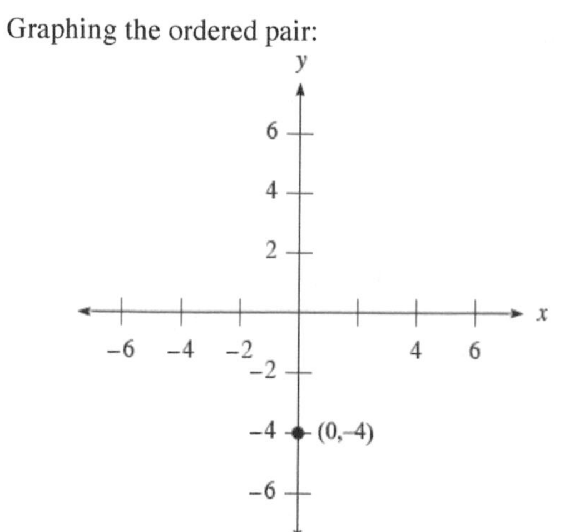

5. Substituting $x = 0$, $y = 0$, $x = 4$, and $y = -6$:

$$\begin{aligned} 3(0)-2y&=6 \\ 0-2y&=6 \\ -2y&=6 \\ y&=-3 \end{aligned} \qquad \begin{aligned} 3x-2(0)&=6 \\ 3x-0&=6 \\ 3x&=6 \\ x&=2 \end{aligned} \qquad \begin{aligned} 3(4)-2y&=6 \\ 12-2y&=6 \\ -2y&=-6 \\ y&=3 \end{aligned} \qquad \begin{aligned} 3x-2(-6)&=6 \\ 3x+12&=6 \\ 3x&=-6 \\ x&=-2 \end{aligned}$$

The ordered pairs are $(0,-3),(2,0),(4,3)$, and $(-2,-6)$.

6. Substituting each ordered pair into the equation:

$$\begin{aligned} (0,7):&\quad -3(0)+7=0+7=7 \\ (2,-1):&\quad -3(2)+7=-6+7=1\neq-1 \\ (4,-5):&\quad -3(4)+7=-12+7=-5 \\ (-5,-3):&\quad -3(-5)+7=15+7=22\neq-3 \end{aligned}$$

The ordered pairs $(0,7)$ and $(4,-5)$ are solutions.

7. Graphing the line:

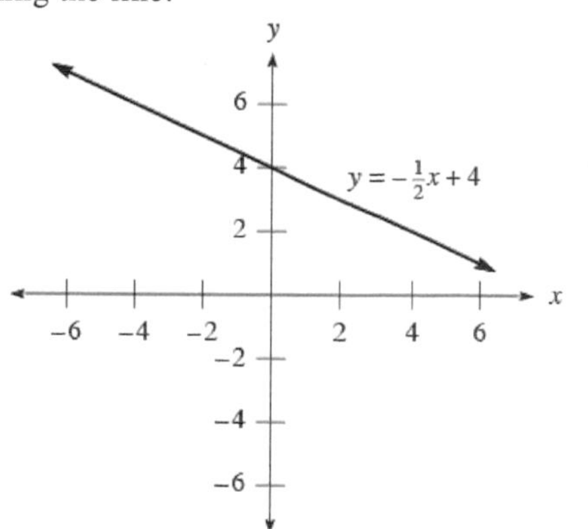

8. Graphing the line:

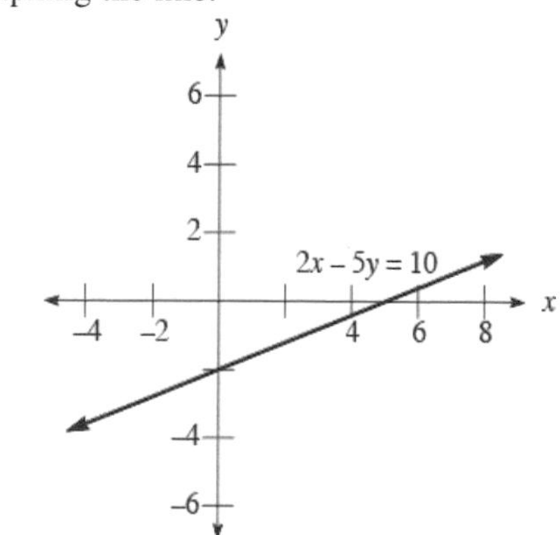

9. Graphing the line:

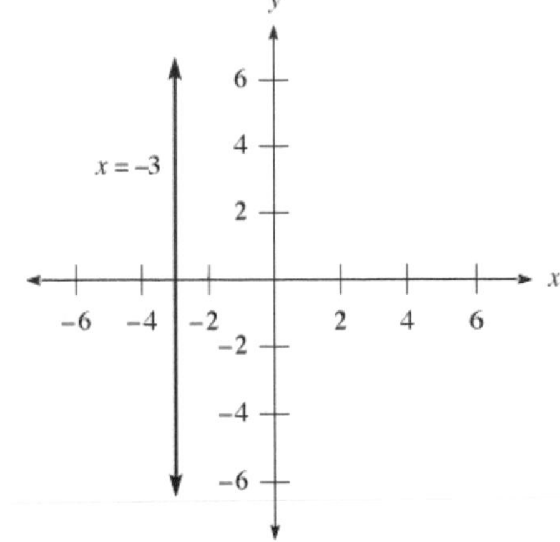

10. Graphing the line:

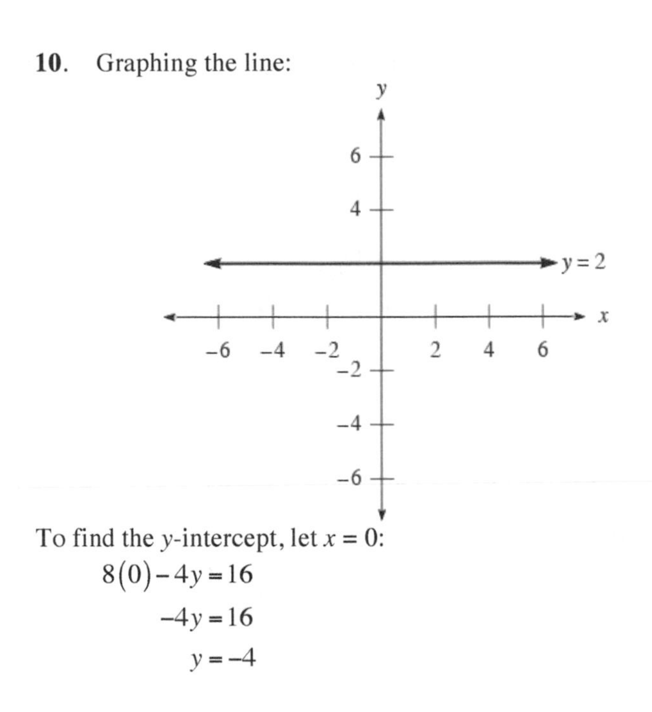

11. To find the x-intercept, let $y = 0$:

$$8x - 4(0) = 16$$
$$8x = 16$$
$$x = 2$$

To find the y-intercept, let $x = 0$:

$$8(0) - 4y = 16$$
$$-4y = 16$$
$$y = -4$$

The x-intercept is 2 and the y-intercept is –4.

12. To find the x-intercept, let $y = 0$:

$$0 = \frac{3}{2}x + 6$$
$$\frac{3}{2}x = -6$$
$$x = -4$$

To find the y-intercept, let $x = 0$:

$$y = \frac{3}{2}(0) + 6$$
$$y = 6$$

The x-intercept is –4 and the y-intercept is 6.

13. There is no x-intercept, and the y-intercept is 3.

14. The x-intercept is –2, and there is no y-intercept.

15. The slope is given by: $m = \dfrac{6-2}{-5-3} = \dfrac{4}{-8} = -\dfrac{1}{2}$

16. The slope is given by: $m = \dfrac{1-9}{7-0} = \dfrac{-8}{7} = -\dfrac{8}{7}$

17. The slope is given by: $m = \dfrac{4-1}{0-1} = \dfrac{3}{-1} = -3$

18. Since the line is vertical, the slope is undefined.

19. Since the line is horizontal, the slope is 0.

20. The slope is $m = -\dfrac{1}{2}$ and the y-intercept is $b = 6$.

21. The slope is $m = 3$ and the y-intercept is $b = 0$.

22. The parallel line slope is $m = \dfrac{3}{7}$.

23. The perpendicular line slope is $m = \dfrac{1}{4}$.

24. Since $m = 3$ and $b = -5$, the equation is $y = 3x - 5$.

25. Using the point-slope formula:

$$y - 1 = -\frac{1}{2}(x - 4)$$
$$y - 1 = -\frac{1}{2}x + 2$$
$$y = -\frac{1}{2}x + 3$$

26. The slope is given by: $m=\frac{2-(-4)}{-6-3}=\frac{6}{-9}=-\frac{2}{3}$

Using the point-slope formula:

$$y-(-4)=-\frac{2}{3}(x-3)$$
$$y+4=-\frac{2}{3}x+2$$
$$y=-\frac{2}{3}x-2$$

27. The slope is given by: $m=\frac{6-0}{-2-3}=-\frac{6}{5}$

Using the point-slope formula:

$$y-0=-\frac{6}{5}(x-3)$$
$$y-0=-\frac{6}{5}x+\frac{18}{5}$$
$$y=-\frac{6}{5}x+\frac{18}{5}$$

28. The parallel slope is $m=\frac{3}{2}$. Using the point-slope formula:

$$y-(-2)=\frac{3}{2}(x-1)$$
$$y+2=\frac{3}{2}x-\frac{3}{2}$$
$$2y+4=3x-3$$
$$-3x+2y=-7$$
$$3x-2y=7$$

29. First solve for y to find the slope:

$$3x+2y=6$$
$$2y=-3x+6$$
$$y=-\frac{3}{2}x+3$$

The perpendicular slope is $m=\frac{2}{3}$. Using the point-slope formula:

$$y-3=\frac{2}{3}(x-(-4))$$
$$y-3=\frac{2}{3}x+\frac{8}{3}$$
$$3y-9=2x+8$$
$$-2x+3y=17$$
$$2x-3y=-17$$

30. Since the slope is 0, the line is horizontal and its equation is $y=-5$.

31. Since the line is vertical, its equation is $x=-2$.

32. Graphing the linear inequality:

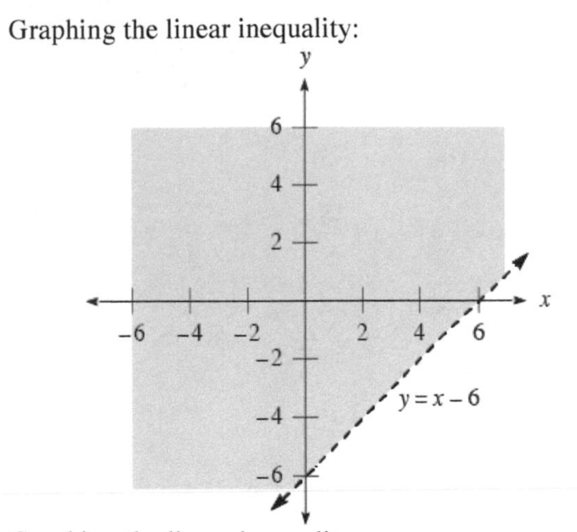

33. Graphing the linear inequality:

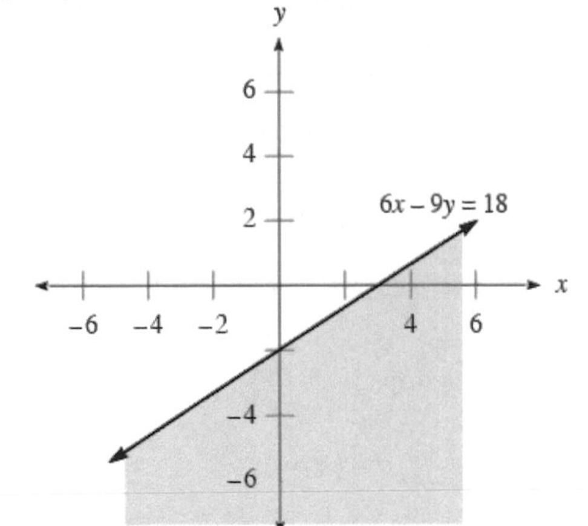

34. Graphing the linear inequality:

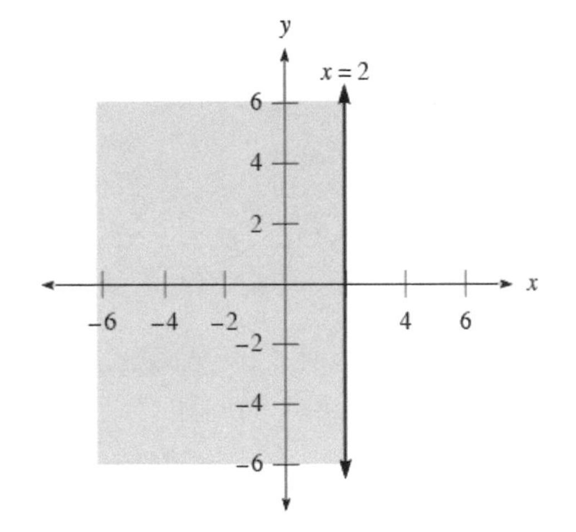

Chapter 4
Systems of Linear Equations

4.1 Solving Linear Systems by Graphing

1. **a.** Substituting $(-1,-2)$ into each equation:

$$x-y=-1-(-2)=-1+2=1$$
$$x+2y=-1-2(-2)=-1+4=3\neq 4$$

So $(-1,-2)$ is not a solution to the system.

b. Substituting $(2,1)$ into each equation:

$$x-y=2-1=1$$
$$x+2y=2+2(1)=2+2=4$$

So $(2,1)$ is a solution to the system.

c. Substituting $(2,2)$ into each equation:

$$x-y=2-2=0\neq 1$$
$$x+2y=2+2(2)=2+4=6\neq 4$$

So $(2,2)$ is not a solution to the system.

3. **a.** Substituting $(5,1)$ into each equation:

$$2x-4y=2(5)-4(1)=10-4=6$$
$$-x+2y=-5+2(1)=-5+2=-3$$

So $(5,1)$ is a solution to the system.

b. Substituting $(-1,-1)$ into each equation:

$$2x-4y=2(-1)-4(-1)=-2+4=2\neq 6$$
$$-x+2y=-(-1)+2(-1)=1-2=-1\neq -3$$

So $(-1,-1)$ is not a solution to the system.

c. Substituting $(3,0)$ into each equation:

$$2x-4y=2(3)-4(0)=6-0=6$$
$$-x+2y=-3+2(0)=-3+0=-3$$

So $(3,0)$ is a solution to the system.

5. Graphing both lines:

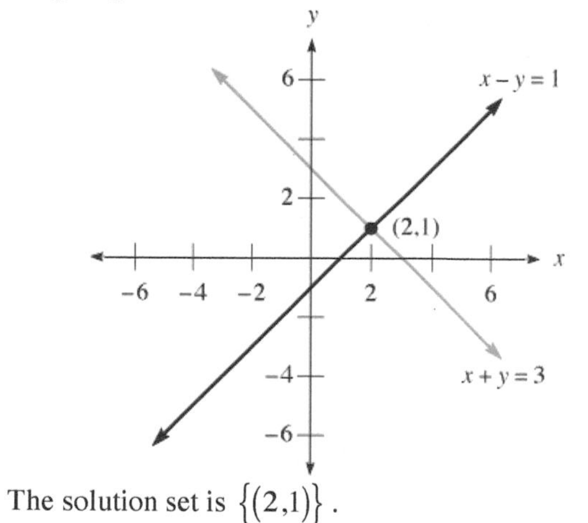

The solution set is $\{(2,1)\}$.

7. Graphing both lines:

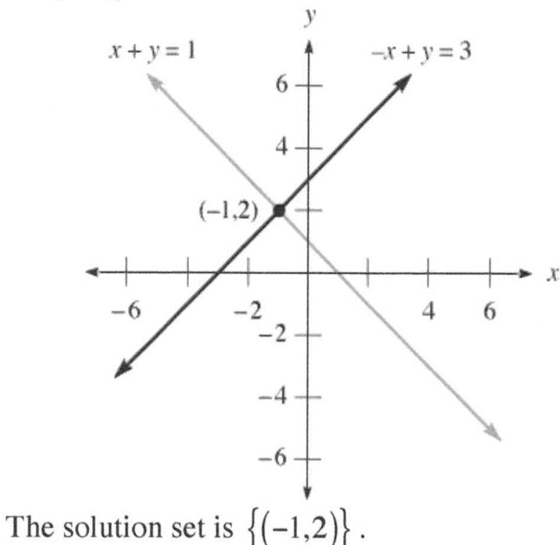

The solution set is $\{(-1,2)\}$.

9. Graphing both lines:

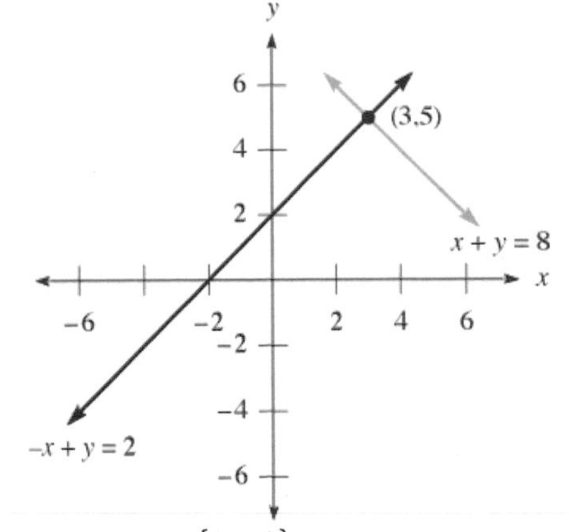

The solution set is $\{(3,5)\}$.

11. Graphing both lines:

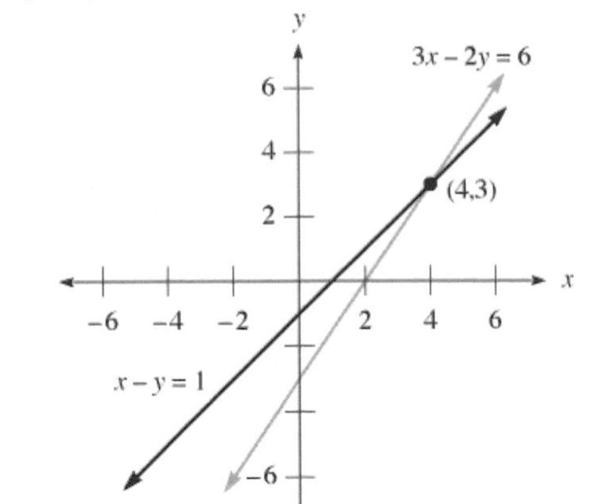

The solution set is $\{(4,3)\}$.

13. Graphing both lines:

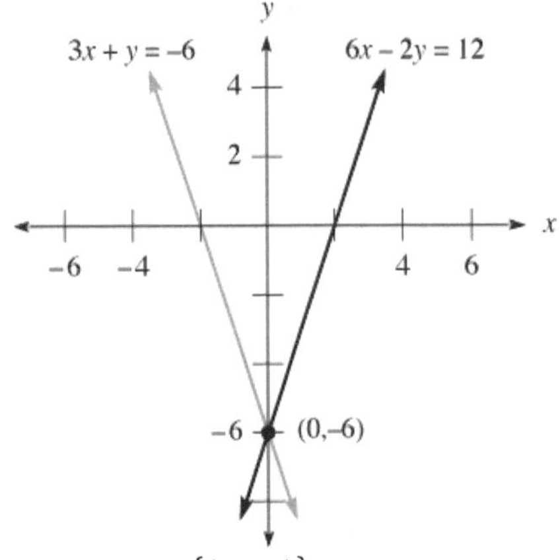

The solution set is $\{(0,-6)\}$.

15. Graphing both lines:

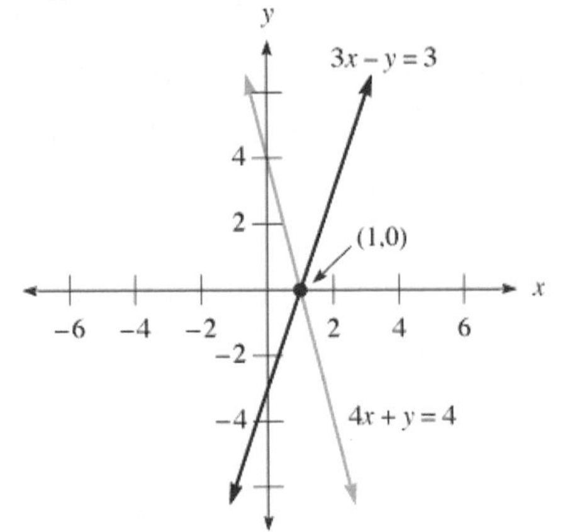

The solution set is $\{(1,0)\}$.

17. Graphing both lines:

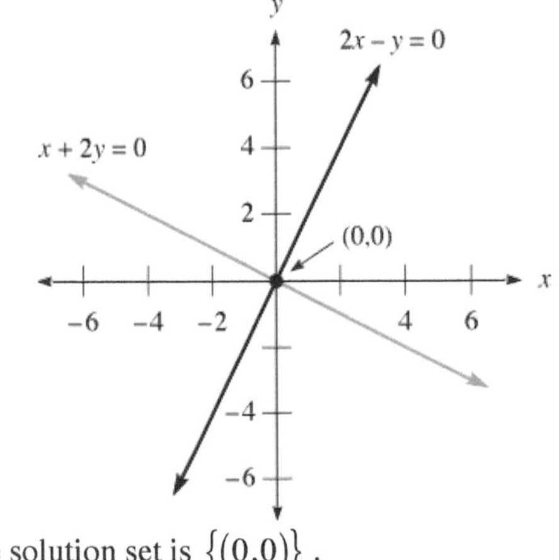

The solution set is $\{(0,0)\}$.

19. Graphing both lines:

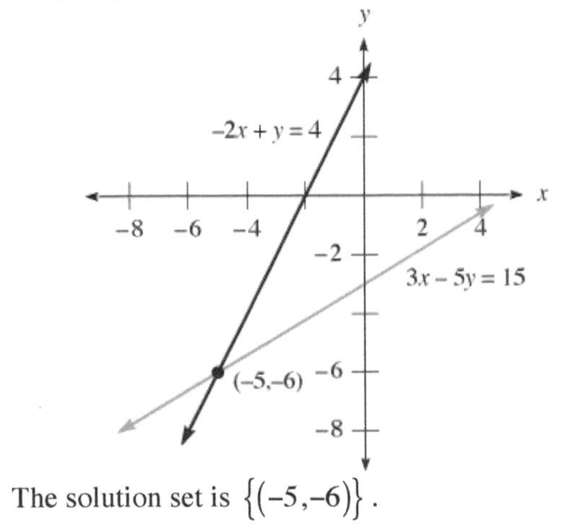

The solution set is $\{(-5,-6)\}$.

21. Graphing both lines:

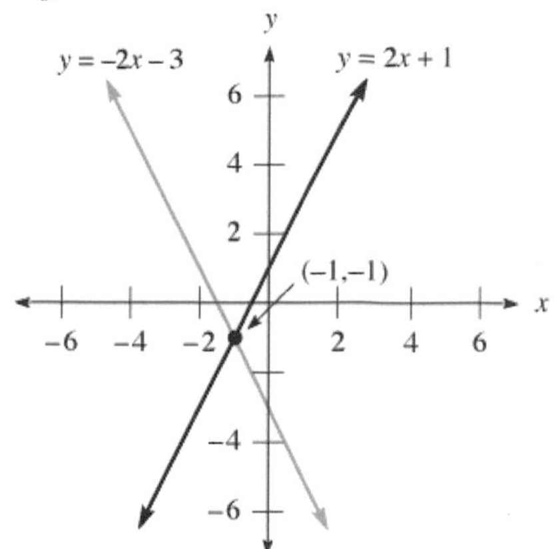

The solution set is $\{(-1,-1)\}$.

23. Graphing both lines:

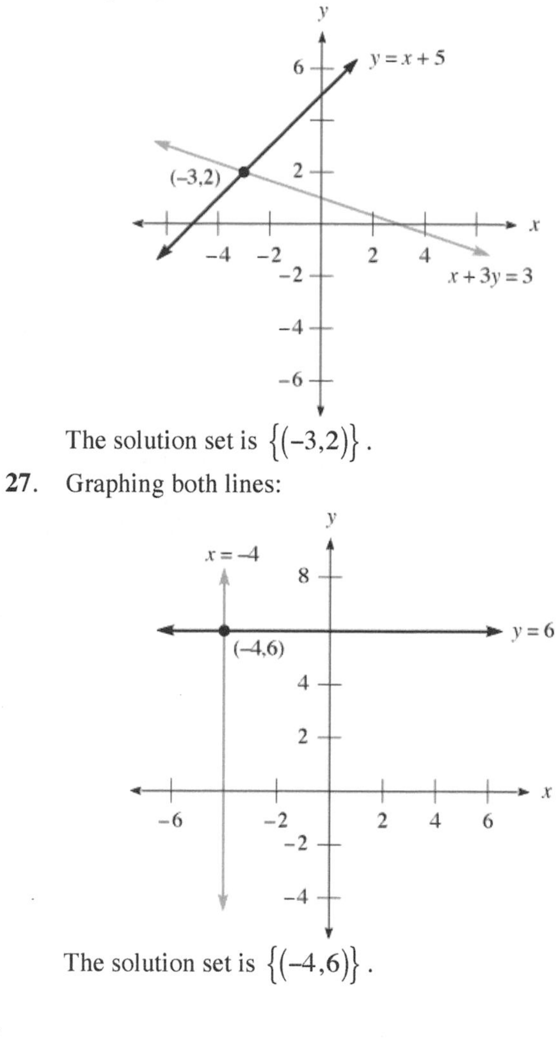

The solution set is $\{(-3,2)\}$.

25. Graphing both lines:

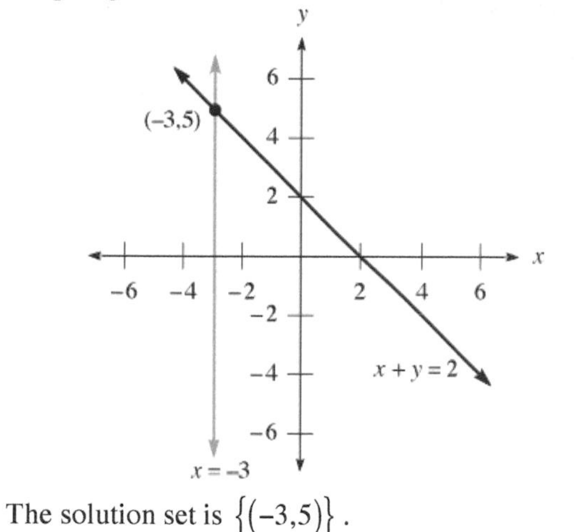

The solution set is $\{(-3,5)\}$.

27. Graphing both lines:

The solution set is $\{(-4,6)\}$.

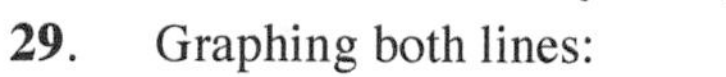

29. Graphing both lines:

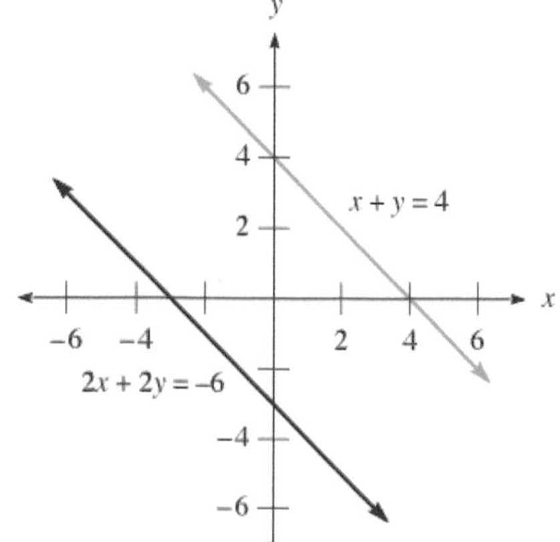

There is no intersection (the lines are parallel). The solution set is $\varnothing$.

31. Graphing both lines:

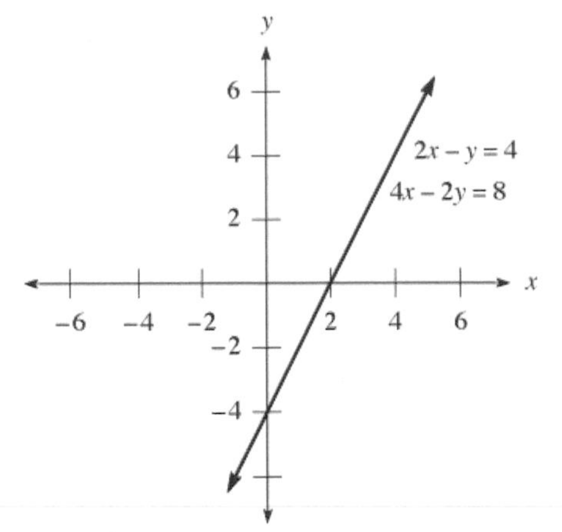

The system is dependent (both lines are the same, they coincide). The solution set is $\{(x,y)\,|\,2x-y=4\}$.

33. Graphing both lines:

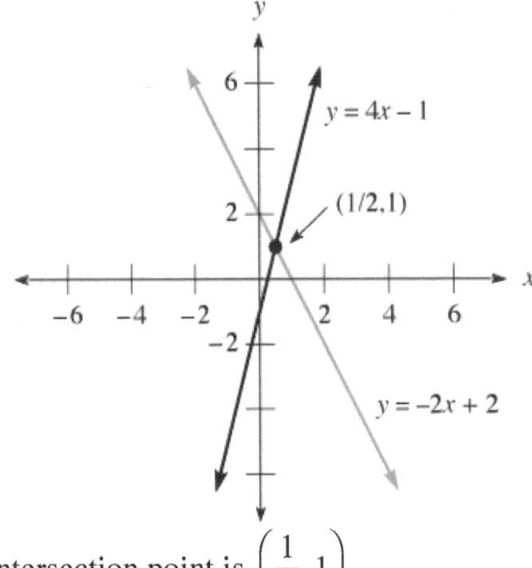

The intersection point is $\left(\frac{1}{2},1\right)$.

35. Graphing both lines:

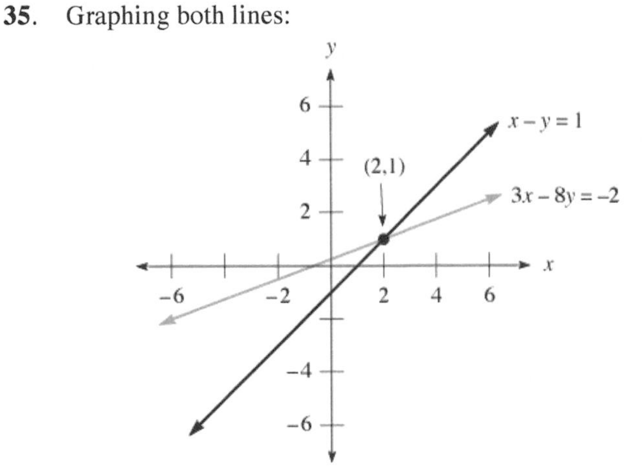

The intersection point is $(2,1)$.

37. Substituting the point $(-3,2)$:

$$a=x-y=2-(-3)=2+3=5$$
$$b=2x+3y=2(2)+3(-3)=4-9=-5$$

39. **a.** If $a = 3$ and $b = 1$, the two equations will be identical and thus have an infinite number of solutions.
b. If $a = 3$ and $b \neq 1$, the two equations will be contradictions, and thus the system will be inconsistent.
c. As long as $a \neq 3$, the lines will not be parallel and thus the solution will be a single ordered pair.

41. **a.** If Jane worked 25 hours, she would earn the same amount at each position.
b. If Jane worked less than 20 hours, she should choose Gigi's since she earns more in that position.
c. If Jane worked more than 30 hours, she should choose Marcy's since she earns more in that position.

43. Simplifying: $(x+y)+(x-y)=x+x+y-y=2x$

45. Simplifying: $3(2x-y)+(x+3y)=6x-3y+x+3y=7x$

47. Simplifying: $-4(3x+5y)+5(5x+4y)=-12x-20y+25x+20y=13x$

49. Simplifying: $6\left(\frac{1}{2}x-\frac{1}{3}y\right)=6\cdot\frac{1}{2}x-6\cdot\frac{1}{3}y=3x-2y$

51. Substituting $x = 3$:

$$3+y=4$$
$$y=1$$

53. Substituting $x = 3$:

$$3+3y=3$$
$$3y=0$$
$$y=0$$

55. Substituting $x = 6$:

$$3(6)+5y=-7$$
$$18+5y=-7$$
$$5y=-25$$
$$y=-5$$

57. **a.** Substituting $(4,-1)$ into each equation:

$$x-y=4-(-1)=4+1=5$$
$$x+y=4+(-1)=4-1=3$$

So $(4,-1)$ is a solution to the system.

b. Substituting $(6,1)$ into each equation:

$$x-y=6-1=5$$
$$x+y=6+1=7\neq 3$$

So $(6,1)$ is not a solution to the system.

c. Substituting $(2,-3)$ into each equation:

$$x-y=2-(-3)=2+3=5$$
$$x+y=2+(-3)=2-3=-1\neq 3$$

So $(2,-3)$ is not a solution to the system.

d. Substituting $(-4,1)$ into each equation:

$$x-y=-4-1=-5\neq 5$$
$$x+y=-4+1=-3\neq 3$$

So $(-4,1)$ is not a solution to the system.

The correct answer is a.

59. Inconsistent systems are a result of parallel lines which do not intersect. The correct answer is b.

4.2 The Elimination Method

1. Adding the two equations:

$$2x=4$$
$$x=2$$

Substituting into the first equation:

$$2+y=3$$
$$y=1$$

The solution is $(2,1)$.

3. Adding the two equations:

$$2y=14$$
$$y=7$$

Substituting into the first equation:

$$x+7=10$$
$$x=3$$

The solution is $(3,7)$.

5. Adding the two equations:

$$-2y=10$$
$$y=-5$$

Substituting into the first equation:

$$x-(-5)=7$$
$$x+5=7$$
$$x=2$$

The solution is $(2,-5)$.

7. Adding the two equations:

$$4x=-4$$
$$x=-1$$

Substituting into the first equation:

$$-1+y=-1$$
$$y=0$$

The solution is $(-1,0)$.

9. Adding the two equations:

$$0=0$$

The lines coincide (the system is dependent). The solution set is $\{(x,y) \mid 3x+2y=1\}$.

11. Multiplying the first equation by 2:

$$6x-2y=8$$
$$2x+2y=24$$

Adding the two equations:

$$8x=32$$
$$x=4$$

Substituting into the first equation:

$$3(4)-y=4$$
$$12-y=4$$
$$-y=-8$$
$$y=8$$

The solution is $(4,8)$.

13. Multiplying the second equation by –3:

$$5x-3y=-2$$
$$-30x+3y=-3$$

Adding the two equations:

$$-25x=-5$$
$$x=\frac{1}{5}$$

Substituting into the first equation:

$$5\left(\frac{1}{5}\right)-3y=-2$$
$$1-3y=-2$$
$$-3y=-3$$
$$y=1$$

The solution is $\left(\frac{1}{5},1\right)$.

15. Multiplying the second equation by 4:

$$11x-4y=11$$
$$20x+4y=20$$

Adding the two equations:

$$31x=31$$
$$x=1$$

Substituting into the second equation:

$$5(1)+y=5$$
$$5+y=5$$
$$y=0$$

The solution is $(1,0)$.

17. Multiplying the second equation by 3:

$$3x-5y=7$$
$$-3x+3y=-3$$

Adding the two equations:

$$-2y=4$$
$$y=-2$$

Substituting into the second equation:

$$-x-2=-1$$
$$-x=1$$
$$x=-1$$

The solution is $(-1,-2)$.

19. Multiplying the first equation by –2:

$$2x+16y=2$$
$$-2x+4y=13$$

Adding the two equations:

$$20y=15$$
$$y=\frac{3}{4}$$

Substituting into the first equation:

$$-x-8\left(\frac{3}{4}\right)=-1$$
$$-x-6=-1$$
$$-x=5$$
$$x=-5$$

The solution is $\left(-5,\frac{3}{4}\right)$.

21. Multiplying the first equation by 2:

$$-6x-2y=14$$
$$6x+7y=11$$

Adding the two equations:

$$5y=25$$
$$y=5$$

Substituting into the first equation:

$$-3x-5=7$$
$$-3x=12$$
$$x=-4$$

The solution is $(-4,5)$.

23. Adding the two equations:

$$8x=-24$$
$$x=-3$$

Substituting into the second equation:

$$2(-3)+y=-16$$
$$-6+y=-16$$
$$y=-10$$

The solution is $(-3,-10)$.

25. Multiplying the second equation by 3:

$$x+3y=9$$
$$6x-3y=12$$

Adding the two equations:

$$7x=21$$
$$x=3$$

Substituting into the first equation:

$$3+3y=9$$
$$3y=6$$
$$y=2$$

The solution is $(3,2)$.

27. Multiplying the second equation by 2:

$$x-6y=3$$
$$8x+6y=42$$

Adding the two equations:

$$5x=0$$
$$x=0$$

Substituting into the second equation:

$$4(5)+3y=21$$
$$20+3y=21$$
$$3y=1$$
$$y=\frac{1}{3}$$

The solution is $\left(5,\frac{1}{3}\right)$.

29. Multiplying the second equation by –3:

$$2x+9y=2$$
$$-15x-9y=24$$

Adding the two equations:

$$-13x=26$$
$$x=-2$$

Substituting into the first equation:

$$2(-2)+9y=2$$
$$-4+9y=2$$
$$9y=6$$
$$y=\frac{2}{3}$$

The solution is $\left(-2,\frac{2}{3}\right)$.

31. To clear each equation of fractions, multiply the first equation by 12 and the second equation by 6:

$$12\left(\frac{1}{3}x+\frac{1}{4}y\right)=12\left(\frac{7}{6}\right)$$
$$4x+3y=14$$

$$6\left(\frac{3}{2}x-\frac{1}{3}y\right)=6\left(\frac{7}{3}\right)$$
$$9x-2y=14$$

The system of equations is:

$$4x+3y=14$$
$$9x-2y=14$$

Multiplying the first equation by 2 and the second equation by 3:

$$8x+6y=28$$
$$27x-6y=42$$

Adding the two equations:

$$35x=70$$
$$x=2$$

Substituting into $4x+3y=14$:

$$4(2)+3y=14$$
$$8+3y=14$$
$$3y=6$$
$$y=2$$

The solution is $(2,2)$.

33. Multiplying the first equation by –2:

$$-6x-4y=2$$
$$6x+4y=0$$

Adding the two equations:

$$0=2$$

Since this statement is false, there is no solution to the system. The system is inconsistent.

35. Multiplying the first equation by 2 and the second equation by 3:

$$22x+12y=34$$
$$15x-12y=3$$

Adding the two equations:

$$37x=37$$
$$x=1$$

Substituting into the second equation:

$$5(1)-4y=1$$
$$5-4y=1$$
$$-4y=-4$$
$$y=1$$

The solution is $(1,1)$.

37. To clear each equation of fractions, multiply the first equation by 6 and the second equation by 6:

$$6\left(\frac{1}{2}x+\frac{1}{6}y\right)=6\left(\frac{1}{3}\right)$$
$$3x+y=2$$

$$6\left(-x-\frac{1}{3}y\right)=6\left(-\frac{1}{6}\right)$$
$$-6x-2y=-1$$

The system of equations is:

$$3x+y=2$$
$$-6x-2y=-1$$

Multiplying the first equation by 2:

$$6x+2y=4$$
$$-6x-2y=-1$$

Adding the two equations:

$$0=3$$

Since this statement is false, there is no solution to the system. The system is inconsistent.

39. Multiplying the second equation by 100 (to eliminate decimals):

$$x+y=22$$
$$5x+10y=170$$

Multiplying the first equation by –5:

$$-5x-5y=-110$$
$$5x+10y=170$$

Adding the two equations:

$$5y=60$$
$$y=12$$

Substituting into the first equation:

$$x+12=22$$
$$x=10$$

The solution is $(10,12)$.

41. Solving the equation:

$$x+(2x-1)=2$$
$$3x-1=2$$
$$3x=3$$
$$x=1$$

43. Solving the equation:

$$2(3y-1)-3y=4$$
$$6y-2-3y=4$$
$$3y-2=4$$
$$3y=6$$
$$y=2$$

45. Solving the equation:

$$\begin{aligned} 4x+2(-2x+4)&=8 \\ 4x-4x+8&=8 \\ 8&=8 \end{aligned}$$

The solution set is all real numbers.

47. Solving for x:

$$\begin{aligned} x-3y&=-1 \\ x&=3y-1 \end{aligned}$$

49. Substituting $x = 1$: $y=2(1)-1=2-1=1$

51. Substituting $y = 2$: $x=3(2)-1=6-1=5$

53. Substituting $x = 13$: $y=1.5(13)+15=19.5+15=34.5$

55. Substituting $x = 12$: $y=0.75(12)+24.95=9+24.95=33.95$

57. Multiplying the first equation by 3:

$$\begin{aligned} 12x-3y&=-9 \\ 2x+3y&=-5 \end{aligned}$$

Adding the two equations:

$$\begin{aligned} 14x&=-14 \\ x&=-1 \end{aligned}$$

Substituting into the second equation:

$$\begin{aligned} 2(-1)+3y&=-5 \\ -2+3y&=-5 \\ 3y&=-3 \\ y&=-1 \end{aligned}$$

The solution is $(-1,-1)$. The correct answer is c.

4.3 The Substitution Method

1. Substituting into the first equation:

$$\begin{aligned} x+(2x-1)&=11 \\ 3x-1&=11 \\ 3x&=12 \\ x&=4 \\ y&=2(4)-1=7 \end{aligned}$$

The solution is $(4,7)$.

3. Substituting into the first equation:

$$\begin{aligned} x+(5x+2)&=20 \\ 6x+2&=20 \\ 6x&=18 \\ x&=3 \\ y&=5(3)+2=17 \end{aligned}$$

The solution is $(3,17)$.

5. Substituting into the first equation:

$$\begin{aligned} -2x+(-4x+8)&=-1 \\ -6x+8&=-1 \\ -6x&=-9 \\ x&=\frac{3}{2} \\ y&=-4\left(\frac{3}{2}\right)+8=-6+8=2 \end{aligned}$$

The solution is $\left(\frac{3}{2},2\right)$.

7. Substituting into the first equation:

$$\begin{aligned} 3(-y+6)-2y&=-2 \\ -3y+18-2y&=-2 \\ -5y+18&=-2 \\ -5y&=-20 \\ y&=4 \\ x&=-4+6=2 \end{aligned}$$

The solution is $(2,4)$.

9. Substituting into the first equation:

$$\begin{aligned} 5x-4(4)&=-16 \\ 5x-16&=-16 \\ 5x&=0 \\ x&=0 \end{aligned}$$

The solution is $(0,4)$.

11. Substituting into the first equation:

$$\begin{aligned} 5x+4(-3x)&=7 \\ 5x-12x&=7 \\ -7x&=7 \\ x&=-1 \\ y&=-3(-1)=3 \end{aligned}$$

The solution is $(-1,3)$.

13. Solving the second equation for x:

$$x - 2y = -1$$
$$x = 2y - 1$$

Substituting into the first equation:

$$(2y-1)+3y=4$$
$$5y-1=4$$
$$5y=5$$
$$y=1$$
$$x=2(1)-1=1$$

The solution is $(1,1)$.

15. Solving the second equation for x:

$$x - 5y = 17$$
$$x = 5y + 17$$

Substituting into the first equation:

$$2(5y+17)+y=1$$
$$10y+34+y=1$$
$$11y+34=1$$
$$11y=-33$$
$$y=-3$$
$$x=5(-3)+17=2$$

The solution is $(2,-3)$.

17. Solving the second equation for x:

$$x - 5y = -5$$
$$x = 5y - 5$$

Substituting into the first equation:

$$3(5y-5)+5y=-3$$
$$15y-15+5y=-3$$
$$20y-15=-3$$
$$20y=12$$
$$y=\frac{3}{5}$$
$$x=5\left(\frac{3}{5}\right)-5=3-5=-2$$

The solution is $\left(-2,\frac{3}{5}\right)$.

19. Solving the second equation for x:

$$x - 3y = -18$$
$$x = 3y - 18$$

Substituting into the first equation:

$$5(3y-18)+3y=0$$
$$15y-90+3y=0$$
$$18y-90=0$$
$$18y=90$$
$$y=5$$
$$x=3(5)-18=-3$$

The solution is $(-3,5)$.

21. Solving the second equation for x:

$$x + 3y = 12$$
$$x = -3y + 12$$

Substituting into the first equation:

$$-3(-3y+12)-9y=7$$
$$9y-36-9y=7$$
$$-36=7$$

Since this statement is false, there is no solution to the system. The system is inconsistent.

23. Solving the second equation for y:

$$x + y = 22$$
$$y = 22 - x$$

Substituting into the first equation:

$$0.05x+0.10(22-x)=1.70$$
$$0.05x+2.2-0.10x=1.70$$
$$-0.05x+2.2=1.7$$
$$-0.05x=-0.5$$
$$x=10$$
$$y=22-10=12$$

The solution is $(10,12)$.

25. Solving the second equation for x:

$$x - y = 2$$
$$x = y + 2$$

Substituting into the first equation:

$$\frac{1}{4}(y+2)+\frac{1}{3}y=-\frac{1}{2}$$
$$12\cdot\frac{1}{4}(y+2)+12\cdot\frac{1}{3}y=12\left(-\frac{1}{2}\right)$$
$$3y+6+4y=-6$$
$$7y+6=-6$$
$$7y=-12$$
$$y=-\frac{12}{7}$$
$$x=-\frac{12}{7}+2=\frac{2}{7}$$

The solution is $\left(\frac{2}{7},-\frac{12}{7}\right)$.

27. Solving the second equation for y:

$$\begin{aligned} 2x-y&=5 \\ -y&=-2x+5 \\ y&=2x-5 \end{aligned}$$

Substituting into the first equation:

$$\begin{aligned} 5x-8(2x-5)&=7 \\ 5x-16x+40&=7 \\ -11x+40&=7 \\ -11x&=-33 \\ x&=3 \\ y&=2(3)-5=1 \end{aligned}$$

The solution is $(3,1)$.

29. Substituting into the first equation:

$$\begin{aligned} 5(y)-6y&=-4 \\ -y&=-4 \\ y&=4 \\ x&=4 \end{aligned}$$

The solution is $(4,4)$.

31. Solving the second equation for x:

$$\begin{aligned} 3x-12y&=-9 \\ 3x&=12y-9 \\ x&=4y-3 \end{aligned}$$

Substituting into the first equation:

$$\begin{aligned} 4(4y-3)+2y&=3 \\ 16y-12+2y&=3 \\ 18y-12&=3 \\ 18y&=15 \\ y&=\frac{5}{6} \\ x&=4\left(\frac{5}{6}\right)-3=\frac{10}{3}-3=\frac{1}{3} \end{aligned}$$

The solution is $\left(\frac{1}{3},\frac{5}{6}\right)$.

33. Substituting $y=3x$ into the first equation:

$$\begin{aligned} -3x+2(3x)&=6 \\ -3x+6x&=6 \\ 3x&=6 \\ x&=2 \\ y&=3(2)=6 \end{aligned}$$

The solution is $(2,6)$.

35. Setting the two equations equal:

$$\begin{aligned} 2x-12&=-x+3 \\ 3x-12&=3 \\ 3x&=15 \\ x&=5 \\ y&=2(5)-12=-2 \end{aligned}$$

The solution is $(5,-2)$.

37. Substituting into the first equation:

$$\begin{aligned} 7x-11(10)&=16 \\ 7x-110&=16 \\ 7x&=126 \\ x&=18 \\ y&=10 \end{aligned}$$

The solution is $(18,10)$.

39. Solving the second equation for x:

$$x-y=2$$
$$x=y+2$$

Substituting into the first equation:

$$-4(y+2)+4y=-8$$
$$-4y-8+4y=-8$$
$$-8=-8$$

Since this statement is true, the system is dependent. The two lines coincide. The solution is $\{(x,y)\mid x-y=2\}$.

41. Multiplying the first equation by 4 and the second equation by –3:

$$12x+28y=8$$
$$-12x-6y=3$$

Adding the two equations:

$$22y=11$$
$$y=\frac{11}{22}=\frac{1}{2}$$

Substituting into the second equation:

$$4x+2\left(\frac{1}{2}\right)=-1$$
$$4x+1=-1$$
$$4x=-2$$
$$x=-\frac{1}{2}$$

The solution is $\left(-\frac{1}{2},\frac{1}{2}\right)$.

43. **a.** Setting the two equations equal:

$$\frac{4.20}{20}x+250=\frac{4.20}{35}x+340$$
$$140\left(\frac{4.20}{20}x+250\right)=140\left(\frac{4.20}{35}x+340\right)$$
$$29.4x+35{,}000=16.8x+47{,}600$$
$$12.6x+35{,}000=47{,}600$$
$$12.6x=12{,}600$$
$$x=1{,}000$$

At 1,000 miles the car and truck cost the same to operate.

b. If Daniel drives more than 1,200 miles, the car will be cheaper to operate.

c. If Daniel drives less than 800 miles, the truck will be cheaper to operate.

d. The graphs appear in the first quadrant only because all quantities are positive.

45. Let x and $5x + 8$ represent the two numbers. The equation is:

$$x+5x+8=26$$
$$6x+8=26$$
$$6x=18$$
$$x=3$$
$$5x+8=23$$

The two numbers are 3 and 23.

47. Let x represent the smaller number and $2x - 6$ represent the larger number. The equation is:

$$2x-6-x=9$$
$$x-6=9$$
$$x=15$$
$$2x-6=24$$

The two numbers are 15 and 24.

49. Let w represent the width and $3w + 5$ represent the length. Using the perimeter formula:

$$\begin{aligned} 2(w)+2(3w+5) &= 58 \\ 2w+6w+10 &= 58 \\ 8w+10 &= 58 \\ 8w &= 48 \\ w &= 6 \\ 3w+5 &= 23 \end{aligned}$$

The width is 6 inches and the length is 23 inches.

51. Completing the table:

	Nickels	Dimes
Number	$x+4$	x
Value (cents)	$5(x+4)$	$10x$

The equation is:

$$\begin{aligned} 5(x+4)+10x &= 170 \\ 5x+20+10x &= 170 \\ 15x+20 &= 170 \\ 15x &= 150 \\ x &= 10 \\ x+4 &= 14 \end{aligned}$$

John has 14 nickels and 10 dimes.

53. Solving the second equation for x:

$$\begin{aligned} x-4y &= -2 \\ x &= 4y-2 \end{aligned}$$

The correct answer is b.

4.4 Applications

1. Let x and y represent the two numbers. The system of equations is:

$$\begin{aligned} x+y &= 25 \\ y &= 5+x \end{aligned}$$

Substituting into the first equation:

$$\begin{aligned} x+(5+x) &= 25 \\ 2x+5 &= 25 \\ 2x &= 20 \\ x &= 10 \\ y &= 5+10=15 \end{aligned}$$

The two numbers are 10 and 15.

3. Let x and y represent the two numbers. The system of equations is:

$$\begin{aligned} x+y &= 15 \\ y &= 4x \end{aligned}$$

Substituting into the first equation:

$$\begin{aligned} x+4x &= 15 \\ 5x &= 15 \\ x &= 3 \\ y &= 4(3)=12 \end{aligned}$$

The two numbers are 3 and 12.

5. Let x represent the larger number and y represent the smaller number. The system of equations is:

$$x - y = 5$$
$$x = 2y + 1$$

Substituting into the first equation:

$$2y + 1 - y = 5$$
$$y + 1 = 5$$
$$y = 4$$
$$x = 2(4) + 1 = 9$$

The two numbers are 4 and 9.

7. Let x and y represent the two numbers. The system of equations is:

$$y = 4x + 5$$
$$x + y = 35$$

Substituting into the second equation:

$$x + 4x + 5 = 35$$
$$5x + 5 = 35$$
$$5x = 30$$
$$x = 6$$
$$y = 4(6) + 5 = 29$$

The two numbers are 6 and 29.

9. Let x represent the amount invested at 6% and y represent the amount invested at 8%. The system of equations is:

$$x + y = 20{,}000$$
$$0.06x + 0.08y = 1{,}380$$

Multiplying the first equation by –0.06:

$$-0.06x - 0.06y = -1{,}200$$
$$0.06x + 0.08y = 1{,}380$$

Adding the two equations:

$$0.02y = 180$$
$$y = 9{,}000$$

Substituting into the first equation:

$$x + 9{,}000 = 20{,}000$$
$$x = 11{,}000$$

Mr. Wilson invested $9,000 at 8% and $11,000 at 6%.

11. Let x represent the amount invested at 5% and y represent the amount invested at 6%. The system of equations is:

$$x = 4y$$
$$0.05x + 0.06y = 520$$

Substituting into the second equation:

$$0.05(4y) + 0.06y = 520$$
$$0.20y + 0.06y = 520$$
$$0.26y = 520$$
$$y = 2{,}000$$
$$x = 4(2{,}000) = 8{,}000$$

She invested $8,000 at 5% and $2,000 at 6%.

13. Let x represent the number of nickels and y represent the number of quarters. The system of equations is:

$$x + y = 14$$
$$0.05x + 0.25y = 2.30$$

Multiplying the first equation by –0.05:

$$-0.05x - 0.05y = -0.7$$
$$0.05x + 0.25y = 2.30$$

Adding the two equations:

$$0.20y = 1.6$$
$$y = 8$$

Substituting into the first equation:

$$x + 8 = 14$$
$$x = 6$$

Ron has 6 nickels and 8 quarters.

15. Let x represent the number of dimes and y represent the number of quarters. The system of equations is:

$$x + y = 21$$
$$0.10x + 0.25y = 3.45$$

Multiplying the first equation by –0.10:

$$-0.10x - 0.10y = -2.10$$
$$0.10x + 0.25y = 3.45$$

Adding the two equations:

$$0.15y = 1.35$$
$$y = 9$$

Substituting into the first equation:

$$x + 9 = 21$$
$$x = 12$$

Tom has 12 dimes and 9 quarters.

17. Let x represent the liters of 50% alcohol solution and y represent the liters of 20% alcohol solution. The system of equations is:

$$x + y = 18$$
$$0.50x + 0.20y = 0.30(18)$$

Multiplying the first equation by –0.20:

$$-0.20x - 0.20y = -3.6$$
$$0.50x + 0.20y = 5.4$$

Adding the two equations:

$$0.30x = 1.8$$
$$x = 6$$

Substituting into the first equation:

$$6 + y = 18$$
$$y = 12$$

The mixture contains 6 liters of 50% alcohol solution and 12 liters of 20% alcohol solution.

19. Let x represent the gallons of 10% disinfectant and y represent the gallons of 7% disinfectant. The system of equations is:

$$x + y = 30$$
$$0.10x + 0.07y = 0.08(30)$$

Multiplying the first equation by –0.07:

$$-0.07x - 0.07y = -2.1$$
$$0.10x + 0.07y = 2.4$$

Adding the two equations:

$$0.03x = 0.3$$
$$x = 10$$

Substituting into the first equation:

$$10 + y = 30$$
$$y = 20$$

The mixture contains 10 gallons of 10% disinfectant and 20 gallons of 7% disinfectant.

21. Let x represent the number of adult tickets and y represent the number of kids tickets. The system of equations is:

$$x + y = 70$$
$$5.50x + 4.00y = 310$$

Multiplying the first equation by –4:

$$-4.00x - 4.00y = -280$$
$$5.50x + 4.00y = 310$$

Adding the two equations:

$$1.5x = 30$$
$$x = 20$$

Substituting into the first equation:

$$20 + y = 70$$
$$y = 50$$

The matinee had 20 adult tickets sold and 50 kids tickets sold.

23. Let x represent the width and y represent the length. The system of equations is:

$$2x + 2y = 96$$
$$y = 2x$$

Substituting into the first equation:

$$2x + 2(2x) = 96$$
$$2x + 4x = 96$$
$$6x = 96$$
$$x = 16$$
$$y = 2(16) = 32$$

The width is 16 feet and the length is 32 feet.

25. Let x represent the number of \$5 chips and y represent the number of \$25 chips. The system of equations is:

$$x + y = 45$$
$$5x + 25y = 465$$

Multiplying the first equation by –5:

$$-5x - 5y = -225$$
$$5x + 25y = 465$$

Adding the two equations:

$$20y = 240$$
$$y = 12$$

Substituting into the first equation:

$$x + 12 = 45$$
$$x = 33$$

The gambler has 33 \$5 chips and 12 \$25 chips.

27. Let x represent the number of shares of \$11 stock and y represent the number of shares of \$20 stock. The system of equations is:

$$x + y = 150$$
$$11x + 20y = 2550$$

Multiplying the first equation by –11:

$$-11x - 11y = -1650$$
$$11x + 20y = 2550$$

Adding the two equations:

$$9y = 900$$
$$y = 100$$

Substituting into the first equation:

$$x + 100 = 150$$
$$x = 50$$

She bought 50 shares at \$11 and 100 shares at \$20.

29. Evaluating: $6(3+4)+5=6(7)+5=42+5=47$

31. Evaluating: $1^2+2^2+3^2=1+4+9=14$

33. Evaluating: $5(6+3\cdot 2)+4+3\cdot 2=5(6+6)+4+3\cdot 2=5(12)+4+6=60+4+6=70$

35. Evaluating: $(1^3+2^3)+[(2\cdot 3)+(4\cdot 5)]=(1+8)+(6+20)=9+26=35$

37. Evaluating: $(2\cdot 3+4+5)\div 3=(6+4+5)\div 3=15\div 3=5$

39. Evaluating: $6\cdot 10^3+5\cdot 10^2+4\cdot 10^1=6{,}000+500+40=6{,}540$

41. Evaluating: $1\cdot 10^3+7\cdot 10^2+6\cdot 10^1+0=1{,}000+700+60+0=1{,}760$

43. Evaluating: $4\cdot 2-1+5\cdot 3-2=8-1+15-2=20$

45. Evaluating: $(2^3+3^2)\cdot 4-5=(8+9)\cdot 4-5=17\cdot 4-5=68-5=63$

47. Evaluating: $2(2^2+3^2)+3(3^2)=2(4+9)+3(9)=2(13)+3(9)=26+27=53$

49. The correct system of equations is a.

51. The correct system of equations is c.

Chapter 4 Test

1. The solution is (0,6).

2. Graphing the two equations:

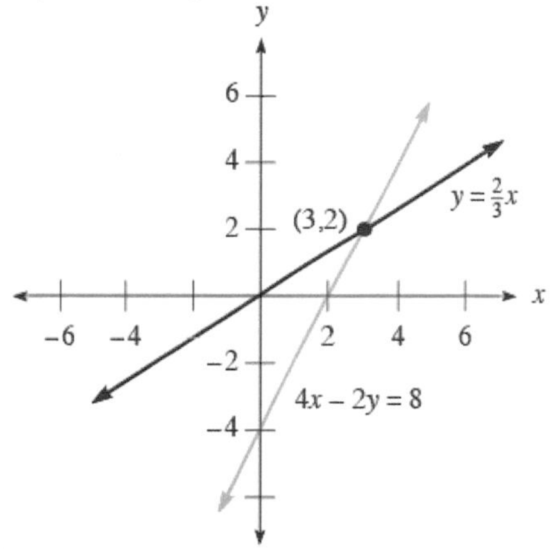

The intersection point is $(3,2)$.

3. Graphing the two equations:

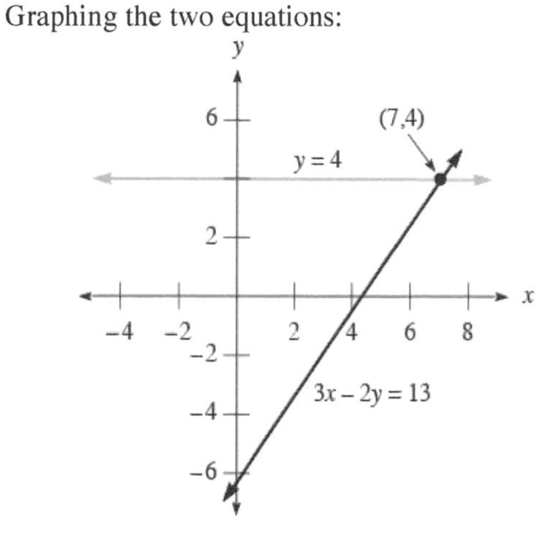

The intersection point is $(7,4)$.

4. Graphing both lines:

The intersection point is $(-2,4)$.

5. Multiplying the first equation by 3:

$$3x-3y=-27$$
$$2x+3y=7$$

Adding the two equations:

$$5x=-20$$
$$x=-4$$

Substituting into the second equation:

$$2(-4)+3y=7$$
$$-8+3y=7$$
$$3y=15$$
$$y=5$$

The solution is $(-4,5)$.

6. Multiplying the first equation by –1:

$$-3x+y=-1$$
$$5x-y=3$$

Adding the two equations:

$$2x=2$$
$$x=1$$

Substituting into the first equation:

$$3(1)-y=1$$
$$3-y=1$$
$$-y=-2$$
$$y=2$$

The solution is $(1,2)$.

7. Multiplying the second equation by –2:

$$2x+3y=-3$$
$$-2x-12y=-24$$

Adding the two equations:

$$-9y=-27$$
$$y=3$$

Substituting into the second equation:

$$x+6(3)=12$$
$$x+18=12$$
$$x=-6$$

The solution is $(-6,3)$.

8. Multiplying the first equation by –2:

$$-4x-6y=-8$$
$$4x+6y=8$$

Adding the two equations: $0=0$

Since this statement is true, the system is dependent. The two lines coincide. The solution set is $\{(x,y)\,|\,2x+3y=4\}$.

9. Substituting into the first equation:

$$3x-(2x-8)=12$$
$$3x-2x+8=12$$
$$x+8=12$$
$$x=4$$

Substituting into the second equation: $y=2(4)-8=8-8=0$. The solution is $(4,0)$.

10. Substituting into the first equation:

$$\begin{aligned} 3(4y-17)-6y&=3 \\ 12y-51-6y&=3 \\ 6y-51&=3 \\ 6y&=54 \\ y&=9 \end{aligned}$$

Substituting into the second equation: $x=4(9)-17=36-17=19$. The solution is $(19,9)$.

11. Solving the second equation for y:

$$\begin{aligned} 3x+y&=-5 \\ y&=-3x-5 \end{aligned}$$

Substituting into the first equation:

$$\begin{aligned} 2x-3(-3x-5)&=-18 \\ 2x+9x+15&=-18 \\ 11x+15&=-18 \\ 11x&=-33 \\ x&=-3 \end{aligned}$$

Substituting into the second equation: $y=-3(-3)-5=9-5=4$. The solution is $(-3,4)$.

12. Solving the second equation for x:

$$\begin{aligned} x-4y&=-1 \\ x&=4y-1 \end{aligned}$$

Substituting into the first equation:

$$\begin{aligned} 2(4y-1)-3y&=13 \\ 8y-2-3y&=13 \\ 5y-2&=13 \\ 5y&=15 \\ y&=3 \end{aligned}$$

Substituting into the second equation: $x=4(3)-1=12-1=11$. The solution is $(11,3)$.

13. Let x and y represent the two numbers. The system of equations is:

$$\begin{aligned} x+y&=18 \\ y&=3x+2 \end{aligned}$$

Substituting into the first equation:

$$\begin{aligned} x+3x+2&=18 \\ 4x+2&=18 \\ 4x&=16 \\ x&=4 \end{aligned}$$

Substituting into the second equation: $y=3(4)+2=12+2=14$. The two numbers are 4 and 14.

14. Let x represent the amount invested at 6% and y represent the amount invested at 7%. The system of equations is:

$$\begin{aligned} x+y&=2000 \\ 0.06x+0.07y&=135.20 \end{aligned}$$

Multiplying the first equation by –0.06:

$$\begin{aligned} -0.06x-0.06y&=-120 \\ 0.06x+0.07y&=135.20 \end{aligned}$$

Adding the two equations:

$$\begin{aligned} 0.01y&=15.20 \\ y&=1520 \end{aligned}$$

Substituting into the first equation:

$$\begin{aligned} x+1520&=2000 \\ x&=480 \end{aligned}$$

Dave should invest \$480 at 6% and \$1,520 at 7%.

15. Let x represent the number of nickels and y represent the number of dimes. The system of equations is:

$$x+y=19$$
$$0.05x+0.10y=1.35$$

Multiplying the first equation by –0.05:

$$-0.05x-0.05y=-0.95$$
$$0.05x+0.10y=1.35$$

Adding the two equations:

$$0.05y=0.40$$
$$y=8$$

Substituting into the first equation:

$$x+8=19$$
$$x=11$$

Maria has 11 nickels and 8 dimes.

16. Let x represent the gallons of 40% antifreeze solution and y represent the gallons of 70% antifreeze solution. The system of equations is:

$$x+y=30$$
$$0.40x+0.70y=0.50(30)$$

Multiplying the first equation by –0.40:

$$-0.40x-0.40y=-12$$
$$0.40x+0.70y=15$$

Adding the two equations:

$$0.30y=3$$
$$y=10$$

Substituting into the first equation:

$$x+10=30$$
$$x=20$$

The solution contains 20 gallons of 40% antifreeze solution and 10 gallons of 70% antifreeze solution.

17. Let w represent the width and l represent the length. The system of equations is:

$$l=2w+15$$
$$2w+2l=198$$

Substituting into the second equation:

$$2w+2(2w+15)=198$$
$$2w+4w+30=198$$
$$6w+30=198$$
$$6w=168$$
$$w=28$$
$$l=2(28)+15=56+15=71$$

The width is 28 feet and the length is 71 feet.

Chapter 5
Exponents and Polynomials

5.1 Multiplication with Exponents and Scientific Notation

1. The base is 4 and the exponent is 2. Evaluating the expression: $4^2 = 4 \cdot 4 = 16$

3. The base is 0.3 and the exponent is 2. Evaluating the expression: $0.3^2 = 0.3 \cdot 0.3 = 0.09$

5. The base is 4 and the exponent is 3. Evaluating the expression: $4^3 = 4 \cdot 4 \cdot 4 = 64$

7. The base is –5 and the exponent is 2. Evaluating the expression: $(-5)^2 = (-5) \cdot (-5) = 25$

9. The base is 2 and the exponent is 3. Evaluating the expression: $-2^3 = -2 \cdot 2 \cdot 2 = -8$

11. The base is 3 and the exponent is 4. Evaluating the expression: $3^4 = 3 \cdot 3 \cdot 3 \cdot 3 = 81$

13. The base is $\frac{2}{3}$ and the exponent is 2. Evaluating the expression: $\left(\frac{2}{3}\right)^2 = \left(\frac{2}{3}\right) \cdot \left(\frac{2}{3}\right) = \frac{4}{9}$

15. The base is $\frac{1}{2}$ and the exponent is 4. Evaluating the expression: $\left(\frac{1}{2}\right)^4 = \left(\frac{1}{2}\right) \cdot \left(\frac{1}{2}\right) \cdot \left(\frac{1}{2}\right) \cdot \left(\frac{1}{2}\right) = \frac{1}{16}$

17. **a.** Completing the table:

Number (x)	1	2	3	4	5	6	7
Square (x^2)	1	4	9	16	25	36	49

b. For numbers larger than 1, the square of the number is larger than the number.

19. Simplifying the expression: $x^4 \cdot x^5 = x^{4+5} = x^9$

21. Simplifying the expression: $y^{10} \cdot y^{20} = y^{10+20} = y^{30}$

23. Simplifying the expression: $2^5 \cdot 2^4 \cdot 2^3 = 2^{5+4+3} = 2^{12}$

25. Simplifying the expression: $x^4 \cdot x^6 \cdot x^8 \cdot x^{10} = x^{4+6+8+10} = x^{28}$

27. Simplifying the expression: $\left(x^2\right)^5 = x^{2 \cdot 5} = x^{10}$

29. Simplifying the expression: $\left(5^4\right)^3 = 5^{4 \cdot 3} = 5^{12}$

31. Simplifying the expression: $\left(y^3\right)^3 = y^{3 \cdot 3} = y^9$

33. Simplifying the expression: $\left(2^5\right)^{10} = 2^{5 \cdot 10} = 2^{50}$

35. Simplifying the expression: $\left(a^3\right)^x = a^{3x}$

37. Simplifying the expression: $\left(b^x\right)^y = b^{xy}$

39. Simplifying the expression: $(4x)^2 = 4^2 \cdot x^2 = 16x^2$

41. Simplifying the expression: $(2y)^5 = 2^5 \cdot y^5 = 32y^5$

43. Simplifying the expression: $(-3x)^4 = (-3)^4 \cdot x^4 = 81x^4$

45. Simplifying the expression: $(0.5ab)^2 = (0.5)^2 \cdot a^2b^2 = 0.25a^2b^2$

47. Simplifying the expression: $(4xyz)^3 = 4^3 \cdot x^3y^3z^3 = 64x^3y^3z^3$

49. Simplifying using properties of exponents: $\left(2x^4\right)^3 = 2^3\left(x^4\right)^3 = 8x^{12}$

51. Simplifying using properties of exponents: $\left(4a^3\right)^2 = 4^2\left(a^3\right)^2 = 16a^6$

53. Simplifying using properties of exponents: $\left(x^2\right)^3\left(x^4\right)^2 = x^6 \cdot x^8 = x^{14}$

55. Simplifying using properties of exponents: $\left(a^3\right)^1\left(a^2\right)^4 = a^3 \cdot a^8 = a^{11}$

57. Simplifying using properties of exponents: $(4x^2y^3)^2 = 4^2x^4y^6 = 16x^4y^6$

59. Simplifying using properties of exponents: $\left(\frac{2}{3}a^4b^5\right)^3 = \left(\frac{2}{3}\right)^3 a^{12}b^{15} = \frac{8}{27}a^{12}b^{15}$

61. Completing the table:

Number (x)	−3	−2	−1	0	1	2	3
Square (x^2)	9	4	1	0	1	4	9

Constructing a line graph:

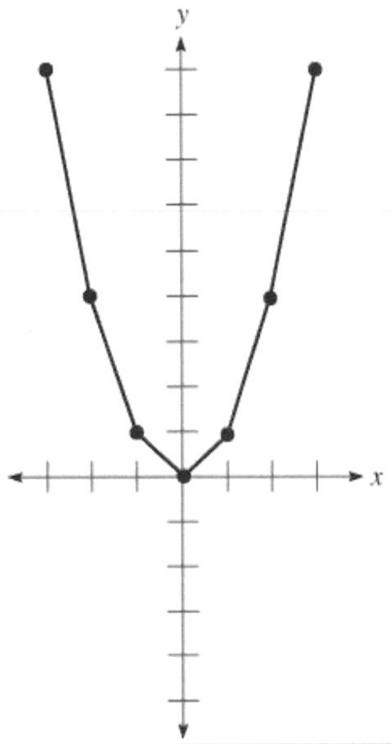

63. Completing the table:

Number (x)	−2.5	−1.5	−0.5	0	0.5	1.5	2.5
Square (x^2)	6.25	2.25	0.25	0	0.25	2.25	6.25

65. Writing in scientific notation: $43{,}200 = 4.32 \times 10^4$

67. Writing in scientific notation: $-570 = -5.7 \times 10^2$

69. Writing in scientific notation: $238{,}000 = 2.38 \times 10^5$

71. Writing in expanded form: $2.49 \times 10^3 = 2{,}490$

73. Writing in expanded form: $-3.52 \times 10^2 = -352$

75. Writing in expanded form: $2.8 \times 10^4 = 28{,}000$

77. The volume is given by: $V = (3 \text{ in.})^3 = 27 \text{ inches}^3$

79. The volume is given by: $V = (2.5 \text{ in.})^3 \approx 15.6 \text{ inches}^3$

81. The volume is given by: $V = (8 \text{ in.})(4.5 \text{ in.})(1 \text{ in.}) = 36 \text{ inches}^3$

83. Possibly, it depends on the actual dimensions of the box.

85. Writing in scientific notation: $650{,}000{,}000$ seconds $= 6.5 \times 10^8$ seconds

87. Writing in expanded form: 7.4×10^5 dollars $= \$740{,}000$

89. Writing in expanded form: 1.8×10^5 dollars $= \$180{,}000$

91. Substitute $c = 8, b = 3.35$, and $s = 3.11$: $d = \pi \cdot 3.11 \cdot 8 \cdot \left(\frac{1}{2} \cdot 3.35\right)^2 \approx 219 \text{ inches}^3$

93. Substitute $c = 6, b = 3.59$, and $s = 2.99$: $d = \pi \cdot 2.99 \cdot 6 \cdot \left(\frac{1}{2} \cdot 3.59\right)^2 \approx 182 \text{ inches}^3$

95. Subtracting: $4 - 7 = 4 + (-7) = -3$

97. Subtracting: $4 - (-7) = 4 + 7 = 11$

99. Subtracting: $15 - 20 = 15 + (-20) = -5$

101. Subtracting: $-15 - (-20) = -15 + 20 = 5$

103. Simplifying: $2(3) - 4 = 6 - 4 = 2$

105. Simplifying: $4(3) - 3(2) = 12 - 6 = 6$

107. Simplifying: $2(5 - 3) = 2(2) = 4$

109. Simplifying: $5 + 4(-2) - 2(-3) = 5 - 8 + 6 = 3$

111. Simplifying: $x^2 \cdot x^4 \cdot x^6 = x^{2+4+6} = x^{12}$. The correct answer is a.

113. Simplifying: $(-2xy)^4 = (-2)^4 \cdot x^4 \cdot y^4 = 16x^4y^4$. The correct answer is a.

5.2 Division with Exponents

1. Writing with positive exponents: $5^{-1} = \frac{1}{5}$

3. Writing with positive exponents: $x^{-1} = \frac{1}{x}$

5. Writing with positive exponents: $3^{-2} = \frac{1}{3^2} = \frac{1}{9}$

7. Writing with positive exponents: $6^{-2} = \frac{1}{6^2} = \frac{1}{36}$

9. Writing with positive exponents: $8^{-2} = \frac{1}{8^2} = \frac{1}{64}$

11. Writing with positive exponents: $5^{-3} = \frac{1}{5^3} = \frac{1}{125}$

13. Writing with positive exponents: $a^{-4} = \frac{1}{a^4}$

15. Writing with positive exponents: $-4^{-2} = -\frac{1}{4^2} = -\frac{1}{16}$

17. Writing with positive exponents: $-x^{-1} = -\frac{1}{x}$

19. Writing with positive exponents: $(-5)^{-3} = \frac{1}{(-5)^3} = \frac{1}{-125} = -\frac{1}{125}$

21. Writing with positive exponents: $(-4)^{-2} = \frac{1}{(-4)^2} = \frac{1}{16}$

23. Writing with positive exponents: $2x^{-3} = 2 \cdot \frac{1}{x^3} = \frac{2}{x^3}$

25. Writing with positive exponents: $(2x)^{-3} = \frac{1}{(2x)^3} = \frac{1}{8x^3}$

27. Writing with positive exponents: $(5y)^{-2} = \frac{1}{(5y)^2} = \frac{1}{25y^2}$

29. Writing with positive exponents: $10^{-2} = \frac{1}{10^2} = \frac{1}{100}$

31. Completing the table:

Number (x)	Square (x^2)	Power of 2 (2^x)
−3	9	$\frac{1}{8}$
−2	4	$\frac{1}{4}$
−1	1	$\frac{1}{2}$
0	0	1
1	1	2
2	4	4
3	9	8

33. Simplifying: $\frac{5^1}{5^3} = 5^{1-3} = 5^{-2} = \frac{1}{5^2} = \frac{1}{25}$

35. Simplifying: $\frac{x^{10}}{x^4} = x^{10-4} = x^6$

37. Simplifying: $\frac{4^3}{4^0} = 4^{3-0} = 4^3 = 64$

39. Simplifying: $\frac{(2x)^7}{(2x)^4} = (2x)^{7-4} = (2x)^3 = 2^3x^3 = 8x^3$

41. Simplifying: $\frac{6^{11}}{6} = \frac{6^{11}}{6^1} = 6^{11-1} = 6^{10} \quad (= 60{,}466{,}176)$

43. Simplifying: $\frac{6}{6^{11}} = \frac{6^1}{6^{11}} = 6^{1-11} = 6^{-10} = \frac{1}{6^{10}} \quad \left(= \frac{1}{60{,}466{,}176}\right)$

45. Simplifying: $\frac{2^{-5}}{2^3} = 2^{-5-3} = 2^{-8} = \frac{1}{2^8} = \frac{1}{256}$

47. Simplifying: $\frac{2^5}{2^{-3}} = 2^{5-(-3)} = 2^{5+3} = 2^8 = 256$

49. Simplifying: $\frac{(3x)^{-5}}{(3x)^{-8}} = (3x)^{-5-(-8)} = (3x)^{-5+8} = (3x)^3 = 3^3x^3 = 27x^3$

51. Simplifying: $(3xy)^4 = 3^4x^4y^4 = 81x^4y^4$

53. Simplifying: $10^0 = 1$

55. Simplifying: $(2a^2b)^1 = 2a^2b$

57. Simplifying: $(7y^3)^{-2} = \frac{1}{(7y^3)^2} = \frac{1}{49y^6}$

59. Simplifying: $x^{-3} \bullet x^{-5} = x^{-3-5} = x^{-8} = \frac{1}{x^8}$

61. Simplifying: $y^7 \bullet y^{-10} = y^{7-10} = y^{-3} = \frac{1}{y^3}$

63. Simplifying: $\frac{(x^2)^3}{x^4} = \frac{x^6}{x^4} = x^{6-4} = x^2$

65. Simplifying: $\frac{(a^4)^3}{(a^3)^2} = \frac{a^{12}}{a^6} = a^{12-6} = a^6$

67. Simplifying: $\frac{y^7}{(y^2)^8} = \frac{y^7}{y^{16}} = y^{7-16} = y^{-9} = \frac{1}{y^9}$

69. Simplifying: $\left(\frac{y^7}{y^2}\right)^8 = (y^{7-2})^8 = (y^5)^8 = y^{40}$

71. Simplifying: $\frac{(x^{-2})^3}{x^{-5}} = \frac{x^{-6}}{x^{-5}} = x^{-6-(-5)} = x^{-6+5} = x^{-1} = \frac{1}{x}$

73. Simplifying: $\left(\frac{x^{-2}}{x^{-5}}\right)^3 = (x^{-2+5})^3 = (x^3)^3 = x^9$

75. Simplifying: $\frac{(a^3)^2(a^4)^5}{(a^5)^2} = \frac{a^6 \bullet a^{20}}{a^{10}} = \frac{a^{26}}{a^{10}} = a^{26-10} = a^{16}$

77. Simplifying: $\frac{(a^{-2})^3(a^4)^2}{(a^{-3})^{-2}} = \frac{a^{-6} \bullet a^8}{a^6} = \frac{a^2}{a^6} = a^{2-6} = a^{-4} = \frac{1}{a^4}$

79. Completing the table:

Number (x)	Power of 2 (2^x)
–3	1/8
–2	1/4
–1	1/2
0	1
1	2
2	4
3	8

Constructing the line graph:

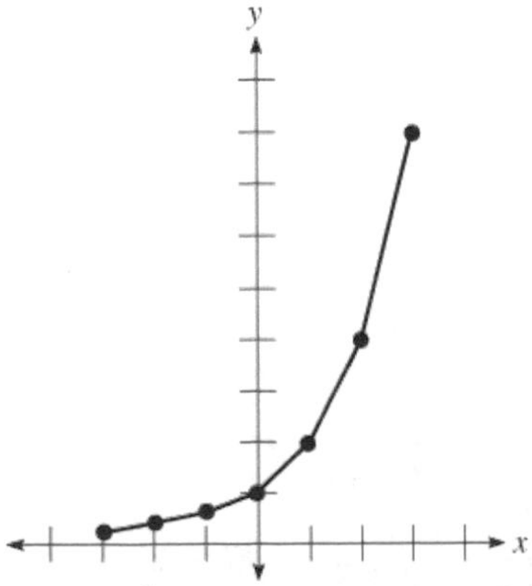

y

x

81. Writing in scientific notation: $0.0048 = 4.8 \times 10^{-3}$

83. Writing in scientific notation: $25 = 2.5 \times 10^1$

85. Writing in scientific notation: $0.25 = 2.5 \times 10^{-1}$

87. Writing in scientific notation: $0.000009 = 9 \times 10^{-6}$

89. Completing the table:

Expanded Form	Scientific Notation $(n \times 10^r)$
0.000357	3.57×10^{-4}
0.00357	3.57×10^{-3}
0.0357	3.57×10^{-2}
0.357	3.57×10^{-1}
3.57	3.57×10^{0}
35.7	3.57×10^{1}
357	3.57×10^{2}
3,570	3.57×10^{3}
35,700	3.57×10^{4}

91. Writing in expanded form: $4.23 \times 10^{-3} = 0.00423$

93. Writing in expanded form: $8 \times 10^{-5} = 0.00008$

95. Writing in expanded form: $4.2 \times 10^{0} = 4.2$

97. Writing in expanded form: $2.4 \times 10^{-1} = 0.24$

99. Writing in expanded form: 2×10^{-3} seconds $= 0.002$ seconds

101. Writing each number in scientific notation:

Craven/Busch: $0.002 = 2 \times 10^{-3}$

Earnhardt/Irvan: $0.005 = 5 \times 10^{-3}$

Harvick/Gordon: $0.006 = 6 \times 10^{-3}$

Kahne/Kenseth: $0.01 = 1 \times 10^{-2}$

Kenseth/Kahne: $0.01 = 1 \times 10^{-2}$

103. Writing in scientific notation: $25 \times 10^{3} = 2.5 \times 10^{4}$

105. Writing in scientific notation: $23.5 \times 10^{4} = 2.35 \times 10^{5}$

107. Writing in scientific notation: $0.82 \times 10^{-3} = 8.2 \times 10^{-4}$

109. The area of the smaller square is $(10 \text{ in.})^2 = 100 \text{ inches}^2$, while the area of the larger square is $(20 \text{ in.})^2 = 400 \text{ inches}^2$. It would take 4 smaller squares to cover the larger square.

111. The area of the smaller square is x^2, while the area of the larger square is $(2x)^2 = 4x^2$. It would take 4 smaller squares to cover the larger square.

113. The volume of the smaller box is $(6 \text{ in.})^3 = 216 \text{ inches}^3$, while the volume of the larger box is $(12 \text{ in.})^3 = 1{,}728 \text{ inches}^3$. Thus 8 smaller boxes will fit inside the larger box ($8 \cdot 216 = 1{,}728$).

115. The volume of the smaller box is x^3, while the volume of the larger box is $(2x)^3 = 8x^3$. Thus 8 smaller boxes will fit inside the larger box.

117. Simplifying: $3(4.5) = 13.5$

119. Simplifying: $\frac{4}{5}(10) = \frac{40}{5} = 8$

121. Simplifying: $6.8(3.9) = 26.52$

123. Simplifying: $-3 + 15 = 12$

125. Simplifying: $x^5 \bullet x^3 = x^{5+3} = x^8$

127. Simplifying: $\frac{x^3}{x^2} = x^{3-2} = x$

129. Simplifying: $\frac{y^3}{y^5} = y^{3-5} = y^{-2} = \frac{1}{y^2}$

131. Writing in expanded form: $3.4 \times 10^{2} = 340$

133. Simplifying: $4^{-3} = \frac{1}{4^3} = \frac{1}{64}$. The correct answer is b.

135. Simplifying: $\left(\frac{x}{3}\right)^4 = \frac{x^4}{3^4} = \frac{x^4}{81}$. The correct answer is a.

137. Writing in scientific notation: $0.00123 = 1.23 \times 10^{-3}$. The correct answer is d.

5.3 Operations with Monomials

1. The coefficient is 7 and the degree is 3.

3. The coefficient is –1 and the degree is 1.

5. The coefficient is $\frac{1}{2}$ and the degree is 2.

7. The coefficient is –4 and the degree is 11.

9. The coefficient is 8 and the degree is 0.

11. Multiplying the monomials: $(3x^4)(4x^3) = 12x^{4+3} = 12x^7$

13. Multiplying the monomials: $(-2y^4)(8y^7) = -16y^{4+7} = -16y^{11}$

15. Multiplying the monomials: $(8x)(4x) = 32x^{1+1} = 32x^2$

17. Multiplying the monomials: $(10a^3)(10a)(2a^2) = 200a^{3+1+2} = 200a^6$

19. Multiplying the monomials: $(6ab^2)(-4a^2b) = -24a^{1+2}b^{2+1} = -24a^3b^3$

21. Multiplying the monomials: $(4x^2y)(3x^3y^3)(2xy^4) = 24x^{2+3+1}y^{1+3+4} = 24x^6y^8$

23. Dividing the monomials: $\frac{15x^3}{5x^2} = \frac{15}{5} \bullet \frac{x^3}{x^2} = 3x$

25. Dividing the monomials: $\frac{18y^9}{3y^{12}} = \frac{18}{3} \bullet \frac{y^9}{y^{12}} = 6 \bullet \frac{1}{y^3} = \frac{6}{y^3}$

27. Dividing the monomials: $\frac{32a^3}{64a^4} = \frac{32}{64} \bullet \frac{a^3}{a^4} = \frac{1}{2} \bullet \frac{1}{a} = \frac{1}{2a}$

29. Dividing the monomials: $\frac{21a^2b^3}{-7ab^5} = \frac{21}{-7} \bullet \frac{a^2}{a} \bullet \frac{b^3}{b^5} = -3 \bullet a \bullet \frac{1}{b^2} = -\frac{3a}{b^2}$

31. Dividing the monomials: $\frac{3x^3y^2z}{27xy^2z^3} = \frac{3}{27} \bullet \frac{x^3}{x} \bullet \frac{y^2}{y^2} \bullet \frac{z}{z^3} = \frac{1}{9} \bullet x^2 \bullet \frac{1}{z^2} = \frac{x^2}{9z^2}$

33. Completing the table:

a	b	ab	$\frac{a}{b}$	$\frac{b}{a}$
10	$5x$	$50x$	$\frac{2}{x}$	$\frac{x}{2}$
$20x^3$	$6x^2$	$120x^5$	$\frac{10x}{3}$	$\frac{3}{10x}$
$25x^5$	$5x^4$	$125x^9$	$5x$	$\frac{1}{5x}$
$3x^{-2}$	$3x^2$	9	$\frac{1}{x^4}$	x^4
$-2y^4$	$8y^7$	$-16y^{11}$	$-\frac{1}{4y^3}$	$-4y^3$

35. Finding the product: $(3\times10^3)(2\times10^5) = 6\times10^8$

37. Finding the product: $(3.5\times10^4)(5\times10^{-6}) = 17.5\times10^{-2} = 1.75\times10^{-1}$

39. Finding the product: $(5.5\times10^{-3})(2.2\times10^{-4}) = 12.1\times10^{-7} = 1.21\times10^{-6}$

41. Finding the quotient: $\frac{8.4\times10^5}{2\times10^2} = 4.2\times10^3$

43. Finding the quotient: $\frac{6\times10^8}{2\times10^{-2}} = 3\times10^{10}$

45. Finding the quotient: $\frac{2.5\times10^{-6}}{5\times10^{-4}} = 0.5\times10^{-2} = 5\times10^{-3}$

47. Combining the monomials: $3x^2 + 5x^2 = (3+5)x^2 = 8x^2$

49. Combining the monomials: $8x^5 - 19x^5 = (8-19)x^5 = -11x^5$

51. Combining the monomials: $2a + a - 3a = (2+1-3)a = 0a = 0$

53. Combining the monomials: $10x^3 - 8x^3 + 2x^3 = (10-8+2)x^3 = 4x^3$

55. Combining the monomials: $20ab^2 - 19ab^2 + 30ab^2 = (20-19+30)ab^2 = 31ab^2$

57. Completing the table:

a	b	ab	$a+b$
$5x$	$3x$	$15x^2$	$8x$
$4x^2$	$2x^2$	$8x^4$	$6x^2$
$3x^3$	$6x^3$	$18x^6$	$9x^3$
$2x^4$	$-3x^4$	$-6x^8$	$-x^4$
x^5	$7x^5$	$7x^{10}$	$8x^5$

59. Simplifying using properties of exponents: $(2x)^3(2x)^4 = (2x)^7 = 2^7x^7 = 128x^7$

61. Simplifying using properties of exponents: $\left(3x^2\right)^3(2x)^4 = 3^3x^6 \cdot 2^4x^4 = 27x^6 \cdot 16x^4 = 432x^{10}$

63. Simplifying using properties of exponents: $\left(4xy^3\right)^2\left(2x^5y\right)^3 = 4^2x^2y^6 \cdot 2^3x^{15}y^3 = 16x^2y^6 \cdot 8x^{15}y^3 = 128x^{17}y^9$

65. Simplifying: $\dfrac{(2x)^5}{(2x)^3} = (2x)^{5-3} = (2x)^2 = 2^2x^2 = 4x^2$

67. Simplifying: $\dfrac{\left(3x^2\right)^3}{(2x)^4} = \dfrac{3^3x^6}{2^4x^4} = \dfrac{27x^6}{16x^4} = \dfrac{27x^2}{16}$

69. Simplifying: $\dfrac{\left(2x^2y\right)^4}{\left(6xy^4\right)^2} = \dfrac{2^4x^8y^4}{6^2x^2y^8} = \dfrac{16x^8y^4}{36x^2y^8} = \dfrac{4x^6}{9y^4}$

71. Simplifying the expression: $\dfrac{\left(3x^2\right)\left(8x^5\right)}{6x^4} = \dfrac{24x^7}{6x^4} = \dfrac{24}{6} \cdot \dfrac{x^7}{x^4} = 4x^3$

73. Simplifying the expression: $\dfrac{\left(9a^2b\right)\left(2a^3b^4\right)}{18a^5b^7} = \dfrac{18a^5b^5}{18a^5b^7} = \dfrac{18}{18} \cdot \dfrac{a^5}{a^5} \cdot \dfrac{b^5}{b^7} = 1 \cdot \dfrac{1}{b^2} = \dfrac{1}{b^2}$

75. Simplifying the expression: $\dfrac{\left(4x^3y^2\right)\left(9x^4y^{10}\right)}{\left(3x^5y\right)\left(2x^6y\right)} = \dfrac{36x^7y^{12}}{6x^{11}y^2} = \dfrac{36}{6} \cdot \dfrac{x^7}{x^{11}} \cdot \dfrac{y^{12}}{y^2} = 6 \cdot \dfrac{1}{x^4} \cdot y^{10} = \dfrac{6y^{10}}{x^4}$

77. Simplifying the expression: $\dfrac{\left(6\times10^8\right)\left(3\times10^5\right)}{9\times10^7} = \dfrac{18\times10^{13}}{9\times10^7} = 2\times10^6$

79. Simplifying the expression: $\dfrac{\left(5\times10^3\right)\left(4\times10^{-5}\right)}{2\times10^{-2}} = \dfrac{20\times10^{-2}}{2\times10^{-2}} = 10 = 1\times10^1$

81. Simplifying the expression: $\dfrac{\left(2.8\times10^{-7}\right)\left(3.6\times10^4\right)}{2.4\times10^3} = \dfrac{10.08\times10^{-3}}{2.4\times10^3} = 4.2\times10^{-6}$

83. Simplifying the expression: $\dfrac{18x^4}{3x} + \dfrac{21x^7}{7x^4} = 6x^3 + 3x^3 = 9x^3$

85. Simplifying the expression: $\dfrac{45a^6}{9a^4} - \dfrac{50a^8}{2a^6} = 5a^2 - 25a^2 = -20a^2$

87. Simplifying the expression: $\dfrac{6x^7y^4}{3x^2y^2} + \dfrac{8x^5y^8}{2y^6} = 2x^5y^2 + 4x^5y^2 = 6x^5y^2$

89. Simplifying: $3-8=-5$

91. Simplifying: $-1+7=6$

93. Simplifying: $3(5)^2+1 = 3(25)+1 = 75+1 = 76$

95. Simplifying: $2x^2+4x^2=6x^2$

97. Simplifying: $-5x+7x=2x$

99. Simplifying: $-(2x+9)=-2x-9$

101. Substituting $x=4$: $2x+3=2(4)+3=8+3=11$

103. The degree is 9. The correct answer is b.

105. Finding the product: $(2.6\times10^{-5})(3.8\times10^{3})=9.88\times10^{-2}$. The correct answer is d.

5.4 Addition and Subtraction of Polynomials

1. This is a trinomial of degree 3.

3. This is a trinomial of degree 3.

5. This is a binomial of degree 1.

7. This is a binomial of degree 2.

9. This is a monomial of degree 2.

11. This is a monomial of degree 0.

13. Writing in standard form: $2+5x^2=5x^2+2$. The degree is 2 and the leading coefficient is 5.

15. Writing in standard form: $3x^2-x^3-6x=-x^3+3x^2-6x$. The degree is 3 and the leading coefficient is -1.

17. Writing in standard form: $6x^2-1+x=6x^2+x-1$. The degree is 2 and the leading coefficient is 6.

19. Evaluating when $x=3$: $4x+2=4(3)+2=12+2=14$

21. Evaluating when $x=-1$: $x^2-9=(-1)^2-9=1-9=-8$

23. Evaluating when $a=2$: $3a^2-2a+4=3(2)^2-2(2)+4=12-4+4=12$

25. Evaluating when $x=3$: $x^2-2x+1=(3)^2-2(3)+1=9-6+1=4$

27. Combining the polynomials: $(2x^2+3x+4)+(3x^2+2x+5)=(2x^2+3x^2)+(3x+2x)+(4+5)=5x^2+5x+9$

29. Combining the polynomials: $(3a^2-4a+1)+(2a^2-5a+6)=(3a^2+2a^2)+(-4a-5a)+(1+6)=5a^2-9a+7$

31. Combining the polynomials: $(x^2+4x)+(2x+8)=x^2+(4x+2x)+8=x^2+6x+8$

33. Combining the polynomials: $(x^2-3x)+(3x-9)=x^2+(-3x+3x)-9=x^2-9$

35. Finding the opposite: $-(10x-5)=-10x+5$

37. Finding the opposite: $-(5x^2+x-2)=-5x^2-x+2$

39. Finding the opposite: $-(3+2x-x^2)=-3-2x+x^2=x^2-2x-3$

41. Combining the polynomials: $(6x^3-4x^2+2x)-(9x^2-6x+3)=6x^3-4x^2+2x-9x^2+6x-3=6x^3-13x^2+8x-3$

43. Combining the polynomials: $(a^2-a-1)-(-a^2+a+1)=a^2-a-1+a^2-a-1=2a^2-2a-2$

45. Combining the polynomials: $(6x^2-3x)-(10x-5)=6x^2-3x-10x+5=6x^2-13x+5$

47. Combining the polynomials: $(3y^2-5y)-(6y-10)=3y^2-5y-6y+10=3y^2-11y+10$

49. Combining the polynomials: $\left(\frac{2}{3}x^2-\frac{1}{5}x-\frac{3}{4}\right)+\left(\frac{4}{3}x^2-\frac{4}{5}x+\frac{7}{4}\right)=\left(\frac{2}{3}x^2+\frac{4}{3}x^2\right)+\left(-\frac{1}{5}x-\frac{4}{5}x\right)+\left(-\frac{3}{4}+\frac{7}{4}\right)=2x^2-x+1$

51. Combining the polynomials:

$$\left(\frac{5}{9}x^3+\frac{1}{3}x^2-2x+1\right)-\left(\frac{2}{3}x^3+x^2+\frac{1}{2}x-\frac{3}{4}\right)=\frac{5}{9}x^3+\frac{1}{3}x^2-2x+1-\frac{2}{3}x^3-x^2-\frac{1}{2}x+\frac{3}{4}=-\frac{1}{9}x^3-\frac{2}{3}x^2-\frac{5}{2}x+\frac{7}{4}$$

53. Combining the polynomials:

$$\begin{aligned}(4y^2-3y+2)+(5y^2+12y-4)-(13y^2-6y+20)&=4y^2-3y+2+5y^2+12y-4-13y^2+6y-20\\&=(4y^2+5y^2-13y^2)+(-3y+12y+6y)+(2-4-20)\\&=-4y^2+15y-22\end{aligned}$$

55. Performing the subtraction:

$$\begin{aligned}(11x^2-10x+13)-(10x^2+23x-50)&=11x^2-10x+13-10x^2-23x+50\\&=(11x^2-10x^2)+(-10x-23x)+(13+50)\\&=x^2-33x+63\end{aligned}$$

57. Performing the subtraction:

$$\begin{aligned}(11y^2+11y+11)-(3y^2+7y-15)&=11y^2+11y+11-3y^2-7y+15\\&=(11y^2-3y^2)+(11y-7y)+(11+15)\\&=8y^2+4y+26\end{aligned}$$

59. Performing the addition:

$$(25x^2-50x+75)+(50x^2-100x-150)=(25x^2+50x^2)+(-50x-100x)+(75-150)=75x^2-150x-75$$

61. Performing the operations:

$$(3x-2)+(11x+5)-(2x+1)=3x-2+11x+5-2x-1=(3x+11x-2x)+(-2+5-1)=12x+2$$

63. Finding the volume of the cylinder and sphere:

$$V_{\text{cylinder}}=\pi(3^2)(6)=54\pi \qquad V_{\text{sphere}}=\frac{4}{3}\pi(3^3)=36\pi$$

Subtracting to find the amount of space to pack: $V=54\pi-36\pi=18\pi\approx 56.52$ inches^3

65. Simplifying: $(-5)(-1)=5$

67. Simplifying: $(-1)(6)=-6$

69. Simplifying: $(5x)(-4x)=-20x^2$

71. Simplifying: $3x(-7)=-21x$

73. Simplifying: $5x+(-3x)=2x$

75. Multiplying: $3(2x-6)=6x-18$

77. The degree of the polynomial is 4. The correct answer is d.

79. Adding the polynomials: $(2x^3-x^2+4)+(3x^2+x-2)=2x^3+2x^2+x+2$. The correct answer is b.

5.5 Multiplication with Polynomials

1. Using the distributive property: $2x(3x+1)=2x(3x)+2x(1)=6x^2+2x$

3. Using the distributive property: $2x^2(3x^2-2x+1)=2x^2(3x^2)-2x^2(2x)+2x^2(1)=6x^4-4x^3+2x^2$

5. Using the distributive property: $2ab(a^2-ab+1)=2ab(a^2)-2ab(ab)+2ab(1)=2a^3b-2a^2b^2+2ab$

7. Using the distributive property: $y^2(3y^2+9y+12)=y^2(3y^2)+y^2(9y)+y^2(12)=3y^4+9y^3+12y^2$

9. Using the distributive property:

$$4x^2y(2x^3y+3x^2y^2+8y^3)=4x^2y(2x^3y)+4x^2y(3x^2y^2)+4x^2y(8y^3)=8x^5y^2+12x^4y^3+32x^2y^4$$

11. Multiplying using the FOIL method: $(x+3)(x+4)=x^2+3x+4x+12=x^2+7x+12$

13. Multiplying using the FOIL method: $(x+6)(x+1)=x^2+6x+1x+6=x^2+7x+6$

15. Multiplying using the FOIL method: $\left(x+\frac{1}{2}\right)\left(x+\frac{3}{2}\right)=x^2+\frac{1}{2}x+\frac{3}{2}x+\frac{3}{4}=x^2+2x+\frac{3}{4}$

17. Multiplying using the FOIL method: $(a+5)(a-3)=a^2+5a-3a-15=a^2+2a-15$

19. Multiplying using the FOIL method: $(x-a)(y+b)=xy+bx-ay-ab$

21. Multiplying using the FOIL method: $(x+6)(x-6)=x^2-6x+6x-36=x^2-36$

23. Multiplying using the FOIL method: $\left(y+\frac{5}{6}\right)\left(y-\frac{5}{6}\right)=y^2-\frac{5}{6}y+\frac{5}{6}y-\frac{25}{36}=y^2-\frac{25}{36}$

25. Multiplying using the FOIL method: $(2x-3)(x-4)=2x^2-8x-3x+12=2x^2-11x+12$

27. Multiplying using the FOIL method: $(a+2)(2a-1)=2a^2-a+4a-2=2a^2+3a-2$

29. Multiplying using the FOIL method: $(2x-5)(3x-2)=6x^2-4x-15x+10=6x^2-19x+10$

31. Multiplying using the FOIL method: $(2x+3)(a+4)=2ax+8x+3a+12$

33. Multiplying using the FOIL method: $(5x-4)(5x+4)=25x^2+20x-20x-16=25x^2-16$

35. Multiplying using the FOIL method: $\left(2x-\frac{1}{2}\right)\left(x+\frac{3}{2}\right)=2x^2+3x-\frac{1}{2}x-\frac{3}{4}=2x^2+\frac{5}{2}x-\frac{3}{4}$

37. Multiplying using the FOIL method: $(1-2a)(3-4a)=3-4a-6a+8a^2=3-10a+8a^2$

39. Multiplying using the column method:

$$\begin{array}{l} a^2-3a+2 \\ \underline{\qquad a-3} \\ a^3-3a^2+2a \\ \underline{\quad -3a^2+9a-6} \\ a^3-6a^2+11a-6 \end{array}$$

41. Multiplying using the column method:

$$\begin{array}{l} x^2-2x+4 \\ \underline{\qquad x+2} \\ x^3-2x^2+4x \\ \underline{\qquad 2x^2-4x+8} \\ x^3+8 \end{array}$$

43. Multiplying using the column method:

$$\begin{array}{l} x^2+8x+9 \\ \underline{\qquad 2x+1} \\ 2x^3+16x^2+18x \\ \underline{\qquad\quad x^2+8x+9} \\ 2x^3+17x^2+26x+9 \end{array}$$

45. Multiplying using the column method:

$$\begin{array}{l} 5x^2+2x+1 \\ \underline{x^2-3x+5} \\ 5x^4+2x^3+x^2 \\ \quad -15x^3-6x^2-3x \\ \underline{\qquad\qquad 25x^2+10x+5} \\ 5x^4-13x^3+20x^2+7x+5 \end{array}$$

47. Multiplying using the column method:

$$\begin{array}{l} 3x^2-5x-2 \\ \underline{2x^2+x-1} \\ 6x^4-10x^3-4x^2 \\ \qquad 3x^3-5x^2-2x \\ \underline{\qquad\quad -3x^2+5x+2} \\ 6x^4-7x^3-12x^2+3x+2 \end{array}$$

49. Multiplying using the column method:

$$\begin{array}{l} a^3+a+2 \\ \underline{a^2-3a+4} \\ a^5+a^3+2a^2 \\ \quad -3a^4-3a^2-6a \\ \underline{\qquad 4a^3+4a+8} \\ a^5-3a^4+5a^3-a^2-2a+8 \end{array}$$

51. Multiplying using the FOIL method: $(x^2+3)(2x^2-5)=2x^4-5x^2+6x^2-15=2x^4+x^2-15$

53. Multiplying using the FOIL method: $(3a^4+2)(2a^2+5)=6a^6+15a^4+4a^2+10$

55. First multiply two polynomials using the FOIL method: $(x+3)(x+4)=x^2+3x+4x+12=x^2+7x+12$

Now using the column method:

$$\begin{array}{l} x^2+7x+12 \\ \underline{\qquad x+5} \\ x^3+7x^2+12x \\ \underline{\qquad 5x^2+35x+60} \\ x^3+12x^2+47x+60 \end{array}$$

57. Simplifying: $(x-3)(x-2)+2=x^2-3x-2x+6+2=x^2-5x+8$

59. Simplifying: $(2x-3)(4x+3)+4=8x^2+6x-12x-9+4=8x^2-6x-5$

61. Simplifying: $(x+4)(x-5)+(-5)(2)=x^2-5x+4x-20-10=x^2-x-30$

63. Simplifying: $2(x-3)+x(x+2)=2x-6+x^2+2x=x^2+4x-6$

65. Simplifying: $3x(x+1)-2x(x-5)=3x^2+3x-2x^2+10x=x^2+13x$

67. Simplifying: $x(x+2)-3=x^2+2x-3$

69. Simplifying: $a(a-3)+6=a^2-3a+6$

71. The product is $(x+2)(x+3)=x^2+5x+6$:

	x	3
x	x^2	$3x$
2	$2x$	6

73. The product is $(x+1)(2x+2)=2x^2+4x+2$:

	x	x	2
x	x^2	x^2	$2x$
1	x	x	2

75. **a.** Adding the expressions: $(2x+5)+(3x-4)=2x+3x+5-4=5x+1$

b. Subtracting the expressions: $(3x-4)-(2x+5)=3x-4-2x-5=x-9$

c. Solving the equation:

$$\begin{aligned} 2x+5&=3x-4\\ -x+5&=-4\\ -x&=-9\\ x&=9 \end{aligned}$$

d. Multiplying the expressions: $(2x+5)(3x-4)=6x^2-8x+15x-20=6x^2+7x-20$

77. Let x represent the width and $2x+5$ represent the length. The area is given by: $A=x(2x+5)=2x^2+5x$

79. Let x and $x+1$ represent the width and length, respectively. The area is given by: $A=x(x+1)=x^2+x$

81. The revenue is: $R=xp=(100-10p)p=100p-10p^2$

83. Simplifying: $13\cdot 13=169$

85. Simplifying: $2(x)(-5)=-10x$

87. Simplifying: $6x+(-6x)=0$

89. Simplifying: $(2x)(-3)+(2x)(3)=-6x+6x=0$

91. Multiplying: $-4(3x-4)=-12x+16$

93. Multiplying: $(x-1)(x+2)=x^2+2x-x-2=x^2+x-2$

95. Multiplying: $(x+3)(x+3)=x^2+3x+3x+9=x^2+6x+9$

97. Using the distributive property: $3x^4(2x^2-5x+4)=3x^4(2x^2)-3x^4(5x)+3x^4(4)=6x^6-15x^5+12x^4$

The correct answer is d.

99. Multiplying using the column method:

$$\begin{array}{r} 2x^2-x-3\\ x+4\\ \hline 2x^3-x^2-3x\\ 8x^2-4x-12\\ \hline 2x^3+7x^2-7x-12 \end{array}$$

The correct answer is c.

5.6 Binomial Squares and Other Special Products

1. Multiplying using the FOIL method: $(x-2)^2=(x-2)(x-2)=x^2-2x-2x+4=x^2-4x+4$

3. Multiplying using the FOIL method: $(a+3)^2=(a+3)(a+3)=a^2+3a+3a+9=a^2+6a+9$

5. Multiplying using the FOIL method: $(x-5)^2=(x-5)(x-5)=x^2-5x-5x+25=x^2-10x+25$

7. Multiplying using the FOIL method: $\left(a-\frac{1}{2}\right)^2=\left(a-\frac{1}{2}\right)\left(a-\frac{1}{2}\right)=a^2-\frac{1}{2}a-\frac{1}{2}a+\frac{1}{4}=a^2-a+\frac{1}{4}$

9. Multiplying using the FOIL method: $(x+10)^2=(x+10)(x+10)=x^2+10x+10x+100=x^2+20x+100$

11. Multiplying using the square of binomial formula: $(a+0.8)^2=a^2+2(a)(0.8)+(0.8)^2=a^2+1.6a+0.64$

13. Multiplying using the square of binomial formula: $(2x-1)^2=(2x)^2-2(2x)(1)+(1)^2=4x^2-4x+1$

15. Multiplying using the square of binomial formula: $(4a+5)^2=(4a)^2+2(4a)(5)+(5)^2=16a^2+40a+25$

17. Multiplying using the square of binomial formula: $(3x-2)^2=(3x)^2-2(3x)(2)+(2)^2=9x^2-12x+4$

19. Multiplying using the square of binomial formula: $(3a+5b)^2=(3a)^2+2(3a)(5b)+(5b)^2=9a^2+30ab+25b^2$

21. Multiplying using the square of binomial formula: $(4x-5y)^2=(4x)^2-2(4x)(5y)+(5y)^2=16x^2-40xy+25y^2$

23. Multiplying using the square of binomial formula: $(x^2+5)^2=(x^2)^2+2(x^2)(5)+(5)^2=x^4+10x^2+25$

25. Multiplying using the square of binomial formula: $(a^3+1)^2=(a^3)^2+2(a^3)(1)+(1)^2=a^6+2a^3+1$

27. Multiplying using the square of binomial formula: $(7m^2+2n)^2=(7m^2)^2+2(7m^2)(2n)+(2n)^2=49m^4+28m^2n+4n^2$

29. Multiplying using the square of binomial formula:

$$(6x^2-10y^2)^2=(6x^2)^2-2(6x^2)(10y^2)+(10y^2)^2=36x^4-120x^2y^2+100y^4$$

31. Completing the table:

x	$(x+3)^2$	x^2+9	x^2+6x+9
1	16	10	16
2	25	13	25
3	36	18	36
4	49	25	49

33. Completing the table:

a	1	3	3	4
b	1	5	4	5
$(a+b)^2$	4	64	49	81
a^2+b^2	2	34	25	41
a^2+ab+b^2	3	49	37	61
$a^2+2ab+b^2$	4	64	49	81

35. Multiplying using the FOIL method: $(a+5)(a-5)=a^2-5a+5a-25=a^2-25$

37. Multiplying using the FOIL method: $(y-1)(y+1)=y^2+y-y-1=y^2-1$

39. Multiplying using the difference of squares formula: $(9+x)(9-x)=(9)^2-(x)^2=81-x^2$

41. Multiplying using the difference of squares formula: $(2x+5)(2x-5)=(2x)^2-(5)^2=4x^2-25$

43. Multiplying using the difference of squares formula: $\left(4x+\frac{1}{3}\right)\left(4x-\frac{1}{3}\right)=(4x)^2-\left(\frac{1}{3}\right)^2=16x^2-\frac{1}{9}$

45. Multiplying using the difference of squares formula: $(2a+7b)(2a-7b)=(2a)^2-(7b)^2=4a^2-49b^2$

47. Multiplying using the difference of squares formula: $(6-7x)(6+7x)=(6)^2-(7x)^2=36-49x^2$

49. Multiplying using the difference of squares formula: $(x^2+3)(x^2-3)=(x^2)^2-(3)^2=x^4-9$

51. Multiplying using the difference of squares formula: $(a^2+4b^2)(a^2-4b^2)=(a^2)^2-(4b^2)^2=a^4-16b^4$

53. Multiplying using the difference of squares formula: $(5y^4-8)(5y^4+8)=(5y^4)^2-(8)^2=25y^8-64$

55. Multiplying and simplifying: $(x+3)(x-3)+(x+5)(x-5)=(x^2-9)+(x^2-25)=2x^2-34$

57. Multiplying and simplifying:

$$(2x+3)^2-(4x-1)^2=(4x^2+12x+9)-(16x^2-8x+1)=4x^2+12x+9-16x^2+8x-1=-12x^2+20x+8$$

59. Multiplying and simplifying:

$$\begin{aligned}(a+1)^2-(a+2)^2+(a+3)^2&=(a^2+2a+1)-(a^2+4a+4)+(a^2+6a+9)\\&=a^2+2a+1-a^2-4a-4+a^2+6a+9\\&=a^2+4a+6\end{aligned}$$

61. Multiplying and simplifying:

$$\begin{aligned}(2x+3)^3&=(2x+3)(2x+3)^2\\&=(2x+3)(4x^2+12x+9)\\&=8x^3+24x^2+18x+12x^2+36x+27\\&=8x^3+36x^2+54x+27\end{aligned}$$

63. Multiplying: $(49)(51)=(50-1)(50+1)=50^2-1^2=2{,}500-1=2{,}499$

65. Evaluating when $x=2$:

$$(x+3)^2=(2+3)^2=(5)^2=25$$

$$x^2+6x+9=(2)^2+6(2)+9=4+12+9=25$$

Both expressions are equal to 25.

67. Let x and $x+1$ represent the two integers. The expression can be written as:

$$(x)^2+(x+1)^2=x^2+(x^2+2x+1)=2x^2+2x+1$$

69. Let x, $x+1$, and $x+2$ represent the three integers. The expression can be written as:

$$(x)^2+(x+1)^2+(x+2)^2=x^2+(x^2+2x+1)+(x^2+4x+4)=3x^2+6x+5$$

71. Verifying the areas: $(a+b)^2=a^2+ab+ab+b^2=a^2+2ab+b^2$

73. Simplifying: $\dfrac{10x^3}{5x}=2x^{3-1}=2x^2$

75. Simplifying: $\dfrac{3x^2}{3}=x^2$

77. Simplifying: $\dfrac{9x^2}{3x}=3x^{2-1}=3x$

79. Simplifying: $\dfrac{24x^3y^2}{8x^2y}=3x^{3-2}y^{2-1}=3xy$

81. Dividing:

```
     146
27)3962
   27
   126
   108
    182
    162
     20
```

The quotient is $146\dfrac{20}{27}$.

83. Multiplying: $(x-3)x = x(x-3) = x^2 - 3x$

85. Multiplying: $2x^2(x-5) = 2x^3 - 10x^2$

87. Subtracting: $(x^2-5x)-(x^2-3x) = x^2-5x-x^2+3x = -2x$

89. Subtracting: $(-2x+8)-(-2x+6) = -2x+8+2x-6 = 2$

91. Multiplying using the square of binomial formula: $(3x-4y)^2 = (3x)^2 - 2(3x)(4y)+(4y)^2 = 9x^2-24xy+16y^2$
The correct answer is b.

5.7 Division with Polynomials

1. Performing the division: $\frac{5x^2-10x}{5x} = \frac{5x^2}{5x} - \frac{10x}{5x} = x-2$

3. Performing the division: $\frac{25x^2y-10xy}{5x} = \frac{25x^2y}{5x} - \frac{10xy}{5x} = 5xy-2y$

5. Performing the division: $\frac{35x^5-30x^4+25x^3}{5x} = \frac{35x^5}{5x} - \frac{30x^4}{5x} + \frac{25x^3}{5x} = 7x^4-6x^3+5x^2$

7. Performing the division: $\frac{8a^2-4a}{-2a} = \frac{8a^2}{-2a} + \frac{-4a}{-2a} = -4a+2$

9. Performing the division: $\frac{12a^3b-6a^2b^2+14ab^3}{-2a} = \frac{12a^3b}{-2a} + \frac{-6a^2b^2}{-2a} + \frac{14ab^3}{-2a} = -6a^2b+3ab^2-7b^3$

11. Performing the division: $\frac{a^2+2ab+b^2}{-2a} = \frac{a^2}{-2a} + \frac{2ab}{-2a} + \frac{b^2}{-2a} = -\frac{a}{2} - b - \frac{b^2}{2a}$

13. Performing the division: $\frac{6x+8y}{2} = \frac{6x}{2} + \frac{8y}{2} = 3x+4y$

15. Performing the division: $\frac{7y-21}{-7} = \frac{7y}{-7} + \frac{-21}{-7} = -y+3$

17. Performing the division: $\frac{10xy-8x}{2x} = \frac{10xy}{2x} - \frac{8x}{2x} = 5y-4$

19. Performing the division: $\frac{x^2y-x^3y^2}{x} = \frac{x^2y}{x} - \frac{x^3y^2}{x} = xy-x^2y^2$

21. Performing the division: $\frac{a^2b^2-ab^2}{-ab^2} = \frac{a^2b^2}{-ab^2} + \frac{-ab^2}{-ab^2} = -a+1$

23. Performing the division: $\frac{x^3-3x^2y+xy^2}{x} = \frac{x^3}{x} - \frac{3x^2y}{x} + \frac{xy^2}{x} = x^2-3xy+y^2$

25. Performing the division: $\frac{10a^2-15a^2b+25a^2b^2}{5a^2} = \frac{10a^2}{5a^2} - \frac{15a^2b}{5a^2} + \frac{25a^2b^2}{5a^2} = 2-3b+5b^2$

27. Performing the division: $\frac{26x^2y^2-13xy}{-13xy} = \frac{26x^2y^2}{-13xy} + \frac{-13xy}{-13xy} = -2xy+1$

29. Performing the division: $\frac{5a^2x-10ax^2+15a^2x^2}{20a^2x^2} = \frac{5a^2x}{20a^2x^2} - \frac{10ax^2}{20a^2x^2} + \frac{15a^2x^2}{20a^2x^2} = \frac{1}{4x} - \frac{1}{2a} + \frac{3}{4}$

31. Performing the division: $\frac{16x^5+8x^2+12x}{12x^3} = \frac{16x^5}{12x^3} + \frac{8x^2}{12x^3} + \frac{12x}{12x^3} = \frac{4x^2}{3} + \frac{2}{3x} + \frac{1}{x^2}$

33. Performing the division: $\frac{9a^{5m}-27a^{3m}}{3a^{2m}} = \frac{9a^{5m}}{3a^{2m}} - \frac{27a^{3m}}{3a^{2m}} = 3a^{5m-2m} - 9a^{3m-2m} = 3a^{3m} - 9a^m$

35. Performing the division: $\frac{10x^{5m}-25x^{3m}+35x^m}{5x^m} = \frac{10x^{5m}}{5x^m} - \frac{25x^{3m}}{5x^m} + \frac{35x^m}{5x^m} = 2x^{5m-m} - 5x^{3m-m} + 7x^{m-m} = 2x^{4m} - 5x^{2m} + 7$

37. Simplifying and then dividing:

$$\frac{2x^3(3x+2)-3x^2(2x-4)}{2x^2}=\frac{6x^4+4x^3-6x^3+12x^2}{2x^2}=\frac{6x^4-2x^3+12x^2}{2x^2}=\frac{6x^4}{2x^2}-\frac{2x^3}{2x^2}+\frac{12x^2}{2x^2}=3x^2-x+6$$

39. Simplifying and then dividing:

$$\frac{(x+2)^2-(x-2)^2}{2x}=\frac{(x^2+4x+4)-(x^2-4x+4)}{2x}=\frac{x^2+4x+4-x^2+4x-4}{2x}=\frac{8x}{2x}=4$$

41. Simplifying and then dividing:

$$\frac{(x+5)^2+(x+5)(x-5)}{2x}=\frac{(x^2+10x+25)+(x^2-25)}{2x}=\frac{2x^2+10x}{2x}=\frac{2x^2}{2x}+\frac{10x}{2x}=x+5$$

43. Using long division:

$$\begin{array}{r}
x-2 \\
x-3\overline{)x^2-5x+6} \\
\underline{x^2-3x} \\
-2x+6 \\
\underline{-2x+6} \\
0
\end{array}$$

The quotient is $x-2$.

45. Using long division:

$$\begin{array}{r}
a+4 \\
a+5\overline{)a^2+9a+20} \\
\underline{a^2+5a} \\
4a+20 \\
\underline{4a+20} \\
0
\end{array}$$

The quotient is $a+4$.

47. Using long division:

$$\begin{array}{r}
x+3 \\
2x-1\overline{)2x^2+5x-3} \\
\underline{2x^2-\ \ x} \\
6x-3 \\
\underline{6x-3} \\
0
\end{array}$$

The quotient is $x+3$.

49. Using long division:

$$\begin{array}{r}
x-3 \\
x^2-6x+9\overline{)x^3-9x^2+27x-27} \\
\underline{x^3-6x^2+\ 9x} \\
-3x^2+18x-27 \\
\underline{-3x^2+18x-27} \\
0
\end{array}$$

The quotient is $x-3$.

51. Using long division:

$$\begin{array}{r}
x^2-x-3 \\
x^2+2\overline{)x^4-x^3-x^2-2x-6} \\
\underline{x^4\quad\ +2x^2} \\
-x^3-3x^2-2x \\
\underline{-x^3\qquad\ -2x} \\
-3x^2\qquad -6 \\
\underline{-3x^2\qquad -6} \\
0
\end{array}$$

The quotient is x^2-x-3.

53. Using long division:

$$\begin{array}{r}
1 \\
x-2\overline{)x+3} \\
\underline{x-2} \\
5
\end{array}$$

The quotient is $1+\frac{5}{x-2}$.

55. Using long division:

$$\begin{array}{r} 3 \\ x+2\overline{)3x+4} \\ \underline{3x+6} \\ -2 \end{array}$$

The quotient is $3+\dfrac{-2}{x+2}$.

57. Using long division:

$$\begin{array}{r} x+2 \\ x+3\overline{)x^2+5x+8} \\ \underline{x^2+3x} \\ 2x+8 \\ \underline{2x+6} \\ 2 \end{array}$$

The quotient is $x+2+\dfrac{2}{x+3}$.

59. Using long division:

$$\begin{array}{r} x+4 \\ x-2\overline{)x^2+2x+1} \\ \underline{x^2-2x} \\ 4x+1 \\ \underline{4x-8} \\ 9 \end{array}$$

The quotient is $x+4+\dfrac{9}{x-2}$.

61. Using long division:

$$\begin{array}{r} x+4 \\ x+1\overline{)x^2+5x-6} \\ \underline{x^2+\ x} \\ 4x-6 \\ \underline{4x+4} \\ -10 \end{array}$$

The quotient is $x+4+\dfrac{-10}{x+1}$.

63. Using long division:

$$\begin{array}{r} x-3 \\ 2x+4\overline{)2x^2-2x+5} \\ \underline{2x^2+4x} \\ -6x+5 \\ \underline{-6x-12} \\ 17 \end{array}$$

The quotient is $x-3+\dfrac{17}{2x+4}$.

65. Using long division:

$$\begin{array}{r} 3a-2 \\ 2a+3\overline{)6a^2+5a+1} \\ \underline{6a^2+9a} \\ -4a+1 \\ \underline{-4a-6} \\ 7 \end{array}$$

The quotient is $3a-2+\dfrac{7}{2a+3}$.

67. Using long division:

$$\begin{array}{r} 2a^2-a-3 \\ 3a-5\overline{)6a^3-13a^2-4a+15} \\ \underline{6a^3-10a^2} \\ -3a^2-4a \\ \underline{-3a^2+5a} \\ -9a+15 \\ \underline{-9a+15} \\ 0 \end{array}$$

The quotient is $2a^2-a-3$.

69. Using long division:

$$\begin{array}{r} 1 \\ x^2-3x-2\overline{)x^2-6x+9} \\ \underline{x^2-3x-2} \\ -3x+11 \end{array}$$

The quotient is $1+\dfrac{-3x+11}{x^2-3x-2}$.

71. Using long division:

$$\begin{array}{r} a-5 \\ 2a^2+a+3\overline{)2a^3-9a^2-5a+4} \\ \underline{2a^3+a^2+3a} \\ -10a^2-8a+4 \\ \underline{-10a^2-5a-15} \\ -3a+19 \end{array}$$

The quotient is $a-5+\dfrac{-3a+19}{2a^2+a+3}$.

73. Using long division:

$$\begin{array}{r} x^2-x+5 \\ x+1\overline{)x^3+0x^2+4x+5} \\ \underline{x^3+x^2} \\ -x^2+4x \\ \underline{-x^2-x} \\ 5x+5 \\ \underline{5x+5} \\ 0 \end{array}$$

The quotient is x^2-x+5.

75. Using long division:

$$\begin{array}{r} x^2+x+1 \\ x-1\overline{)x^3+0x^2+0x-1} \\ \underline{x^3-x^2} \\ x^2+0x \\ \underline{x^2-x} \\ x-1 \\ \underline{x-1} \\ 0 \end{array}$$

The quotient is x^2+x+1.

77. Using long division:

$$\begin{array}{r} x^2+2x+4 \\ x-2\overline{)x^3+0x^2+0x-8} \\ \underline{x^3-2x^2} \\ 2x^2+0x \\ \underline{2x^2-4x} \\ 4x-8 \\ \underline{4x-8} \\ 0 \end{array}$$

The quotient is x^2+2x+4.

79. Using long division:

$$\begin{array}{r} 1 \\ a^2+1\overline{)a^2+3a+2} \\ \underline{a^2+0a+1} \\ 3a+1 \end{array}$$

The quotient is $1+\dfrac{3a+1}{a^2+1}$.

81. Using long division:

$$\begin{array}{r} a+3 \\ a^2-2\overline{)a^3+3a^2+0a+1} \\ \underline{a^3+0a^2-2a} \\ 3a^2+2a+1 \\ \underline{3a^2+0a-6} \\ 2a+7 \end{array}$$

The quotient is $a+3+\dfrac{2a+7}{a^2-2}$.

83. Using long division:

$$\begin{array}{r} 2a^2+3 \\ 2a^2-1\overline{)4a^4+4a^2-2} \\ \underline{4a^4-2a^2} \\ 6a^2-2 \\ \underline{6a^2-3} \\ 1 \end{array}$$

The quotient is $2a^2+3+\dfrac{1}{2a^2-1}$.

85. Evaluating when $x = 2$: $2x+3=2(2)+3=4+3=7$

Evaluating when $x = 2$: $\dfrac{10x+15}{5}=\dfrac{10(2)+15}{5}=\dfrac{20+15}{5}=\dfrac{35}{5}=7$

87. Evaluating when $x = 10$: $\frac{3x+8}{2} = \frac{3(10)+8}{2} = \frac{30+8}{2} = \frac{38}{2} = 19$

Evaluating when $x = 10$: $3x+4 = 3(10)+4 = 30+4 = 34$

89. Dividing: $\frac{\$5,894}{12} \approx \$491.17$ per month

91. Dividing: $\frac{\$3,977}{12} \approx \$331.42$ per month

93. Simplifying the expression: $\left(5x^3\right)^2\left(2x^6\right)^3 = 25x^6 \cdot 8x^{18} = 200x^{24}$

95. Simplifying the expression: $\frac{x^4}{x^{-3}} = x^{4-(-3)} = x^{4+3} = x^7$

97. Simplifying the expression: $\left(2\times10^{-4}\right)\left(4\times10^5\right) = 8\times10^1 = 80$

99. Simplifying the expression: $20ab^2 - 16ab^2 + 6ab^2 = 10ab^2$

101. Multiplying using the distributive property: $2x^2\left(3x^2+3x-1\right) = 2x^2\left(3x^2\right)+2x^2(3x)-2x^2(1) = 6x^4+6x^3-2x^2$

103. Multiplying using the square of binomial formula: $(3y-5)^2 = (3y)^2 - 2(3y)(5)+(5)^2 = 9y^2-30y+25$

105. Multiplying using the difference of squares formula: $\left(2a^2+7\right)\left(2a^2-7\right) = \left(2a^2\right)^2-(7)^2 = 4a^4-49$

107. Dividing the monomial: $\frac{12x^3y^2-3xy^3}{6x^2y^2} = \frac{12x^3y^2}{6x^2y^2} - \frac{3xy^3}{6x^2y^2} = 2x - \frac{y}{2x}$. The correct answer is d.

Chapter 5 Test

1. Simplifying the expression: $(-2)^5 = (-2)(-2)(-2)(-2)(-2) = -32$

2. Simplifying the expression: $-4^2 = -(4\cdot4) = -16$

3. Simplifying the expression: $x^9 \cdot x^{14} = x^{9+14} = x^{23}$

4. Simplifying the expression: $\left(4x^2y^3\right)^2 = 4^2x^4y^6 = 16x^4y^6$

5. Simplifying the expression: $4^{-2} = \frac{1}{4^2} = \frac{1}{16}$

6. Simplifying the expression: $\left(4a^5b^3\right)^0 = 1$

7. Simplifying the expression: $\frac{x^{-4}}{x^{-7}} = x^{-4-(-7)} = x^{-4+7} = x^3$

8. Simplifying the expression: $\left(\frac{x}{3}\right)^3 = \frac{x^3}{3^3} = \frac{x^3}{27}$

9. Simplifying the expression: $\frac{\left(x^{-3}\right)^2\left(x^{-5}\right)^{-3}}{\left(x^{-3}\right)^{-4}} = \frac{x^{-6}x^{15}}{x^{12}} = \frac{x^9}{x^{12}} = x^{9-12} = x^{-3} = \frac{1}{x^3}$

10. Writing in scientific notation: $0.04307 = 4.307\times10^{-2}$

11. Writing in expanded form: $7.63\times10^6 = 7,630,000$

12. Simplifying the expression: $\left(6a^2b\right)\left(-4ab^3\right) = -24a^{2+1}b^{1+3} = -24a^3b^4$

13. Simplifying the expression: $\frac{17x^2y^5z^3}{51x^4y^2z} = \frac{17}{51}\cdot\frac{x^2}{x^4}\cdot\frac{y^5}{y^2}\cdot\frac{z^3}{z} = \frac{1}{3}\cdot\frac{1}{x^2}\cdot y^3\cdot z^2 = \frac{y^3z^2}{3x^2}$

14. Simplifying the expression: $\frac{(3a^3b)(4a^2b^5)}{24a^2b^4}=\frac{12a^5b^6}{24a^2b^4}=\frac{a^3b^2}{2}$

15. Simplifying the expression: $\frac{28x^4}{4x}+\frac{30x^7}{6x^4}=7x^3+5x^3=12x^3$

16. Simplifying the expression: $\frac{(1.1\times10^5)(3\times10^{-2})}{4.4\times10^{-5}}=\frac{3.3\times10^3}{4.4\times10^{-5}}=0.75\times10^8=7.5\times10^7$

17. Performing the operations: $(9x^2-2x)+(7x+4)=9x^2-2x+7x+4=9x^2+5x+4$

18. Performing the operations: $(4x^2+5x-6)-(2x^2-x-4)=4x^2+5x-6-2x^2+x+4=2x^2+6x-2$

19. Performing the operations: $(7x+3)-(2x+7)=7x+3-2x-7=5x-4$

20. Evaluating when $a=-3$: $3a^2+4a+6=3(-3)^2+4(-3)+6=27-12+6=21$

21. Multiplying using the distributive property: $3x^2(5x^2-2x+4)=3x^2(5x^2)-3x^2(2x)+3x^2(4)=15x^4-6x^3+12x^2$

22. Multiplying using the FOIL method: $\left(x+\frac{1}{4}\right)\left(x-\frac{1}{3}\right)=x^2-\frac{1}{3}x+\frac{1}{4}x-\frac{1}{12}=x^2-\frac{1}{12}x-\frac{1}{12}$

23. Multiplying using the FOIL method: $(2x-3)(5x+6)=10x^2+12x-15x-18=10x^2-3x-18$

24. Multiplying using the column method:

$$\begin{array}{r} x^2-4x+16 \\ x+4 \\ \hline x^3-4x^2+16x \\ 4x^2-16x+64 \\ \hline x^3+64 \end{array}$$

25. Multiplying using the square of binomial formula: $(x-6)^2=(x)^2-2(x)(6)+(6)^2=x^2-12x+36$

26. Multiplying using the square of binomial formula: $(2a+4b)^2=(2a)^2+2(2a)(4b)+(4b)^2=4a^2+16ab+16b^2$

27. Multiplying using the difference of squares formula: $(3x-6)(3x+6)=(3x)^2-(6)^2=9x^2-36$

28. Multiplying using the difference of squares formula: $(x^2-4)(x^2+4)=(x^2)^2-(4)^2=x^4-16$

29. Dividing the monomial: $\frac{18x^3-36x^2+6x}{6x}=\frac{18x^3}{6x}-\frac{36x^2}{6x}+\frac{6x}{6x}=3x^2-6x+1$

30. Using long division:

$$\begin{array}{r} 3x-1 \\ 3x-1\overline{)9x^2-6x-4} \\ \underline{9x^2-3x} \\ -3x-4 \\ \underline{-3x+1} \\ -5 \end{array}$$

The quotient is $3x-1+\frac{-5}{3x-1}$.

31. Using long division:

$$\begin{array}{r} 4x+3 \\ x^2+2\overline{)4x^3+3x^2+0x+1} \\ \underline{4x^3+0x^2+8x} \\ 3x^2-8x+1 \\ \underline{3x^2+0x+6} \\ -8x-5 \end{array}$$

The quotient is $4x+3+\frac{-8x-5}{x^2+2}$.

32. Using the volume formula: $V=(3.2 \text{ in.})^3\approx 32.77 \text{ inches}^3$

33. Let w represent the width, $3w$ represent the length, and $\frac{1}{3}w$ represent the height. The volume is given by:

$$V=(w)(3w)\left(\frac{1}{3}w\right)=w^3$$

Chapter 6
Factoring

6.1 The Greatest Common Factor and Factoring by Grouping

1. The greatest common factor is 3.

3. The greatest common factor is $2x$.

5. The greatest common factor is $4a^2b$.

7. The greatest common factor is $2x+1$.

9. Factoring out the greatest common factor: $15x+25=5(3x+5)$

11. Factoring out the greatest common factor: $6a+9=3(2a+3)$

13. Factoring out the greatest common factor: $4x-8y=4(x-2y)$

15. Factoring out the greatest common factor: $3x^2-6x+9=3(x^2-2x+3)$

17. Factoring out the greatest common factor: $3a^2-3a+60=3(a^2-a+20)$

19. Factoring out the greatest common factor: $24y^2-52y+24=4(6y^2-13y+6)$

21. Factoring out the greatest common factor: $9x^2-8x^3=x^2(9-8x)$

23. Factoring out the greatest common factor: $13a^2-26a^3=13a^2(1-2a)$

25. Factoring out the greatest common factor: $21x^2y-28xy^2=7xy(3x-4y)$

27. Factoring out the greatest common factor: $22a^2b^2-11ab^2=11ab^2(2a-1)$

29. Factoring out the greatest common factor: $7x^3+21x^2-28x=7x(x^2+3x-4)$

31. Factoring out the greatest common factor: $121y^4-11x^4=11(11y^4-x^4)$

33. Factoring out the greatest common factor: $100x^4-50x^3+25x^2=25x^2(4x^2-2x+1)$

35. Factoring out the greatest common factor: $8a^2+16b^2+32c^2=8(a^2+2b^2+4c^2)$

37. Factoring out the greatest common factor: $4a^2b-16ab^2+32a^2b^2=4ab(a-4b+8ab)$

39. Factoring out the greatest common factor: $121a^3b^2-22a^2b^3+33a^3b^3=11a^2b^2(11a-2b+3ab)$

41. Factoring out the greatest common factor: $12x^2y^3-72x^5y^3-36x^4y^4=12x^2y^3(1-6x^3-3x^2y)$

43. Factoring by grouping: $xy+5x+3y+15=x(y+5)+3(y+5)=(y+5)(x+3)$

45. Factoring by grouping: $xy+6x+2y+12=x(y+6)+2(y+6)=(y+6)(x+2)$

47. Factoring by grouping: $ab+7a-3b-21=a(b+7)-3(b+7)=(b+7)(a-3)$

49. Factoring by grouping: $ax-bx+ay-by=x(a-b)+y(a-b)=(a-b)(x+y)$

51. Factoring by grouping: $2ax+5a-2x-5=a(2x+5)-1(2x+5)=(2x+5)(a-1)$

53. Factoring by grouping: $27by-6y+9b-2=3y(9b-2)+1(9b-2)=(9b-2)(3y+1)$

55. Factoring by grouping: $3xb-4b-6x+8=b(3x-4)-2(3x-4)=(3x-4)(b-2)$

57. Factoring by grouping: $x^2+2a+2x+ax=x^2+ax+2x+2a=x(x+a)+2(x+a)=(x+a)(x+2)$

59. Factoring by grouping: $x^2+ab-ax-bx=x^2-ax-bx+ab=x(x-a)-b(x-a)=(x-a)(x-b)$

61. Factoring by grouping: $ax+ay+bx+by+cx+cy=a(x+y)+b(x+y)+c(x+y)=(x+y)(a+b+c)$

63. Factoring by grouping: $6x^2+9x+4x+6=3x(2x+3)+2(2x+3)=(2x+3)(3x+2)$

65. Factoring by grouping: $20x^2-2x+50x-5=2x(10x-1)+5(10x-1)=(10x-1)(2x+5)$

67. Factoring by grouping: $20x^2+4x+25x+5=4x(5x+1)+5(5x+1)=(5x+1)(4x+5)$

69. Factoring by grouping: $x^3+2x^2+3x+6=x^2(x+2)+3(x+2)=(x+2)(x^2+3)$

71. Factoring by grouping: $6x^3-4x^2+15x-10=2x^2(3x-2)+5(3x-2)=(3x-2)(2x^2+5)$

73. Its greatest common factor is $3\bullet 2=6$.

75. The correct factoring is: $12x^2+6x+3=3(4x^2+2x+1)$

77. Factoring: $1{,}000+1{,}000r=1{,}000(1+r)$

Evaluating when $r=0.12$: $1{,}000(1+0.12)=1{,}000(1.12)=\$1{,}120$

79. **a.** Factoring: $A=1{,}000{,}000+1{,}000{,}000r=1{,}000{,}000(1+r)$

b. Evaluating when $r=0.30$: $A=1{,}000{,}000(1+0.30)=1{,}300{,}000$

81. Multiplying using the FOIL method: $(x-7)(x+2)=x^2+2x-7x-14=x^2-5x-14$

83. Multiplying using the FOIL method: $(x-3)(x+2)=x^2+2x-3x-6=x^2-x-6$

85. Multiplying using the column method:

$$\begin{array}{r} x^2-3x+9 \\ \underline{x+3} \\ x^3-3x^2+9x \\ \underline{3x^2-9x+27} \\ x^3+27 \end{array}$$

87. Multiplying using the column method:

$$\begin{array}{r} x^2+4x-3 \\ \underline{2x+1} \\ 2x^3+8x^2-6x \\ \underline{x^2+4x-3} \\ 2x^3+9x^2-2x-3 \end{array}$$

89. Multiplying: $3x^4(6x^3-4x^2+2x)=3x^4\bullet 6x^3-3x^4\bullet 4x^2+3x^4\bullet 2x=18x^7-12x^6+6x^5$

91. Multiplying: $\left(x+\frac{1}{3}\right)\left(x+\frac{2}{3}\right)=x^2+\frac{2}{3}x+\frac{1}{3}x+\frac{2}{9}=x^2+x+\frac{2}{9}$

93. Multiplying: $(6x+4y)(2x-3y)=12x^2-18xy+8xy-12y^2=12x^2-10xy-12y^2$

95. Multiplying: $(9a+1)(9a-1)=81a^2-9a+9a-1=81a^2-1$

97. Multiplying: $(x-9)(x-9)=x^2-9x-9x+81=x^2-18x+81$

99. Multiplying: $(x+2)(x^2-2x+4)=x(x^2-2x+4)+2(x^2-2x+4)=x^3-2x^2+4x+2x^2-4x+8=x^3+8$

101. The greatest common factor is $3a^2b$. The correct answer is a.

103. Factoring by grouping: $2ax+6x-5a-15=2x(a+3)-5(a+3)=(a+3)(2x-5)$. The correct answer is c.

6.2 Factoring Trinomials

1. Factoring the trinomial: $x^2+7x+12=(x+3)(x+4)$

3. Factoring the trinomial: $x^2+3x+2=(x+2)(x+1)$

5. Factoring the trinomial: $a^2+10a+21=(a+7)(a+3)$

7. Factoring the trinomial: $x^2-7x+10=(x-5)(x-2)$

9. Factoring the trinomial: $y^2-10y+21=(y-7)(y-3)$

11. Factoring the trinomial: $x^2-x-12=(x-4)(x+3)$

13. Factoring the trinomial: $y^2+y-12=(y+4)(y-3)$

15. Factoring the trinomial: $x^2+5x-14=(x+7)(x-2)$

17. Factoring the trinomial: $r^2-8r-9=(r-9)(r+1)$

19. Factoring the trinomial: $x^2-x-30=(x-6)(x+5)$

21. Factoring the trinomial: $a^2+15a+56=(a+7)(a+8)$

23. Factoring the trinomial: $y^2-y-42=(y-7)(y+6)$

25. Factoring the trinomial: $x^2+13x+42=(x+7)(x+6)$

27. Factoring the trinomial: $x^2+5xy+6y^2=(x+2y)(x+3y)$

29. Factoring the trinomial: $x^2-9xy+20y^2=(x-4y)(x-5y)$

31. Factoring the trinomial: $a^2+2ab-8b^2=(a+4b)(a-2b)$

33. Factoring the trinomial: $a^2-10ab+25b^2=(a-5b)(a-5b)=(a-5b)^2$

35. Factoring the trinomial: $a^2+10ab+25b^2=(a+5b)(a+5b)=(a+5b)^2$

37. Factoring the trinomial: $x^2+2xa-48a^2=(x+8a)(x-6a)$

39. Factoring the trinomial: $x^2-5xb-36b^2=(x-9b)(x+4b)$

41. Factoring the trinomial: $2x^2+6x+4=2(x^2+3x+2)=2(x+2)(x+1)$

43. Factoring the trinomial: $3a^2-3a-60=3(a^2-a-20)=3(a-5)(a+4)$

45. Factoring the trinomial: $100x^2-500x+600=100(x^2-5x+6)=100(x-3)(x-2)$

47. Factoring the trinomial: $100p^2-1{,}300p+4{,}000=100(p^2-13p+40)=100(p-8)(p-5)$

49. Factoring the trinomial: $x^4-x^3-12x^2=x^2(x^2-x-12)=x^2(x-4)(x+3)$

51. Factoring the trinomial: $2r^3+4r^2-30r=2r(r^2+2r-15)=2r(r+5)(r-3)$

53. Factoring the trinomial: $2y^4-6y^3-8y^2=2y^2(y^2-3y-4)=2y^2(y-4)(y+1)$

55. Factoring the trinomial: $x^5+4x^4+4x^3=x^3(x^2+4x+4)=x^3(x+2)(x+2)=x^3(x+2)^2$

57. Factoring the trinomial: $3y^4-12y^3-15y^2=3y^2(y^2-4y-5)=3y^2(y-5)(y+1)$

59. Factoring the trinomial: $4x^4-52x^3+144x^2=4x^2(x^2-13x+36)=4x^2(x-9)(x-4)$

61. Factoring the trinomial: $-a^2-11a-30=-1(a^2+11a+30)=-1(a+6)(a+5)$

63. Factoring the trinomial: $56-x-x^2=-1(x^2+x-56)=-1(x+8)(x-7)$

65. Factoring the trinomial: $x^4-5x^2+6=(x^2-2)(x^2-3)$

67. Factoring the trinomial: $x^2-80x-2{,}000=(x-100)(x+20)$

69. Factoring the trinomial: $x^2-x+\frac{1}{4}=\left(x-\frac{1}{2}\right)\left(x-\frac{1}{2}\right)=\left(x-\frac{1}{2}\right)^2$

71. Factoring the trinomial: $x^2+0.6x+0.08=(x+0.4)(x+0.2)$

73. We can use long division to find the other factor:

$$\begin{array}{r} x+16 \\ x+8\overline{)x^2+24x+128} \\ \underline{x^2+\ \ 8x} \\ 16x+128 \\ \underline{16x+128} \\ 0 \end{array}$$

The other factor is $x+16$.

75. Using FOIL to multiply out the factors: $(4x+3)(x-1)=4x^2-4x+3x-3=4x^2-x-3$

77. Multiplying using the FOIL method: $(6a+1)(a+2)=6a^2+12a+a+2=6a^2+13a+2$

79. Multiplying using the FOIL method: $(3a+2)(2a+1)=6a^2+3a+4a+2=6a^2+7a+2$

81. Multiplying using the FOIL method: $(6a+2)(a+1)=6a^2+6a+2a+2=6a^2+8a+2$

83. Factoring the trinomial: $x^2+8x+12=(x+6)(x+2)$. The correct answer is b.

85. Factoring the trinomial: $a^2-5a-6=(a-6)(a+1)$. The correct answer is c.

6.3 More on Factoring Trinomials

1. Factoring the trinomial: $2x^2+7x+3=(2x+1)(x+3)$

3. Factoring the trinomial: $2a^2-a-3=(2a-3)(a+1)$

5. Factoring the trinomial: $3x^2+2x-5=(3x+5)(x-1)$

7. Factoring the trinomial: $3y^2-14y-5=(3y+1)(y-5)$

9. Factoring the trinomial: $6x^2+13x+6=(3x+2)(2x+3)$

11. Factoring the trinomial: $4x^2-12xy+9y^2=(2x-3y)(2x-3y)=(2x-3y)^2$

13. Factoring the trinomial: $4y^2-11y-3=(4y+1)(y-3)$

15. Factoring the trinomial: $20x^2-41x+20=(4x-5)(5x-4)$

17. Factoring the trinomial: $20a^2+48ab-5b^2=(10a-b)(2a+5b)$

19. Factoring the trinomial: $20x^2-21x-5=(4x-5)(5x+1)$

21. Factoring the trinomial: $12m^2+16m-3=(6m-1)(2m+3)$

23. Factoring the trinomial: $20x^2+37x+15=(4x+5)(5x+3)$

25. Factoring the trinomial: $12a^2-25ab+12b^2=(3a-4b)(4a-3b)$

27. Factoring the trinomial: $3x^2-xy-14y^2=(3x-7y)(x+2y)$

29. Factoring the trinomial: $14x^2+29x-15=(2x+5)(7x-3)$

31. Factoring the trinomial: $6x^2-43x+55=(3x-5)(2x-11)$

33. Factoring the trinomial: $15t^2-67t+38=(5t-19)(3t-2)$

35. Factoring the trinomial: $4x^2+2x-6=2(2x^2+x-3)=2(2x+3)(x-1)$

37. Factoring the trinomial: $24a^2-50a+24=2(12a^2-25a+12)=2(4a-3)(3a-4)$

39. Factoring the trinomial: $10+13x-3x^2=-1(3x^2-13x-10)=-1(3x+2)(x-5)$

41. Factoring the trinomial: $-12x^2+10x+8=-2(6x^2-5x-4)=-2(3x-4)(2x+1)$

43. Factoring the trinomial: $10x^3-23x^2+12x=x(10x^2-23x+12)=x(5x-4)(2x-3)$

45. Factoring the trinomial: $6x^4-11x^3-10x^2=x^2(6x^2-11x-10)=x^2(3x+2)(2x-5)$

47. Factoring the trinomial: $10a^3-6a^2-4a=2a(5a^2-3a-2)=2a(5a+2)(a-1)$

49. Factoring the trinomial: $15x^3-102x^2-21x=3x(5x^2-34x-7)=3x(5x+1)(x-7)$

51. Factoring the trinomial: $35y^3-60y^2-20y=5y(7y^2-12y-4)=5y(7y+2)(y-2)$

53. Factoring the trinomial: $15a^4-2a^3-a^2=a^2(15a^2-2a-1)=a^2(5a+1)(3a-1)$

55. Factoring the trinomial: $24x^2y-6xy-45y=3y(8x^2-2x-15)=3y(4x+5)(2x-3)$

57. Factoring the trinomial: $12x^2y-34xy^2+14y^3=2y(6x^2-17xy+7y^2)=2y(2x-y)(3x-7y)$

59. Evaluating each expression when $x=2$:

$$2x^2+7x+3=2(2)^2+7(2)+3=8+14+3=25$$

$$(2x+1)(x+3)=(2\cdot 2+1)(2+3)=(5)(5)=25$$

Both expressions equal 25.

61. Multiplying using the difference of squares formula: $(2x+3)(2x-3)=(2x)^2-(3)^2=4x^2-9$

63. Multiplying using the difference of squares formula: $(x+3)(x-3)(x^2+9)=(x^2-9)(x^2+9)=(x^2)^2-(9)^2=x^4-81$

65. a. Factoring: $h=8+62t-16t^2=-2(8t^2-31t-4)=-2(t-4)(8t+1)$

b. Completing the table:

Time t (seconds)	0	1	2	3	4
Height h (feet)	8	54	68	50	0

67. a. Factoring: $V=x(99-40x+4x^2)=x(11-2x)(9-2x)$

b. Since $2x$ is cut from each side, the original box had dimensions of 11 inches by 9 inches.

69. Multiplying: $(x+3)(x-3)=x^2-(3)^2=x^2-9$

71. Multiplying: $(2x-3y)(2x+3y)=(2x)^2-(3y)^2=4x^2-9y^2$

73. Multiplying: $(x^2+4)(x+2)(x-2)=(x^2+4)(x^2-4)=(x^2)^2-(4)^2=x^4-16$

75. Multiplying: $(x+3)^2=x^2+2(x)(3)+(3)^2=x^2+6x+9$

77. Multiplying: $(2x+3)^2=(2x)^2+2(2x)(3)+(3)^2=4x^2+12x+9$

79. Multiplying: $(4x-2y)^2=(4x)^2-2(4x)(2y)+(2y)^2=16x^2-16xy+4y^2$

81. a. Multiplying: $1^3=1$

b. Multiplying: $2^3=8$

c. Multiplying: $3^3=27$

d. Multiplying: $4^3=64$

e. Multiplying: $5^3=125$

83. a. Multiplying: $x(x^2-2x+4)=x^3-2x^2+4x$

b. Multiplying: $2(x^2-2x+4)=2x^2-4x+8$

c. Multiplying: $(x+2)(x^2-2x+4)=x(x^2-2x+4)+2(x^2-2x+4)=x^3-2x^2+4x+2x^2-4x+8=x^3+8$

85. a. Multiplying: $x(x^2-3x+9)=x^3-3x^2+9x$

b. Multiplying: $3(x^2-3x+9)=3x^2-9x+27$

c. Multiplying: $(x+3)(x^2-3x+9)=x(x^2-3x+9)+3(x^2-3x+9)=x^3-3x^2+9x+3x^2-9x+27=x^3+27$

87. Factoring the trinomial: $4x^2-5x-6=(4x+3)(x-2)$. The correct answer is d.

89. Factoring the trinomial: $h=8+28t-16t^2=-16t^2+28t+8=-4(4t^2-7t-2)=-4(4t+1)(t-2)$

The correct answer is b.

6.4 Special Factoring Patterns

1. Factoring the trinomial: $x^2-2x+1=(x-1)(x-1)=(x-1)^2$
3. Factoring the trinomial: $x^2+2x+1=(x+1)(x+1)=(x+1)^2$
5. Factoring the trinomial: $a^2-10a+25=(a-5)(a-5)=(a-5)^2$
7. Factoring the trinomial: $y^2+4y+4=(y+2)(y+2)=(y+2)^2$
9. Factoring the trinomial: $x^2-4x+4=(x-2)(x-2)=(x-2)^2$
11. Factoring the trinomial: $m^2-12m+36=(m-6)(m-6)=(m-6)^2$
13. Factoring the trinomial: $4a^2+12a+9=(2a+3)(2a+3)=(2a+3)^2$
15. Factoring the trinomial: $49x^2-14x+1=(7x-1)(7x-1)=(7x-1)^2$
17. Factoring the trinomial: $9y^2-30y+25=(3y-5)(3y-5)=(3y-5)^2$
19. Factoring the trinomial: $x^2+10xy+25y^2=(x+5y)(x+5y)=(x+5y)^2$
21. Factoring the trinomial: $9a^2+6ab+b^2=(3a+b)(3a+b)=(3a+b)^2$
23. Factoring the trinomial: $3a^2+18a+27=3(a^2+6a+9)=3(a+3)(a+3)=3(a+3)^2$
25. Factoring the trinomial: $2x^2+20xy+50y^2=2(x^2+10xy+25y^2)=2(x+5y)(x+5y)=2(x+5y)^2$
27. Factoring the trinomial: $5x^3+30x^2y+45xy^2=5x(x^2+6xy+9y^2)=5x(x+3y)(x+3y)=5x(x+3y)^2$
29. Factoring the binomial: $x^2-9=(x+3)(x-3)$
31. Factoring the binomial: $a^2-36=(a+6)(a-6)$
33. Factoring the binomial: $x^2-49=(x+7)(x-7)$
35. Factoring the binomial: $4a^2-16=4(a^2-4)=4(a+2)(a-2)$
37. The expression $9x^2+25$ cannot be factored.
39. Factoring the binomial: $25x^2-169=(5x+13)(5x-13)$
41. Factoring the binomial: $9a^2-16b^2=(3a+4b)(3a-4b)$
43. Factoring the binomial: $9-m^2=(3+m)(3-m)$
45. Factoring the binomial: $25-4x^2=(5+2x)(5-2x)$
47. Factoring the binomial: $2x^2-18=2(x^2-9)=2(x+3)(x-3)$
49. Factoring: $x^3-y^3=(x-y)(x^2+xy+y^2)$
51. Factoring: $a^3+8=(a+2)(a^2-2a+4)$
53. Factoring: $27+x^3=(3+x)(9-3x+x^2)$
55. Factoring: $y^3-1=(y-1)(y^2+y+1)$
57. Factoring: $64-y^3=(4-y)(16+4y+y^2)$
59. Factoring: $125h^3-t^3=(5h-t)(25h^2+5ht+t^2)$
61. Factoring: $x^3-216=(x-6)(x^2+6x+36)$
63. Factoring: $2y^3-54=2(y^3-27)=2(y-3)(y^2+3y+9)$
65. Factoring: $64+27a^3=(4+3a)(16-12a+9a^2)$
67. Factoring: $8x^3-27y^3=(2x-3y)(4x^2+6xy+9y^2)$
69. Factoring the binomial: $32a^2-128=32(a^2-4)=32(a+2)(a-2)$
71. Factoring the binomial: $8x^2y-18y=2y(4x^2-9)=2y(2x+3)(2x-3)$
73. Factoring: $2a^3-128b^3=2(a^3-64b^3)=2(a-4b)(a^2+4ab+16b^2)$
75. Factoring: $2x^3+432y^3=2(x^3+216y^3)=2(x+6y)(x^2-6xy+36y^2)$
77. Factoring: $10a^3-640b^3=10(a^3-64b^3)=10(a-4b)(a^2+4ab+16b^2)$

79. Factoring: $10r^3-1250=10(r^3-125)=10(r-5)(r^2+5r+25)$

81. Factoring: $t^3+\frac{1}{27}=\left(t+\frac{1}{3}\right)\left(t^2-\frac{1}{3}t+\frac{1}{9}\right)$

83. Factoring: $27x^3-\frac{1}{27}=\left(3x-\frac{1}{3}\right)\left(9x^2+x+\frac{1}{9}\right)$

85. Factoring: $64a^3+125b^3=(4a+5b)(16a^2-20ab+25b^2)$

87. Factoring: $\frac{1}{8}x^3-\frac{1}{27}y^3=\left(\frac{1}{2}x-\frac{1}{3}y\right)\left(\frac{1}{4}x^2+\frac{1}{6}xy+\frac{1}{9}y^2\right)$

89. Factoring the binomial: $a^4-b^4=(a^2+b^2)(a^2-b^2)=(a^2+b^2)(a+b)(a-b)$

91. Factoring the binomial: $16m^4-81=(4m^2+9)(4m^2-9)=(4m^2+9)(2m+3)(2m-3)$

93. Factoring the binomial: $3x^3y-75xy^3=3xy(x^2-25y^2)=3xy(x+5y)(x-5y)$

95. Factoring: $a^6-b^6=(a^3+b^3)(a^3-b^3)=(a+b)(a^2-ab+b^2)(a-b)(a^2+ab+b^2)$

97. Factoring: $64x^6-y^6=(8x^3+y^3)(8x^3-y^3)=(2x+y)(4x^2-2xy+y^2)(2x-y)(4x^2+2xy+y^2)$

99. Factoring: $x^6-(5y)^6=\left(x^3+(5y)^3\right)\left(x^3-(5y)^3\right)=(x+5y)(x^2-5xy+25y^2)(x-5y)(x^2+5xy+25y^2)$

101. Factoring by grouping: $x^2+6x+9-y^2=(x+3)^2-y^2=(x+3+y)(x+3-y)$

103. Factoring by grouping: $x^2+2xy+y^2-9=(x+y)^2-9=(x+y+3)(x+y-3)$

105. Since $(x+7)^2=x^2+14x+49$, the value is $b=14$.

107. Since $(x+5)^2=x^2+10x+25$, the value is $c=25$.

109. **a.** Subtracting the area of the missing square, the area is $A=x^2-16$.

b. Factoring: $A=x^2-16=(x+4)(x-4)$

c. Make a single rectangle of dimensions $x-4$ by $x+4$.

111. Subtracting the area of the missing square: $A=a^2-b^2=(a+b)(a-b)$

113. Multiplying: $2x^3(x+2)(x-2)=2x^3(x^2-4)=2x^5-8x^3$

115. Multiplying: $3x^2(x-3)^2=3x^2(x^2-6x+9)=3x^4-18x^3+27x^2$

117. Multiplying: $y(y^2+25)=y^3+25y$

119. Multiplying: $(5a-2)(3a+1)=15a^2+5a-6a-2=15a^2-a-2$

121. Multiplying: $4x^2(x-5)(x+2)=4x^2(x^2-3x-10)=4x^4-12x^3-40x^2$

123. Multiplying: $2ab^3(b^2-4b+1)=2ab^5-8ab^4+2ab^3$

125. Factoring: $9x^2-48x+64=(3x-8)^2$. The correct answer is c.

127. The expression $4y^2+25$ cannot be factored. The correct answer is b.

6.5 Factoring: A General Review

1. Factoring the polynomial: $x^2-81=(x+9)(x-9)$
3. Factoring the polynomial: $x^2+2x-15=(x+5)(x-3)$
5. Factoring the polynomial: $x^2+6x+9=(x+3)(x+3)=(x+3)^2$
7. Factoring the polynomial: $y^2-10y+25=(y-5)(y-5)=(y-5)^2$
9. Factoring the polynomial: $2a^3b+6a^2b+2ab=2ab(a^2+3a+1)$
11. The polynomial x^2+x+1 cannot be factored.
13. Factoring the polynomial: $12a^2-75=3(4a^2-25)=3(2a+5)(2a-5)$
15. Factoring the polynomial: $9x^2-12xy+4y^2=(3x-2y)(3x-2y)=(3x-2y)^2$
17. Factoring the polynomial: $4x^3+16xy^2=4x(x^2+4y^2)$
19. Factoring the polynomial: $2y^3+20y^2+50y=2y(y^2+10y+25)=2y(y+5)(y+5)=2y(y+5)^2$
21. Factoring the polynomial: $a^6+4a^4b^2=a^4(a^2+4b^2)$
23. Factoring the polynomial: $xy+3x+4y+12=x(y+3)+4(y+3)=(y+3)(x+4)$
25. Factoring the polynomial: $x^4-16=(x^2+4)(x^2-4)=(x^2+4)(x+2)(x-2)$
27. Factoring the polynomial: $xy-5x+2y-10=x(y-5)+2(y-5)=(y-5)(x+2)$
29. Factoring the polynomial: $5a^2+10ab+5b^2=5(a^2+2ab+b^2)=5(a+b)(a+b)=5(a+b)^2$
31. Factoring the polynomial: $64+x^3=(4+x)(16-4x+x^2)$
33. Factoring the polynomial: $3x^2+15xy+18y^2=3(x^2+5xy+6y^2)=3(x+2y)(x+3y)$
35. Factoring the polynomial: $2x^2+15x-38=(2x+19)(x-2)$
37. Factoring the polynomial: $100x^2-300x+200=100(x^2-3x+2)=100(x-2)(x-1)$
39. Factoring the polynomial: $x^2-64=(x+8)(x-8)$
41. Factoring the polynomial: $x^2+3x+ax+3a=x(x+3)+a(x+3)=(x+3)(x+a)$
43. Factoring the polynomial: $49a^7-9a^5=a^5(49a^2-9)=a^5(7a+3)(7a-3)$
45. The polynomial $49x^2+9y^2$ cannot be factored.
47. Factoring the polynomial: $25a^3+20a^2+3a=a(25a^2+20a+3)=a(5a+3)(5a+1)$
49. Factoring the polynomial: $xa-xb+ay-by=x(a-b)+y(a-b)=(a-b)(x+y)$
51. Factoring the polynomial: $48a^4b-3a^2b=3a^2b(16a^2-1)=3a^2b(4a+1)(4a-1)$
53. Factoring the polynomial: $5x^5-40x^2=5x^2(x^3-8)=5x^2(x-2)(x^2+2x+4)$
55. Factoring the polynomial: $3x^2+35xy-82y^2=(3x+41y)(x-2y)$
57. Factoring the polynomial: $16x^5-44x^4+30x^3=2x^3(8x^2-22x+15)=2x^3(2x-3)(4x-5)$
59. Factoring the polynomial: $2x^2+2ax+3x+3a=2x(x+a)+3(x+a)=(x+a)(2x+3)$
61. Factoring the polynomial: $y^4-1=(y^2+1)(y^2-1)=(y^2+1)(y+1)(y-1)$
63. Factoring the polynomial:

$$12x^4y^2+36x^3y^3+27x^2y^4=3x^2y^2(4x^2+12xy+9y^2)=3x^2y^2(2x+3y)(2x+3y)=3x^2y^2(2x+3y)^2$$

65. Solving the equation:

$$3x-6=9$$
$$3x=15$$
$$x=5$$

67. Solving the equation:

$$2x+3=0$$
$$2x=-3$$
$$x=-\frac{3}{2}$$

69. Solving the equation:

$$4x+3=0$$
$$4x=-3$$
$$x=-\frac{3}{4}$$

71. This would be factored using trial and error. The correct answer is c.

73. This would be factored using a greatest common factor. The correct answer is a.

6.6 Solving Equations by Factoring

1. Setting each factor equal to 0:

$$x+2=0 \qquad x-1=0$$
$$x=-2 \qquad x=1$$

The solutions are –2 and 1.

3. Setting each factor equal to 0:

$$a-4=0 \qquad a-5=0$$
$$a=4 \qquad a=5$$

The solutions are 4 and 5.

5. Setting each factor equal to 0:

$$x=0 \qquad x+1=0 \qquad x-3=0$$
$$x=-1 \qquad x=3$$

The solutions are 0, –1 and 3.

7. Setting each factor equal to 0:

$$3x+2=0 \qquad 2x+3=0$$
$$3x=-2 \qquad 2x=-3$$
$$x=-\frac{2}{3} \qquad x=-\frac{3}{2}$$

The solutions are $-\frac{2}{3}$ and $-\frac{3}{2}$.

9. Setting each factor equal to 0:

$$m=0 \qquad 3m+4=0 \qquad 3m-4=0$$
$$3m=-4 \qquad 3m=4$$
$$m=-\frac{4}{3} \qquad m=\frac{4}{3}$$

The solutions are 0, $-\frac{4}{3}$ and $\frac{4}{3}$.

11. Setting each factor equal to 0:

$$2y=0 \qquad 3y+1=0 \qquad 5y+3=0$$
$$y=0 \qquad 3y=-1 \qquad 5y=-3$$
$$y=-\frac{1}{3} \qquad y=-\frac{3}{5}$$

The solutions are 0, $-\frac{1}{3}$ and $-\frac{3}{5}$.

13. Solving by factoring:

$$x^2+3x+2=0$$
$$(x+2)(x+1)=0$$
$$x=-2,-1$$

15. Solving by factoring:

$$x^2-9x+20=0$$
$$(x-4)(x-5)=0$$
$$x=4,5$$

17. Solving by factoring:

$$a^2-2a-24=0$$
$$(a-6)(a+4)=0$$
$$a=6,-4$$

19. Solving by factoring:

$$100x^2-500x+600=0$$
$$100(x^2-5x+6)=0$$
$$100(x-2)(x-3)=0$$
$$x=2,3$$

21. Solving by factoring:

$$x^2=-6x-9$$
$$x^2+6x+9=0$$
$$(x+3)^2=0$$
$$x+3=0$$
$$x=-3$$

23. Solving by factoring:

$$a^2-16=0$$
$$(a+4)(a-4)=0$$
$$a=-4,4$$

25. Solving by factoring:

$$2x^2+5x-12=0$$
$$(2x-3)(x+4)=0$$
$$x=\frac{3}{2},-4$$

27. Solving by factoring:

$$9x^2+12x+4=0$$
$$(3x+2)^2=0$$
$$3x+2=0$$
$$x=-\frac{2}{3}$$

29. Solving by factoring:

$$a^2+25=10a$$
$$a^2-10a+25=0$$
$$(a-5)^2=0$$
$$a-5=0$$
$$a=5$$

31. Solving by factoring:

$$0=20+3x-2x^2$$
$$2x^2-3x-20=0$$
$$(2x+5)(x-4)=0$$
$$x=-\frac{5}{2},4$$

33. Solving by factoring:

$$3m^2=20-7m$$
$$3m^2+7m-20=0$$
$$(3m-5)(m+4)=0$$
$$m=\frac{5}{3},-4$$

35. Solving by factoring:

$$4x^2-49=0$$
$$(2x+7)(2x-7)=0$$
$$x=-\frac{7}{2},\frac{7}{2}$$

37. Solving by factoring:

$$x^2+6x=0$$
$$x(x+6)=0$$
$$x=0,-6$$

39. Solving by factoring:

$$3x-x^2=0$$
$$x(3-x)=0$$
$$x=0,3$$

41. Solving by factoring:

$$2x^2=8x$$
$$2x^2-8x=0$$
$$2x(x-4)=0$$
$$x=0,4$$

43. Solving by factoring:

$$3x^2=15x$$
$$3x^2-15x=0$$
$$3x(x-5)=0$$
$$x=0,5$$

45. Solving by factoring:

$$\begin{aligned} 1{,}400 &= 400 + 700x - 100x^2 \\ 100x^2 - 700x + 1{,}000 &= 0 \\ 100(x^2 - 7x + 10) &= 0 \\ 100(x-5)(x-2) &= 0 \\ x &= 2,5 \end{aligned}$$

47. Solving by factoring:

$$\begin{aligned} 6x^2 &= -5x + 4 \\ 6x^2 + 5x - 4 &= 0 \\ (3x+4)(2x-1) &= 0 \\ x &= -\frac{4}{3}, \frac{1}{2} \end{aligned}$$

49. Solving by factoring:

$$\begin{aligned} x(2x-3) &= 20 \\ 2x^2 - 3x &= 20 \\ 2x^2 - 3x - 20 &= 0 \\ (2x+5)(x-4) &= 0 \\ x &= -\frac{5}{2}, 4 \end{aligned}$$

51. Solving by factoring:

$$\begin{aligned} t(t+2) &= 80 \\ t^2 + 2t &= 80 \\ t^2 + 2t - 80 &= 0 \\ (t+10)(t-8) &= 0 \\ t &= -10, 8 \end{aligned}$$

53. Solving by factoring:

$$\begin{aligned} 4{,}000 &= (1{,}300 - 100p)p \\ 4{,}000 &= 1{,}300p - 100p^2 \\ 100p^2 - 1{,}300p + 4{,}000 &= 0 \\ 100(p^2 - 13p + 40) &= 0 \\ 100(p-8)(p-5) &= 0 \\ p &= 5,8 \end{aligned}$$

55. Solving by factoring:

$$\begin{aligned} x(14-x) &= 48 \\ 14x - x^2 &= 48 \\ -x^2 + 14x - 48 &= 0 \\ x^2 - 14x + 48 &= 0 \\ (x-6)(x-8) &= 0 \\ x &= 6,8 \end{aligned}$$

57. Solving by factoring:

$$\begin{aligned} (x+5)^2 &= 2x + 9 \\ x^2 + 10x + 25 &= 2x + 9 \\ x^2 + 8x + 16 &= 0 \\ (x+4)^2 &= 0 \\ x + 4 &= 0 \\ x &= -4 \end{aligned}$$

59. Solving by factoring:

$$\begin{aligned} (y-6)^2 &= y - 4 \\ y^2 - 12y + 36 &= y - 4 \\ y^2 - 13y + 40 &= 0 \\ (y-5)(y-8) &= 0 \\ y &= 5,8 \end{aligned}$$

61. Solving by factoring:

$$\begin{aligned} 10^2 &= (x+2)^2 + x^2 \\ 100 &= x^2 + 4x + 4 + x^2 \\ 100 &= 2x^2 + 4x + 4 \\ 0 &= 2x^2 + 4x - 96 \\ 0 &= 2(x^2 + 2x - 48) \\ 0 &= 2(x+8)(x-6) \\ x &= -8, 6 \end{aligned}$$

63. Solving by factoring:

$$\begin{aligned} 2x^3 + 11x^2 + 12x &= 0 \\ x(2x^2 + 11x + 12) &= 0 \\ x(2x+3)(x+4) &= 0 \\ x &= 0, -\frac{3}{2}, -4 \end{aligned}$$

65. Solving by factoring:

$$4y^3-2y^2-30y=0$$
$$2y(2y^2-y-15)=0$$
$$2y(2y+5)(y-3)=0$$
$$y=0,-\frac{5}{2},3$$

67. Solving by factoring:

$$8x^3+16x^2=10x$$
$$8x^3+16x^2-10x=0$$
$$2x(4x^2+8x-5)=0$$
$$2x(2x-1)(2x+5)=0$$
$$x=0,\frac{1}{2},-\frac{5}{2}$$

69. Solving by factoring:

$$20a^3=-18a^2+18a$$
$$20a^3+18a^2-18a=0$$
$$2a(10a^2+9a-9)=0$$
$$2a(5a-3)(2a+3)=0$$
$$a=0,\frac{3}{5},-\frac{3}{2}$$

71. Solving by factoring:

$$16t^2-32t+12=0$$
$$4(4t^2-8t+3)=0$$
$$4(2t-1)(2t-3)=0$$
$$t=\frac{1}{2},\frac{3}{2}$$

73. Solving the equation:

$$(a-5)(a+4)=-2a$$
$$a^2-a-20=-2a$$
$$a^2+a-20=0$$
$$(a+5)(a-4)=0$$
$$a=-5,4$$

75. Solving the equation:

$$3x(x+1)-2x(x-5)=-42$$
$$3x^2+3x-2x^2+10x=-42$$
$$x^2+13x+42=0$$
$$(x+7)(x+6)=0$$
$$x=-7,-6$$

77. Solving the equation:

$$2x(x+3)=x(x+2)-3$$
$$2x^2+6x=x^2+2x-3$$
$$x^2+4x+3=0$$
$$(x+3)(x+1)=0$$
$$x=-3,-1$$

79. Solving the equation:

$$a(a-3)+6=2a$$
$$a^2-3a+6=2a$$
$$a^2-5a+6=0$$
$$(a-2)(a-3)=0$$
$$a=2,3$$

81. Solving the equation:

$$15(x+20)+15x=2x(x+20)$$
$$15x+300+15x=2x^2+40x$$
$$30x+300=2x^2+40x$$
$$0=2x^2+10x-300$$
$$0=2(x^2+5x-150)$$
$$0=2(x+15)(x-10)$$
$$x=-15,10$$

83. Solving the equation:

$$15=a(a+2)$$
$$15=a^2+2a$$
$$0=a^2+2a-15$$
$$0=(a+5)(a-3)$$
$$a=-5,3$$

85. Solving by factoring:

$$x^3+3x^2-4x-12=0$$
$$x^2(x+3)-4(x+3)=0$$
$$(x+3)(x^2-4)=0$$
$$(x+3)(x+2)(x-2)=0$$
$$x=-3,-2,2$$

87. Solving by factoring:

$$x^3+x^2-16x-16=0$$
$$x^2(x+1)-16(x+1)=0$$
$$(x+1)(x^2-16)=0$$
$$(x+1)(x+4)(x-4)=0$$
$$x=-1,-4,4$$

89. **a.** Solving by factoring:

$$2x^2+7x-4=0$$
$$(2x-1)(x+4)=0$$
$$x=\frac{1}{2},-4$$

b. Factoring: $2x^2+7x-4=(2x-1)(x+4)$

c. Solving by factoring:

$$2x^2+7x-4=-7$$
$$2x^2+7x+3=0$$
$$(2x+1)(x+3)=0$$
$$x=-\frac{1}{2},-3$$

d. Solving the equation:

$$2x+7x-4=0$$
$$9x-4=0$$
$$9x=4$$
$$x=\frac{4}{9}$$

91. Let x and $x + 1$ represent the two consecutive integers. The equation is: $x(x+1)=72$

93. Let x and $x + 2$ represent the two consecutive odd integers. The equation is: $x(x+2)=99$

95. Let x and $x + 2$ represent the two consecutive even integers. The equation is: $x(x+2)=5[x+(x+2)]-10$

97. Let x represent the cost of the suit and $5x$ represent the cost of the bicycle. The equation is:

$$x+5x=90$$
$$6x=90$$
$$x=15$$
$$5x=75$$

The suit costs $15 and the bicycle costs $75.

99. Let x represent the cost of the lot and $4x$ represent the cost of the house. The equation is:

$$x+4x=3000$$
$$5x=3000$$
$$x=600$$
$$4x=2400$$

The lot costs $600 and the house costs $2,400.

101. Setting each factor equal to 0:

$$5x+2=0$$
$$5x=-2$$
$$x=-\frac{2}{5}$$

$$x-3=0$$
$$x=3$$

The solutions are $-\frac{2}{5}$ and 3. The correct answer is b.

6.7 Applications

1. Let x and $x + 2$ represent the two integers. The equation is:

$$\begin{aligned} x(x+2) &= 80 \\ x^2+2x &= 80 \\ x^2+2x-80 &= 0 \\ (x+10)(x-8) &= 0 \\ x &= -10, 8 \\ x+2 &= -8, 10 \end{aligned}$$

The two numbers are either –10 and –8, or 8 and 10.

3. Let x and $x + 2$ represent the two integers. The equation is:

$$\begin{aligned} x(x+2) &= 99 \\ x^2+2x &= 99 \\ x^2+2x-99 &= 0 \\ (x+11)(x-9) &= 0 \\ x &= -11, 9 \\ x+2 &= -9, 11 \end{aligned}$$

The two numbers are either –11 and –9, or 9 and 11.

5. Let x and $x + 2$ represent the two integers. The equation is:

$$\begin{aligned} x(x+2) &= 5(x+x+2)-10 \\ x^2+2x &= 5(2x+2)-10 \\ x^2+2x &= 10x+10-10 \\ x^2+2x &= 10x \\ x^2-8x &= 0 \\ x(x-8) &= 0 \\ x &= 0, 8 \\ x+2 &= 2, 10 \end{aligned}$$

The two numbers are either 0 and 2, or 8 and 10.

7. Let x and $14 - x$ represent the two numbers. The equation is:

$$\begin{aligned} x(14-x) &= 48 \\ 14x-x^2 &= 48 \\ 0 &= x^2-14x+48 \\ 0 &= (x-8)(x-6) \\ x &= 8, 6 \\ 14-x &= 6, 8 \end{aligned}$$

The two numbers are 6 and 8.

9. Let x and $5x + 2$ represent the two numbers. The equation is:

$$x(5x+2)=24$$
$$5x^2+2x=24$$
$$5x^2+2x-24=0$$
$$(5x+12)(x-2)=0$$
$$x=-\frac{12}{5},2$$
$$5x+2=-10,12$$

The two numbers are either $-\frac{12}{5}$ and -10, or 2 and 12.

11. Let x and $4x$ represent the two numbers. The equation is:

$$x(4x)=4(x+4x)$$
$$4x^2=4(5x)$$
$$4x^2=20x$$
$$4x^2-20x=0$$
$$4x(x-5)=0$$
$$x=0,5$$
$$4x=0,20$$

The two numbers are either 0 and 0, or 5 and 20.

13. Let w represent the width and $w + 1$ represent the length. The equation is:

$$w(w+1)=12$$
$$w^2+w=12$$
$$w^2+w-12=0$$
$$(w+4)(w-3)=0$$
$$w=3 \quad (w=-4 \text{ is impossible})$$
$$w+1=4$$

The width is 3 inches and the length is 4 inches.

15. Let b represent the base and $2b$ represent the height. The equation is:

$$\frac{1}{2}b(2b)=9$$
$$b^2=9$$
$$b^2-9=0$$
$$(b+3)(b-3)=0$$
$$b=3 \quad (b=-3 \text{ is impossible})$$

The base is 3 inches.

17. Let x and $x + 2$ represent the two legs. The equation is:

$$\begin{aligned} x^2+(x+2)^2&=10^2 \\ x^2+x^2+4x+4&=100 \\ 2x^2+4x+4&=100 \\ 2x^2+4x-96&=0 \\ 2(x^2+2x-48)&=0 \\ 2(x+8)(x-6)&=0 \\ x&=6 \quad (x=-8 \text{ is impossible}) \\ x+2&=8 \end{aligned}$$

The legs are 6 inches and 8 inches.

19. Let x represent the longer leg and $x + 1$ represent the hypotenuse. The equation is:

$$\begin{aligned} 5^2+x^2&=(x+1)^2 \\ 25+x^2&=x^2+2x+1 \\ 25&=2x+1 \\ 24&=2x \\ x&=12 \end{aligned}$$

The longer leg is 12 meters.

21. Setting $C = \$1,400$:

$$\begin{aligned} 1400&=400+700x-100x^2 \\ 100x^2-700x+1000&=0 \\ 100(x^2-7x+10)&=0 \\ 100(x-5)(x-2)&=0 \\ x&=2,5 \end{aligned}$$

The company can manufacture either 200 items or 500 items.

23. The revenue is given by: $R=xp=(1{,}700-100p)p$. Setting $R = \$7,000$:

$$\begin{aligned} 7{,}000&=(1{,}700-100p)p \\ 7{,}000&=1{,}700p-100p^2 \\ 100p^2-1{,}700p+7{,}000&=0 \\ 100(p^2-17p+70)&=0 \\ 100(p-7)(p-10)&=0 \\ p&=7,10 \end{aligned}$$

The calculators should be sold for either \$7 or \$10.

25. **a.** Let x represent the distance from the base to the wall, and $2x + 2$ represent the height on the wall. Using the Pythagorean theorem:

$$\begin{aligned} x^2+(2x+2)^2&=13^2 \\ x^2+4x^2+8x+4&=169 \\ 5x^2+8x-165&=0 \\ (5x+33)(x-5)&=0 \\ x&=5 \quad \left(x=-\frac{33}{5} \text{ is impossible}\right) \end{aligned}$$

The base of the ladder is 5 feet from the wall.

b. Since $2x+2=2\cdot 5+2=12$, the ladder reaches a height of 12 feet.

27. **a.** Finding when $h = 0$:

$$0 = -16t^2 + 396t + 100$$
$$16t^2 - 396t - 100 = 0$$
$$4t^2 - 99t - 25 = 0$$
$$(4t+1)(t-25) = 0$$
$$t = 25 \quad \left(t = -\frac{1}{4} \text{ is impossible}\right)$$

The bullet will land on the ground after 25 seconds.

b. Completing the table:

t (seconds)	h (feet)
0	100
5	1,680
10	2,460
15	2,440
20	1,620
25	0

29. Substituting $x = -2$, $x = 0$, and $x = 2$:

$$y = \frac{1}{2}(-2) + 3 = -1 + 3 = 2 \qquad y = \frac{1}{2}(0) + 3 = 0 + 3 = 3 \qquad y = \frac{1}{2}(2) + 3 = 1 + 3 = 4$$

The ordered pairs are $(-2,2)$, $(0,3)$, and $(2,4)$.

31. Graphing the line:

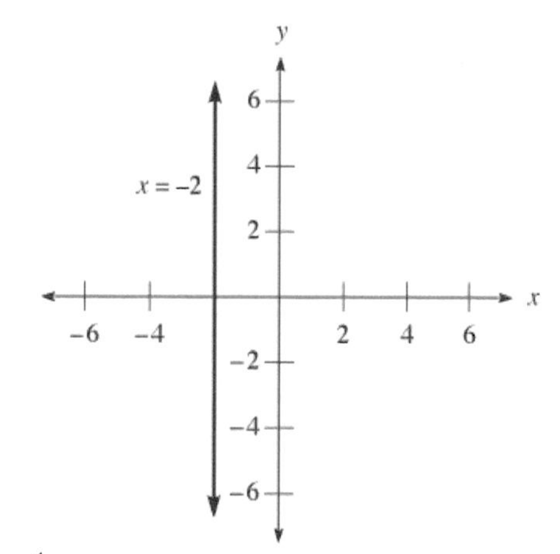

33. Computing the slope: $m = \dfrac{1-5}{0-2} = \dfrac{-4}{-2} = 2$

35. Using the point-slope formula:

$$y - 1 = \frac{1}{2}(x - (-2))$$
$$y - 1 = \frac{1}{2}(x + 2)$$
$$y - 1 = \frac{1}{2}x + 1$$
$$y = \frac{1}{2}x + 2$$

37. Computing the slope: $m=\frac{1-5}{0-2}=\frac{-4}{-2}=2$. Using the point-slope formula:

$$\begin{aligned} y-1&=2(x-0)\\ y-1&=2x\\ y&=2x+1 \end{aligned}$$

39. Let x and $2x-1$ represent the two numbers. The equation is:

$$\begin{aligned} x(2x-1)&=120\\ 2x^2-x&=120\\ 2x^2-x-120&=0\\ (2x+15)(x-8)&=0\\ x&=-\frac{15}{2},8\\ 2x-1&=-16,15 \end{aligned}$$

The two numbers are either $-\frac{15}{2}$ and –16, or 8 and 15. The correct answer is a.

41. Let x and $x+7$ represent the two legs. The equation is: $x^2+(x+7)^2=17^2$, or $x^2+(x+7)^2=289$

The correct answer is d.

Chapter 6 Test

1. Factoring the polynomial: $6x+18=6(x+3)$

2. Factoring the polynomial: $12a^2b-24ab+8ab^2=4ab(3a-6+2b)$

3. Factoring the polynomial: $x^2+3ax-2bx-6ab=x(x+3a)-2b(x+3a)=(x+3a)(x-2b)$

4. Factoring the polynomial: $15y-5xy-12+4x=5y(3-x)-4(3-x)=(3-x)(5y-4)$

5. Factoring the polynomial: $x^2+x-12=(x+4)(x-3)$

6. Factoring the polynomial: $x^2-4x-21=(x-7)(x+3)$

7. Factoring the polynomial: $x^2-25=(x+5)(x-5)$

8. Factoring the polynomial: $x^4-16=(x^2+4)(x^2-4)=(x^2+4)(x+2)(x-2)$

9. The polynomial x^2+36 cannot be factored.

10. Factoring the polynomial: $18x^2-32y^2=2(9x^2-16y^2)=2(3x+4y)(3x-4y)$

11. Factoring the polynomial: $x^3+4x^2-3x-12=x^2(x+4)-3(x+4)=(x+4)(x^2-3)$

12. Factoring the polynomial: $x^2+bx-3x-3b=x(x+b)-3(x+b)=(x+b)(x-3)$

13. Factoring the polynomial: $4x^2-6x-10=2(2x^2-3x-5)=2(2x-5)(x+1)$

14. Factoring the polynomial: $4n^2+13n-12=(4n-3)(n+4)$

15. Factoring the polynomial: $12c^2+c-6=(3c-2)(4c+3)$

16. Factoring the polynomial: $12x^3+12x^2-9x=3x(4x^2+4x-3)=3x(2x-1)(2x+3)$

17. Factoring the polynomial: $x^3+125y^3=(x+5y)(x^2-5xy+25y^2)$

18. Factoring the polynomial: $54b^3-128=2(27b^3-64)=2(3b-4)(9b^2+12b+16)$

19. Solving the equation by factoring:

$$\begin{aligned} x^2-2x-15&=0 \\ (x+3)(x-5)&=0 \\ x&=-3,5 \end{aligned}$$

20. Solving the equation by factoring:

$$\begin{aligned} x^2-7x+12&=0 \\ (x-3)(x-4)&=0 \\ x&=3,4 \end{aligned}$$

21. Solving the equation by factoring:

$$\begin{aligned} x^2-25&=0 \\ (x+5)(x-5)&=0 \\ x&=-5,5 \end{aligned}$$

22. Solving the equation by factoring:

$$\begin{aligned} x^2&=5x+14 \\ x^2-5x-14&=0 \\ (x+2)(x-7)&=0 \\ x&=-2,7 \end{aligned}$$

23. Solving the equation by factoring:

$$\begin{aligned} x^2+x&=30 \\ x^2+x-30&=0 \\ (x+6)(x-5)&=0 \\ x&=-6,5 \end{aligned}$$

24. Solving the equation by factoring:

$$\begin{aligned} y^3&=9y \\ y^3-9y&=0 \\ y\left(y^2-9\right)&=0 \\ y(y+9)(y-9)&=0 \\ y&=-3,0,3 \end{aligned}$$

25. Solving the equation by factoring:

$$\begin{aligned} 2x^2&=-5x+12 \\ 2x^2+5x-12&=0 \\ (2x-3)(x+4)&=0 \\ x&=-4,\frac{3}{2} \end{aligned}$$

26. Solving the equation by factoring:

$$\begin{aligned} 15x^3-65x^2-150x&=0 \\ 5x\left(3x^2-13x-30\right)&=0 \\ 5x(3x+5)(x-6)&=0 \\ x&=-\frac{5}{3},0,6 \end{aligned}$$

27. Let x and $18-x$ represent the numbers. The equation is:

$$\begin{aligned} x(18-x)&=72 \\ 18x-x^2&=72 \\ 0&=x^2-18x+72 \\ 0&=(x-6)(x-12) \\ x&=6,12 \\ 18-x&=12,6 \end{aligned}$$

The numbers are 6 and 12.

28. Let x and $x+2$ represent the two integers. The equation is:

$$\begin{aligned} x(x+2)&=x+x+2+14 \\ x^2+2x&=2x+16 \\ x^2-16&=0 \\ (x+4)(x-4)&=0 \\ x&=-4,4 \\ x+2&=-2,6 \end{aligned}$$

The integers are either –4 and –2, or 4 and 6.

29. Let w represent the width and $3w + 1$ represent the length. The equation is:

$$w(3w+1)=52$$
$$3w^2+w=52$$
$$3w^2+w-52=0$$
$$(3w+13)(w-4)=0$$
$$w=4 \qquad \left(w=-\frac{13}{3} \text{ is impossible}\right)$$
$$3w+1=13$$

The width is 4 feet and the length is 13 feet.

30. Let x and $x + 2$ represent the two legs. The equation is:

$$x^2+(x+2)^2=10^2$$
$$x^2+x^2+4x+4=100$$
$$2x^2+4x-96=0$$
$$2(x^2+2x-48)=0$$
$$2(x+8)(x-6)=0$$
$$x=6 \qquad (x=-8 \text{ is impossible})$$
$$x+2=8$$

The two legs are 6 feet and 8 feet in length.

31. Setting C = \$700:

$$100+500x-100x^2=700$$
$$-100x^2+500x-600=0$$
$$-100(x^2-5x+6)=0$$
$$-100(x-2)(x-3)=0$$
$$x=2,3$$

The company can manufacture either 200 items or 300 items.

32. The revenue is given by: $R=xp=(800-100p)p$. Setting R = \$1,800:

$$(800-100p)p=1{,}500$$
$$800p-100p^2=1{,}500$$
$$-100p^2+800p-1{,}500=0$$
$$-100(p^2-8p+15)=0$$
$$-100(p-5)(p-3)=0$$
$$p=3,5$$

The manufacturer should sell the items at either \$3 or \$5.

Chapter 7
Rational Expressions

7.1 Reducing Rational Expressions to Lowest Terms

1. Evaluating for $x = 2$: $\frac{x-3}{x+4}=\frac{2-3}{2+4}=\frac{-1}{6}=-\frac{1}{6}$

Evaluating for $x = 3$: $\frac{x-3}{x+4}=\frac{3-3}{3+4}=\frac{0}{7}=0$

Evaluating for $x = -4$: $\frac{x-3}{x+4}=\frac{-4-3}{-4+4}=\frac{-7}{0}$, which is undefined

3. Evaluating for $x = -3$: $\frac{2x+1}{x^2+x-2}=\frac{2(-3)+1}{(-3)^2+(-3)-2}=\frac{-6+1}{9-3-2}=\frac{-5}{4}=-\frac{5}{4}$

Evaluating for $x = 0$: $\frac{2x+1}{x^2+x-2}=\frac{2(0)+1}{(0)^2+0-2}=\frac{0+1}{0+0-2}=\frac{1}{-2}=-\frac{1}{2}$

Evaluating for $x = 1$: $\frac{2x+1}{x^2+x-2}=\frac{2(1)+1}{(1)^2+1-2}=\frac{2+1}{1+1-2}=\frac{3}{0}$, which is undefined

5. The denominator is 0 when $x = 0$.

7. The denominator is 0 when $x = -2$ or $x = 3$.

9. Setting the denominator equal to 0:

$$3x^2-2x-1=0$$
$$(3x+1)(x-1)=0$$
$$x=-\frac{1}{3},1$$

11. The denominator is always positive, so the rational expression is defined for all real numbers.

13. Reducing the rational expression: $\frac{5}{5x-10}=\frac{5}{5(x-2)}=\frac{1}{x-2}$. The variable restriction is $x \neq 2$.

15. Reducing the rational expression: $\frac{a-3}{a^2-9}=\frac{1(a-3)}{(a+3)(a-3)}=\frac{1}{a+3}$. The variable restriction is $a \neq -3,3$.

17. Reducing the rational expression: $\frac{x+5}{x^2-25}=\frac{1(x+5)}{(x+5)(x-5)}=\frac{1}{x-5}$. The variable restriction is $x \neq -5,5$.

19. Reducing the rational expression: $\frac{2x^2-8}{4}=\frac{2(x^2-4)}{4}=\frac{2(x+2)(x-2)}{4}=\frac{(x+2)(x-2)}{2}$

There are no variable restrictions.

21. Reducing the rational expression: $\frac{2x-10}{3x-6}=\frac{2(x-5)}{3(x-2)}$. The variable restriction is $x \neq 2$.

23. Reducing the rational expression: $\frac{10a+20}{5a+10}=\frac{10(a+2)}{5(a+2)}=\frac{2}{1}=2$. The variable restriction is $a\neq -2$.

25. Reducing the rational expression: $\frac{5x^2-5}{4x+4}=\frac{5(x^2-1)}{4(x+1)}=\frac{5(x+1)(x-1)}{4(x+1)}=\frac{5(x-1)}{4}$

27. Reducing the rational expression: $\frac{x-3}{x^2-6x+9}=\frac{1(x-3)}{(x-3)^2}=\frac{1}{x-3}$

29. Reducing the rational expression: $\frac{3x+15}{3x^2+24x+45}=\frac{3(x+5)}{3(x^2+8x+15)}=\frac{3(x+5)}{3(x+5)(x+3)}=\frac{1}{x+3}$

31. Reducing the rational expression: $\frac{a^2-3a}{a^3-8a^2+15a}=\frac{a(a-3)}{a(a^2-8a+15)}=\frac{a(a-3)}{a(a-3)(a-5)}=\frac{1}{a-5}$

33. Reducing the rational expression: $\frac{3x-2}{9x^2-4}=\frac{1(3x-2)}{(3x+2)(3x-2)}=\frac{1}{3x+2}$

35. Reducing the rational expression: $\frac{x^2+8x+15}{x^2+5x+6}=\frac{(x+5)(x+3)}{(x+2)(x+3)}=\frac{x+5}{x+2}$

37. Reducing the rational expression: $\frac{2m^3-2m^2-12m}{m^2-5m+6}=\frac{2m(m^2-m-6)}{(m-2)(m-3)}=\frac{2m(m-3)(m+2)}{(m-2)(m-3)}=\frac{2m(m+2)}{m-2}$

39. Reducing the rational expression: $\frac{x^3+3x^2-4x}{x^3-16x}=\frac{x(x^2+3x-4)}{x(x^2-16)}=\frac{x(x+4)(x-1)}{x(x+4)(x-4)}=\frac{x-1}{x-4}$

41. Reducing the rational expression: $\frac{4x^3-10x^2+6x}{2x^3+x^2-3x}=\frac{2x(2x^2-5x+3)}{x(2x^2+x-3)}=\frac{2x(2x-3)(x-1)}{x(2x+3)(x-1)}=\frac{2(2x-3)}{2x+3}$

43. Reducing the rational expression: $\frac{4x^2-12x+9}{4x^2-9}=\frac{(2x-3)^2}{(2x+3)(2x-3)}=\frac{2x-3}{2x+3}$

45. Reducing the rational expression: $\frac{x+3}{x^4-81}=\frac{x+3}{(x^2+9)(x^2-9)}=\frac{x+3}{(x^2+9)(x+3)(x-3)}=\frac{1}{(x^2+9)(x-3)}$

47. Reducing the rational expression: $\frac{3x^2+x-10}{x^4-16}=\frac{(3x-5)(x+2)}{(x^2+4)(x^2-4)}=\frac{(3x-5)(x+2)}{(x^2+4)(x+2)(x-2)}=\frac{3x-5}{(x^2+4)(x-2)}$

49. Reducing the rational expression: $\frac{42x^3-20x^2-48x}{6x^2-5x-4}=\frac{2x(21x^2-10x-24)}{(3x-4)(2x+1)}=\frac{2x(7x+6)(3x-4)}{(3x-4)(2x+1)}=\frac{2x(7x+6)}{2x+1}$

51. Reducing the rational expression: $\frac{x^3-y^3}{x^2-y^2}=\frac{(x-y)(x^2+xy+y^2)}{(x+y)(x-y)}=\frac{x^2+xy+y^2}{x+y}$

53. Reducing the rational expression: $\frac{x^3+8}{x^2-4}=\frac{(x+2)(x^2-2x+4)}{(x+2)(x-2)}=\frac{x^2-2x+4}{x-2}$

55. Reducing the rational expression: $\frac{x^3+8}{x^2+x-2}=\frac{(x+2)(x^2-2x+4)}{(x+2)(x-1)}=\frac{x^2-2x+4}{x-1}$

57. Reducing the rational expression: $\frac{xy+3x+2y+6}{xy+3x+5y+15}=\frac{x(y+3)+2(y+3)}{x(y+3)+5(y+3)}=\frac{(y+3)(x+2)}{(y+3)(x+5)}=\frac{x+2}{x+5}$

59. Reducing the rational expression: $\frac{x^2-3x+ax-3a}{x^2-3x+bx-3b}=\frac{x(x-3)+a(x-3)}{x(x-3)+b(x-3)}=\frac{(x-3)(x+a)}{(x-3)(x+b)}=\frac{x+a}{x+b}$

61. **a.** Adding: $(x^2-4x)+(4x-16)=x^2-4x+4x-16=x^2-16$

b. Subtracting: $(x^2-4x)-(4x-16)=x^2-4x-4x+16=x^2-8x+16$

c. Multiplying: $(x^2-4x)(4x-16)=4x^3-16x^2-16x^2+64x=4x^3-32x^2+64x$

d. Reducing: $\dfrac{x^2-4x}{4x-16}=\dfrac{x(x-4)}{4(x-4)}=\dfrac{x}{4}$

63. Writing as a ratio: $\dfrac{8}{6}=\dfrac{4}{3}$

65. Writing as a ratio: $\dfrac{200}{250}=\dfrac{4}{5}$

67. Writing as a ratio: $\dfrac{32}{4}=\dfrac{8}{1}$

69. Completing the table:

Checks Written x	Total Cost (dollars) $2.00+0.15x$	Cost per Check (dollars) $\dfrac{2.00+0.15x}{x}$
0	2.00	undefined
5	2.75	0.55
10	3.50	0.35
15	4.25	0.28
20	5.00	0.25

71. The average speed is: $\dfrac{122 \text{ miles}}{3 \text{ hours}} \approx 40.7$ miles/hour

73. Find the strikeouts per inning by dividing by 9:

Chris Sale: $\dfrac{11.82 \text{ strikeouts}}{9 \text{ innings}} \approx 1.31$ strikeouts/inning

Clayton Kershaw: $\dfrac{11.64 \text{ strikeouts}}{9 \text{ innings}} \approx 1.29$ strikeouts/inning

Max Scherzer: $\dfrac{10.86 \text{ strikeouts}}{9 \text{ innings}} \approx 1.21$ strikeouts/inning

Chris Archer: $\dfrac{10.70 \text{ strikeouts}}{9 \text{ innings}} \approx 1.19$ strikeouts/inning

75. Finding the average speed: $\dfrac{3 \text{ miles}}{24 \text{ minutes}} = 0.125$ miles/minute

77. Simplifying: $\dfrac{3}{4}\cdot\dfrac{10}{21}=\dfrac{3\cdot2\cdot5}{2\cdot2\cdot3\cdot7}=\dfrac{5}{14}$

79. Simplifying: $\dfrac{4}{5}\div\dfrac{8}{9}=\dfrac{4}{5}\cdot\dfrac{9}{8}=\dfrac{2\cdot2\cdot9}{5\cdot2\cdot2\cdot2}=\dfrac{9}{10}$

81. Factoring: $x^2-9=(x+3)(x-3)$

83. Factoring: $3x-9=3(x-3)$

85. Factoring: $x^2-x-20=(x-5)(x+4)$

87. Factoring: $a^2+5a=a(a+5)$

89. Simplifying: $\dfrac{a(a+5)(a-5)(a+4)}{a^2+5a}=\dfrac{a(a+5)(a-5)(a+4)}{a(a+5)}=(a-5)(a+4)$

91. Multiplying: $\dfrac{5603}{11}\cdot\dfrac{1}{2580}\cdot\dfrac{60}{1}\approx 11.8$

93. Evaluating for $x=-1$: $\dfrac{2x-3}{x^2-4}=\dfrac{2(-1)-3}{(-1)^2-4}=\dfrac{-2-3}{1-4}=\dfrac{-5}{-3}=\dfrac{5}{3}$. The correct answer is d.

95. Reducing the rational expression: $\frac{x^2-4x}{x^2-16}=\frac{x(x-4)}{(x+4)(x-4)}=\frac{x}{x+4}$. The correct answer is b.

7.2 Multiplication and Division of Rational Expressions

1. Simplifying the expression: $\frac{x+y}{3}\bullet\frac{6}{x+y}=\frac{6(x+y)}{3(x+y)}=2$

3. Simplifying the expression: $\frac{2x+10}{x^2}\bullet\frac{x^3}{4x+20}=\frac{2(x+5)}{x^2}\bullet\frac{x^3}{4(x+5)}=\frac{2x^3(x+5)}{4x^2(x+5)}=\frac{x}{2}$

5. Simplifying the expression: $\frac{9}{2a-8}\div\frac{3}{a-4}=\frac{9}{2a-8}\bullet\frac{a-4}{3}=\frac{9}{2(a-4)}\bullet\frac{a-4}{3}=\frac{9(a-4)}{6(a-4)}=\frac{3}{2}$

7. Simplifying the expression: $\frac{x+1}{x^2-9}\div\frac{2x+2}{x+3}=\frac{x+1}{x^2-9}\bullet\frac{x+3}{2x+2}=\frac{x+1}{(x+3)(x-3)}\bullet\frac{x+3}{2(x+1)}=\frac{(x+1)(x+3)}{2(x+3)(x-3)(x+1)}=\frac{1}{2(x-3)}$

9. Simplifying the expression: $\frac{a^2+5a}{7a}\bullet\frac{4a^2}{a^2+4a}=\frac{a(a+5)}{7a}\bullet\frac{4a^2}{a(a+4)}=\frac{4a^3(a+5)}{7a^2(a+4)}=\frac{4a(a+5)}{7(a+4)}$

11. Simplifying the expression:

$$\frac{y^2-5y+6}{2y+4}\div\frac{2y-6}{y+2}=\frac{y^2-5y+6}{2y+4}\bullet\frac{y+2}{2y-6}=\frac{(y-2)(y-3)}{2(y+2)}\bullet\frac{y+2}{2(y-3)}=\frac{(y-2)(y-3)(y+2)}{4(y+2)(y-3)}=\frac{y-2}{4}$$

13. Simplifying the expression: $\frac{2x-8}{x^2-4}\bullet\frac{x^2+6x+8}{x-4}=\frac{2(x-4)}{(x+2)(x-2)}\bullet\frac{(x+4)(x+2)}{x-4}=\frac{2(x-4)(x+4)(x+2)}{(x+2)(x-2)(x-4)}=\frac{2(x+4)}{x-2}$

15. Simplifying the expression:

$$\frac{x-1}{6+x-x^2}\bullet\frac{x^2+5x+6}{x^2-1}=\frac{x-1}{-(x-3)(x+2)}\bullet\frac{(x+2)(x+3)}{(x+1)(x-1)}=-\frac{(x-1)(x+2)(x+3)}{(x-3)(x+2)(x+1)(x-1)}=-\frac{x+3}{(x-3)(x+1)}$$

17. Simplifying the expression:

$$\frac{a^2+10a+25}{a+5}\div\frac{a^2-25}{a-5}=\frac{a^2+10a+25}{a+5}\bullet\frac{a-5}{a^2-25}=\frac{(a+5)^2}{a+5}\bullet\frac{a-5}{(a+5)(a-5)}=\frac{(a+5)^2(a-5)}{(a+5)^2(a-5)}=1$$

19. Simplifying the expression:

$$\begin{aligned}\frac{y^3-5y^2}{y^4+3y^3+2y^2}\div\frac{y^2-5y+6}{y^2-2y-3}&=\frac{y^3-5y^2}{y^4+3y^3+2y^2}\bullet\frac{y^2-2y-3}{y^2-5y+6}\\&=\frac{y^2(y-5)}{y^2(y+2)(y+1)}\bullet\frac{(y-3)(y+1)}{(y-2)(y-3)}\\&=\frac{y^2(y-5)(y-3)(y+1)}{y^2(y+2)(y+1)(y-2)(y-3)}\\&=\frac{y-5}{(y+2)(y-2)}\end{aligned}$$

21. Simplifying the expression:

$$\frac{2x^2+17x+21}{x^2+2x-35}\bullet\frac{25-x^2}{2x^2-7x-15}=\frac{(2x+3)(x+7)}{(x+7)(x-5)}\bullet\frac{-(x+5)(x-5)}{(2x+3)(x-5)}=-\frac{(2x+3)(x+7)(x+5)(x-5)}{(x+7)(x-5)^2(2x+3)}=-\frac{x+5}{x-5}$$

23. Simplifying the expression:

$$\begin{aligned}\frac{2x^2+10x+12}{4x^2+24x+32}\bullet\frac{2x^2+18x+40}{x^2+8x+15}&=\frac{2(x^2+5x+6)}{4(x^2+6x+8)}\bullet\frac{2(x^2+9x+20)}{x^2+8x+15}\\&=\frac{2(x+2)(x+3)}{4(x+4)(x+2)}\bullet\frac{2(x+5)(x+4)}{(x+5)(x+3)}\\&=\frac{4(x+2)(x+3)(x+4)(x+5)}{4(x+2)(x+3)(x+4)(x+5)}\\&=1\end{aligned}$$

25. Simplifying the expression:

$$\begin{aligned}\frac{2a^2+7a+3}{a^2-16}\div\frac{4a^2+8a+3}{2a^2-5a-12}&=\frac{2a^2+7a+3}{a^2-16}\bullet\frac{2a^2-5a-12}{4a^2+8a+3}\\&=\frac{(2a+1)(a+3)}{(a+4)(a-4)}\bullet\frac{(2a+3)(a-4)}{(2a+1)(2a+3)}\\&=\frac{(2a+1)(a+3)(2a+3)(a-4)}{(a+4)(a-4)(2a+1)(2a+3)}\\&=\frac{a+3}{a+4}\end{aligned}$$

27. Simplifying the expression:

$$\begin{aligned}\frac{4y^2-12y+9}{36-y^2}\div\frac{2y^2-5y+3}{y^2+5y-6}&=\frac{4y^2-12y+9}{-(y^2-36)}\bullet\frac{y^2+5y-6}{2y^2-5y+3}\\&=\frac{(2y-3)^2}{-(y+6)(y-6)}\bullet\frac{(y+6)(y-1)}{(2y-3)(y-1)}\\&=-\frac{(2y-3)^2(y+6)(y-1)}{(y+6)(y-6)(2y-3)(y-1)}\\&=-\frac{2y-3}{y-6}\end{aligned}$$

29. Simplifying the expression:

$$\begin{aligned}\frac{x^2-1}{6x^2+18x+12}\bullet\frac{7x^2+17x+6}{x+1}\bullet\frac{6x+30}{7x^2-11x-6}&=\frac{(x+1)(x-1)}{6(x+2)(x+1)}\bullet\frac{(7x+3)(x+2)}{(x+1)}\bullet\frac{6(x+5)}{(7x+3)(x-2)}\\&=\frac{6(x+1)(x-1)(7x+3)(x+2)(x+5)}{6(x+2)(x+1)^2(7x+3)(x-2)}\\&=\frac{(x-1)(x+5)}{(x+1)(x-2)}\end{aligned}$$

31. Simplifying the expression:

$$\begin{aligned}\frac{18x^3+21x^2-60x}{21x^2-25x-4}\bullet\frac{28x^2-17x-3}{16x^3+28x^2-30x}&=\frac{3x(6x^2+7x-20)}{21x^2-25x-4}\bullet\frac{28x^2-17x-3}{2x(8x^2+14x-15)}\\&=\frac{3x(3x-4)(2x+5)}{(7x+1)(3x-4)}\bullet\frac{(7x+1)(4x-3)}{2x(4x-3)(2x+5)}\\&=\frac{3x(3x-4)(2x+5)(7x+1)(4x-3)}{2x(7x+1)(3x-4)(4x-3)(2x+5)}\\&=\frac{3}{2}\end{aligned}$$

33. **a.** Simplifying: $\frac{9-1}{27-1}=\frac{8}{26}=\frac{4}{13}$

b. Reducing: $\frac{x^2-1}{x^3-1}=\frac{(x+1)(x-1)}{(x-1)(x^2+x+1)}=\frac{x+1}{x^2+x+1}$

c. Multiplying: $\frac{x^2-1}{x^3-1}\cdot\frac{x-2}{x+1}=\frac{(x+1)(x-1)}{(x-1)(x^2+x+1)}\cdot\frac{x-2}{x+1}=\frac{x-2}{x^2+x+1}$

d. Dividing: $\frac{x^2-1}{x^3-1}\div\frac{x-1}{x^2+x+1}=\frac{(x+1)(x-1)}{(x-1)(x^2+x+1)}\cdot\frac{x^2+x+1}{x-1}=\frac{x+1}{x-1}$

35. Simplifying the expression: $(x^2-9)\left(\frac{2}{x+3}\right)=\frac{(x+3)(x-3)}{1}\cdot\frac{2}{x+3}=\frac{2(x+3)(x-3)}{x+3}=2(x-3)$

37. Simplifying the expression: $(x^2-x-6)\left(\frac{x+1}{x-3}\right)=\frac{(x-3)(x+2)}{1}\cdot\frac{x+1}{x-3}=\frac{(x-3)(x+2)(x+1)}{x-3}=(x+2)(x+1)$

39. Simplifying the expression: $(x^2-4x-5)\left(\frac{-2x}{x+1}\right)=\frac{(x-5)(x+1)}{1}\cdot\frac{-2x}{x+1}=\frac{-2x(x-5)(x+1)}{x+1}=-2x(x-5)$

41. Simplifying the expression:

$$\frac{x^2-9}{x^2-3x}\cdot\frac{2x+10}{xy+5x+3y+15}=\frac{(x+3)(x-3)}{x(x-3)}\cdot\frac{2(x+5)}{x(y+5)+3(y+5)}=\frac{2(x+3)(x-3)(x+5)}{x(x-3)(y+5)(x+3)}=\frac{2(x+5)}{x(y+5)}$$

43. Simplifying the expression:

$$\frac{2x^2+4x}{x^2-y^2}\cdot\frac{x^2+3x+xy+3y}{x^2+5x+6}=\frac{2x(x+2)}{(x+y)(x-y)}\cdot\frac{x(x+3)+y(x+3)}{(x+2)(x+3)}=\frac{2x(x+2)(x+3)(x+y)}{(x+y)(x-y)(x+2)(x+3)}=\frac{2x}{x-y}$$

45. Simplifying the expression:

$$\begin{aligned}\frac{x^3-3x^2+4x-12}{x^4-16}\cdot\frac{3x^2+5x-2}{3x^2-10x+3}&=\frac{x^2(x-3)+4(x-3)}{(x^2+4)(x^2-4)}\cdot\frac{(3x-1)(x+2)}{(3x-1)(x-3)}\\&=\frac{(x-3)(x^2+4)}{(x^2+4)(x+2)(x-2)}\cdot\frac{(3x-1)(x+2)}{(3x-1)(x-3)}\\&=\frac{(x-3)(x^2+4)(3x-1)(x+2)}{(x^2+4)(x+2)(x-2)(3x-1)(x-3)}\\&=\frac{1}{x-2}\end{aligned}$$

47. Simplifying the expression: $\left(1-\frac{1}{2}\right)\left(1-\frac{1}{3}\right)\left(1-\frac{1}{4}\right)\left(1-\frac{1}{5}\right)=\left(\frac{2}{2}-\frac{1}{2}\right)\left(\frac{3}{3}-\frac{1}{3}\right)\left(\frac{4}{4}-\frac{1}{4}\right)\left(\frac{5}{5}-\frac{1}{5}\right)=\frac{1}{2}\cdot\frac{2}{3}\cdot\frac{3}{4}\cdot\frac{4}{5}=\frac{1}{5}$

49. Simplifying the expression: $\left(1-\frac{1}{2}\right)\left(1-\frac{1}{3}\right)\left(1-\frac{1}{4}\right)\cdots\left(1-\frac{1}{99}\right)\left(1-\frac{1}{100}\right)=\frac{1}{2}\cdot\frac{2}{3}\cdot\frac{3}{4}\cdots\frac{98}{99}\cdot\frac{99}{100}=\frac{1}{100}$

51. Converting the units: $14{,}494\text{ feet}\cdot\frac{1\text{ mile}}{5{,}280\text{ feet}}\approx 2.7\text{ miles}$

53. Converting the units: $\frac{1{,}088\text{ feet}}{1\text{ second}}\cdot\frac{3{,}600\text{ seconds}}{1\text{ hour}}\cdot\frac{1\text{ mile}}{5{,}280\text{ feet}}\approx 742\text{ miles per hour}$

55. Converting the units: $\frac{785\text{ feet}}{20\text{ minutes}}\cdot\frac{60\text{ minutes}}{1\text{ hour}}\cdot\frac{1\text{ mile}}{5{,}280\text{ feet}}\approx 0.45\text{ miles per hour}$

57. Converting the units: $\frac{518\text{ feet}}{40\text{ seconds}}\cdot\frac{3{,}600\text{ seconds}}{1\text{ hour}}\cdot\frac{1\text{ mile}}{5{,}280\text{ feet}}\approx 8.8\text{ miles per hour}$

59. Adding the fractions: $\frac{1}{5}+\frac{3}{5}=\frac{4}{5}$

61. Adding the fractions: $\frac{1}{10}+\frac{3}{14}=\frac{1}{10}\cdot\frac{7}{7}+\frac{3}{14}\cdot\frac{5}{5}=\frac{7}{70}+\frac{15}{70}=\frac{22}{70}=\frac{11}{35}$

63. Subtracting the fractions: $\frac{1}{10}-\frac{3}{14}=\frac{1}{10}\cdot\frac{7}{7}-\frac{3}{14}\cdot\frac{5}{5}=\frac{7}{70}-\frac{15}{70}=-\frac{8}{70}=-\frac{4}{35}$

65. Multiplying: $2(x-3)=2x-6$

67. Multiplying: $(x+4)(x-5)=x^2-5x+4x-20=x^2-x-20$

69. Reducing the fraction: $\frac{x+3}{x^2-9}=\frac{x+3}{(x+3)(x-3)}=\frac{1}{x-3}$

71. Reducing the fraction: $\frac{x^2-x-30}{2(x+5)(x-5)}=\frac{(x+5)(x-6)}{2(x+5)(x-5)}=\frac{x-6}{2(x-5)}$

73. Simplifying: $(x+4)(x-5)-10=x^2-5x+4x-20-10=x^2-x-30$

75. Simplifying the expression: $\frac{x^2-x-6}{2x-10}\cdot\frac{6x-30}{x^2+3x+2}=\frac{(x-3)(x+2)}{2(x-5)}\cdot\frac{6(x-5)}{(x+2)(x+1)}=\frac{6(x-3)(x+2)(x-5)}{2(x+2)(x+1)(x-5)}=\frac{3(x-3)}{x+1}$

The correct answer is b.

77. Converting the units: $\frac{10\text{ miles}}{1\text{ hour}}\cdot\frac{1\text{ hour}}{60\text{ minutes}}\cdot\frac{5{,}280\text{ feet}}{1\text{ mile}}=880$ feet per minute . The correct answer is b.

7.3 Addition and Subtraction of Rational Expressions

1. Combining the fractions: $\frac{3}{x}+\frac{4}{x}=\frac{7}{x}$

3. Combining the fractions: $\frac{9}{a}-\frac{5}{a}=\frac{4}{a}$

5. Combining the fractions: $\frac{1}{x+1}+\frac{x}{x+1}=\frac{1+x}{x+1}=1$

7. Combining the fractions: $\frac{y^2}{y-1}-\frac{1}{y-1}=\frac{y^2-1}{y-1}=\frac{(y+1)(y-1)}{y-1}=y+1$

9. Combining the fractions: $\frac{x^2}{x+2}+\frac{4x+4}{x+2}=\frac{x^2+4x+4}{x+2}=\frac{(x+2)^2}{x+2}=x+2$

11. Combining the fractions: $\frac{x^2}{x-2}-\frac{4x-4}{x-2}=\frac{x^2-4x+4}{x-2}=\frac{(x-2)^2}{x-2}=x-2$

13. Combining the fractions: $\frac{x+2}{x+6}-\frac{x-4}{x+6}=\frac{x+2-x+4}{x+6}=\frac{6}{x+6}$

15. The least common denominator is $5x$.

17. The least common denominator is $x(x-3)$.

19. The least common denominator is $2y^2(y+4)$.

21. Factoring each denominator:

$$a^2-6a+8=(a-2)(a-4)$$
$$a^2+a-6=(a-2)(a+3)$$

The least common denominator is $(a-2)(a-4)(a+3)$.

23. Factoring each denominator:

$$x^2+6x+9=(x+3)(x+3)=(x+3)^2$$

$$x^2+5x+6=(x+3)(x+2)$$

The least common denominator is $(x+3)^2(x+2)$.

25. Combining the fractions: $\frac{y}{2}-\frac{2}{y}=\frac{y\cdot y}{2\cdot y}-\frac{2\cdot 2}{y\cdot 2}=\frac{y^2}{2y}-\frac{4}{2y}=\frac{y^2-4}{2y}=\frac{(y+2)(y-2)}{2y}$

27. Combining the fractions: $\frac{1}{2}+\frac{a}{3}=\frac{1\cdot 3}{2\cdot 3}+\frac{a\cdot 2}{3\cdot 2}=\frac{3}{6}+\frac{2a}{6}=\frac{2a+3}{6}$

29. Combining the fractions: $\frac{x}{x+1}+\frac{3}{4}=\frac{x\cdot 4}{(x+1)\cdot 4}+\frac{3\cdot(x+1)}{4\cdot(x+1)}=\frac{4x}{4(x+1)}+\frac{3x+3}{4(x+1)}=\frac{4x+3x+3}{4(x+1)}=\frac{7x+3}{4(x+1)}$

31. Combining the fractions: $\frac{x+1}{x-2}-\frac{4x+7}{5x-10}=\frac{(x+1)\cdot 5}{(x-2)\cdot 5}-\frac{4x+7}{5(x-2)}=\frac{5x+5}{5(x-2)}-\frac{4x+7}{5(x-2)}=\frac{5x+5-4x-7}{5(x-2)}=\frac{x-2}{5(x-2)}=\frac{1}{5}$

33. Combining the fractions: $\frac{4x-2}{3x+12}-\frac{x-2}{x+4}=\frac{4x-2}{3(x+4)}-\frac{(x-2)\cdot 3}{(x+4)\cdot 3}=\frac{4x-2}{3(x+4)}-\frac{3x-6}{3(x+4)}=\frac{4x-2-3x+6}{3(x+4)}=\frac{x+4}{3(x+4)}=\frac{1}{3}$

35. Combining the fractions: $\frac{6}{x(x-2)}+\frac{3}{x}=\frac{6}{x(x-2)}+\frac{3\cdot(x-2)}{x\cdot(x-2)}=\frac{6}{x(x-2)}+\frac{3x-6}{x(x-2)}=\frac{6+3x-6}{x(x-2)}=\frac{3x}{x(x-2)}=\frac{3}{x-2}$

37. Combining the fractions: $\frac{4}{a}-\frac{12}{a^2+3a}=\frac{4\cdot(a+3)}{a\cdot(a+3)}-\frac{12}{a(a+3)}=\frac{4a+12}{a(a+3)}-\frac{12}{a(a+3)}=\frac{4a+12-12}{a(a+3)}=\frac{4a}{a(a+3)}=\frac{4}{a+3}$

39. Combining the fractions:

$$\begin{aligned}\frac{2}{x+5}-\frac{10}{x^2-25}&=\frac{2\cdot(x-5)}{(x+5)\cdot(x-5)}-\frac{10}{(x+5)(x-5)}\\&=\frac{2x-10}{(x+5)(x-5)}-\frac{10}{(x+5)(x-5)}\\&=\frac{2x-10-10}{(x+5)(x-5)}\\&=\frac{2x-20}{(x+5)(x-5)}\\&=\frac{2(x-10)}{(x+5)(x-5)}\end{aligned}$$

41. Combining the fractions:

$$\begin{aligned}\frac{x-4}{x-3}+\frac{6}{x^2-9}&=\frac{(x-4)\cdot(x+3)}{(x-3)\cdot(x+3)}+\frac{6}{(x+3)(x-3)}\\&=\frac{x^2-x-12}{(x+3)(x-3)}+\frac{6}{(x+3)(x-3)}\\&=\frac{x^2-x-12+6}{(x+3)(x-3)}\\&=\frac{x^2-x-6}{(x+3)(x-3)}\\&=\frac{(x-3)(x+2)}{(x+3)(x-3)}\\&=\frac{x+2}{x+3}\end{aligned}$$

43. Combining the fractions:

$$\begin{aligned}\frac{a-4}{a-3}+\frac{5}{a^2-a-6}&=\frac{(a-4)\cdot(a+2)}{(a-3)\cdot(a+2)}+\frac{5}{(a-3)(a+2)}\\&=\frac{a^2-2a-8}{(a-3)(a+2)}+\frac{5}{(a-3)(a+2)}\\&=\frac{a^2-2a-8+5}{(a-3)(a+2)}\\&=\frac{a^2-2a-3}{(a-3)(a+2)}\\&=\frac{(a-3)(a+1)}{(a-3)(a+2)}\\&=\frac{a+1}{a+2}\end{aligned}$$

45. Combining the fractions:

$$\begin{aligned}\frac{8}{x^2-16}-\frac{7}{x^2-x-12}&=\frac{8}{(x+4)(x-4)}-\frac{7}{(x-4)(x+3)}\\&=\frac{8(x+3)}{(x+4)(x-4)(x+3)}-\frac{7(x+4)}{(x+4)(x-4)(x+3)}\\&=\frac{8x+24}{(x+4)(x-4)(x+3)}-\frac{7x+28}{(x+4)(x-4)(x+3)}\\&=\frac{8x+24-7x-28}{(x+4)(x-4)(x+3)}\\&=\frac{x-4}{(x+4)(x-4)(x+3)}\\&=\frac{1}{(x+4)(x+3)}\end{aligned}$$

47. Combining the fractions:

$$\begin{aligned}\frac{4y}{y^2+6y+5}-\frac{3y}{y^2+5y+4}&=\frac{4y}{(y+5)(y+1)}-\frac{3y}{(y+4)(y+1)}\\&=\frac{4y(y+4)}{(y+5)(y+1)(y+4)}-\frac{3y(y+5)}{(y+5)(y+1)(y+4)}\\&=\frac{4y^2+16y}{(y+5)(y+1)(y+4)}-\frac{3y^2+15y}{(y+5)(y+1)(y+4)}\\&=\frac{4y^2+16y-3y^2-15y}{(y+5)(y+1)(y+4)}\\&=\frac{y^2+y}{(y+5)(y+1)(y+4)}\\&=\frac{y(y+1)}{(y+5)(y+1)(y+4)}\\&=\frac{y}{(y+5)(y+4)}\end{aligned}$$

49. Combining the fractions:

$$\frac{4x+1}{x^2+5x+4}-\frac{x+3}{x^2+4x+3}=\frac{4x+1}{(x+4)(x+1)}-\frac{x+3}{(x+3)(x+1)}$$

$$=\frac{(4x+1)(x+3)}{(x+4)(x+1)(x+3)}-\frac{(x+3)(x+4)}{(x+4)(x+1)(x+3)}$$

$$=\frac{4x^2+13x+3}{(x+4)(x+1)(x+3)}-\frac{x^2+7x+12}{(x+4)(x+1)(x+3)}$$

$$=\frac{4x^2+13x+3-x^2-7x-12}{(x+4)(x+1)(x+3)}$$

$$=\frac{3x^2+6x-9}{(x+4)(x+1)(x+3)}$$

$$=\frac{3(x+3)(x-1)}{(x+4)(x+1)(x+3)}$$

$$=\frac{3(x-1)}{(x+4)(x+1)}$$

51. Combining the fractions:

$$\frac{x-3}{x^2+4x+4}-\frac{x+1}{x^2-4}=\frac{x-3}{(x+2)^2}-\frac{x+1}{(x+2)(x-2)}$$

$$=\frac{(x-3)(x-2)}{(x+2)^2(x-2)}-\frac{(x+1)(x+2)}{(x+2)^2(x-2)}$$

$$=\frac{x^2-5x+6}{(x+2)^2(x-2)}-\frac{x^2+3x+2}{(x+2)^2(x-2)}$$

$$=\frac{x^2-5x+6-x^2-3x-2}{(x+2)^2(x-2)}$$

$$=\frac{-8x+4}{(x+2)^2(x-2)}$$

$$=-\frac{4(2x-1)}{(x+2)^2(x-2)}$$

53. Combining the fractions:

$$\frac{1}{x}+\frac{x}{3x+9}-\frac{3}{x^2+3x}=\frac{1}{x}+\frac{x}{3(x+3)}-\frac{3}{x(x+3)}$$
$$=\frac{1\cdot 3(x+3)}{x\cdot 3(x+3)}+\frac{x\cdot x}{3(x+3)\cdot x}-\frac{3\cdot 3}{x(x+3)\cdot 3}$$
$$=\frac{3x+9}{3x(x+3)}+\frac{x^2}{3x(x+3)}-\frac{9}{3x(x+3)}$$
$$=\frac{3x+9+x^2-9}{3x(x+3)}$$
$$=\frac{x^2+3x}{3x(x+3)}$$
$$=\frac{x(x+3)}{3x(x+3)}$$
$$=\frac{1}{3}$$

55. **a.** Multiplying: $\frac{4}{9}\cdot\frac{1}{6}=\frac{2\cdot 2}{3\cdot 3}\cdot\frac{1}{2\cdot 3}=\frac{2}{3\cdot 3\cdot 3}=\frac{2}{27}$

b. Dividing: $\frac{4}{9}\div\frac{1}{6}=\frac{4}{9}\cdot\frac{6}{1}=\frac{2\cdot 2}{3\cdot 3}\cdot\frac{2\cdot 3}{1}=\frac{2\cdot 2\cdot 2}{3}=\frac{8}{3}$

c. Adding: $\frac{4}{9}+\frac{1}{6}=\frac{4}{9}\cdot\frac{2}{2}+\frac{1}{6}\cdot\frac{3}{3}=\frac{8}{18}+\frac{3}{18}=\frac{11}{18}$

d. Multiplying: $\frac{x+2}{x-2}\cdot\frac{3x+10}{x^2-4}=\frac{x+2}{x-2}\cdot\frac{3x+10}{(x+2)(x-2)}=\frac{3x+10}{(x-2)^2}$

e. Dividing: $\frac{x+2}{x-2}\div\frac{3x+10}{x^2-4}=\frac{x+2}{x-2}\cdot\frac{(x+2)(x-2)}{3x+10}=\frac{(x+2)^2}{3x+10}$

f. Subtracting:

$$\frac{x+2}{x-2}-\frac{3x+10}{x^2-4}=\frac{x+2}{x-2}\cdot\frac{x+2}{x+2}-\frac{3x+10}{(x+2)(x-2)}$$
$$=\frac{x^2+4x+4}{(x+2)(x-2)}-\frac{3x+10}{(x+2)(x-2)}$$
$$=\frac{x^2+x-6}{(x+2)(x-2)}$$
$$=\frac{(x+3)(x-2)}{(x+2)(x-2)}$$
$$=\frac{x+3}{x+2}$$

57. Completing the table:

Number x	Reciprocal $\frac{1}{x}$	Sum $1+\frac{1}{x}$	Quotient $\frac{x+1}{x}$
1	1	2	2
2	$\frac{1}{2}$	$\frac{3}{2}$	$\frac{3}{2}$
3	$\frac{1}{3}$	$\frac{4}{3}$	$\frac{4}{3}$
4	$\frac{1}{4}$	$\frac{5}{4}$	$\frac{5}{4}$

59. Combining the fractions: $1+\frac{1}{x+2}=\frac{1\bullet(x+2)}{1\bullet(x+2)}+\frac{1}{x+2}=\frac{x+2}{x+2}+\frac{1}{x+2}=\frac{x+2+1}{x+2}=\frac{x+3}{x+2}$

61. Combining the fractions: $1-\frac{1}{x+3}=\frac{1\bullet(x+3)}{1\bullet(x+3)}-\frac{1}{x+3}=\frac{x+3}{x+3}-\frac{1}{x+3}=\frac{x+3-1}{x+3}=\frac{x+2}{x+3}$

63. Simplifying each parenthesis first:

$$1-\frac{1}{x}=\frac{x}{x}-\frac{1}{x}=\frac{x-1}{x}$$
$$1-\frac{1}{x+1}=\frac{x+1}{x+1}-\frac{1}{x+1}=\frac{x}{x+1}$$
$$1-\frac{1}{x+2}=\frac{x+2}{x+2}-\frac{1}{x+2}=\frac{x+1}{x+2}$$

Now performing the multiplication: $\left(1-\frac{1}{x}\right)\left(1-\frac{1}{x+1}\right)\left(1-\frac{1}{x+2}\right)=\frac{x-1}{x}\bullet\frac{x}{x+1}\bullet\frac{x+1}{x+2}=\frac{x-1}{x+2}$

65. Simplifying each parenthesis first:

$$1+\frac{1}{x+3}=\frac{x+3}{x+3}+\frac{1}{x+3}=\frac{x+4}{x+3}$$
$$1+\frac{1}{x+2}=\frac{x+2}{x+2}+\frac{1}{x+2}=\frac{x+3}{x+2}$$
$$1+\frac{1}{x+1}=\frac{x+1}{x+1}+\frac{1}{x+1}=\frac{x+2}{x+1}$$

Now performing the multiplication: $\left(1+\frac{1}{x+3}\right)\left(1+\frac{1}{x+2}\right)\left(1+\frac{1}{x+1}\right)=\frac{x+4}{x+3}\bullet\frac{x+3}{x+2}\bullet\frac{x+2}{x+1}=\frac{x+4}{x+1}$

67. Writing the expression: $x+\frac{2}{x}=\frac{x\bullet x}{1\bullet x}+\frac{2}{x}=\frac{x^2}{x}+\frac{2}{x}=\frac{x^2+2}{x}$

69. Writing the expression: $\frac{1}{x}+\frac{1}{2x}=\frac{1\bullet 2}{x\bullet 2}+\frac{1}{2x}=\frac{2}{2x}+\frac{1}{2x}=\frac{3}{2x}$

71. Simplifying: $\frac{1}{2}\div\frac{2}{3}=\frac{1}{2}\bullet\frac{3}{2}=\frac{3}{4}$

73. Simplifying: $1+\frac{1}{2}=\frac{2}{2}+\frac{1}{2}=\frac{3}{2}$

75. Simplifying: $y^5\bullet\frac{2x^3}{y^2}=\frac{2x^3y^5}{y^2}=2x^3y^3$

77. Simplifying: $\frac{2x^3}{y^2}\bullet\frac{y^5}{4x}=\frac{2x^3y^5}{4xy^2}=\frac{x^2y^3}{2}$

79. Factoring: $x^2y+x=x(xy+1)$

81. Reducing the fraction: $\frac{2x^3y^2}{4x}=\frac{x^2y^2}{2}$

83. Reducing the fraction: $\frac{x^2-4}{x^2-x-6}=\frac{(x+2)(x-2)}{(x+2)(x-3)}=\frac{x-2}{x-3}$

85. Adding the fractions: $\frac{x^2-4x}{x-5}+\frac{x-10}{x-5}=\frac{x^2-3x-10}{x-5}=\frac{(x-5)(x+2)}{x-5}=x+2$. The correct answer is c.

87. Factoring each denominator:

$$x^3+4x^2=x^2(x+4)$$
$$x^2-x=x(x-1)$$

The least common denominator is $x^2(x+4)(x-1)$. The correct answer is c.

7.4 Complex Fractions

1. Simplifying the complex fraction: $\dfrac{3/4}{1/8}=\dfrac{3/4\bullet 8}{1/8\bullet 8}=\dfrac{6}{1}=6$

3. Simplifying the complex fraction: $\dfrac{2/3}{4}=\dfrac{2/3\bullet 3}{4\bullet 3}=\dfrac{2}{12}=\dfrac{1}{6}$

5. Simplifying the complex fraction: $\dfrac{\frac{x^2}{y}}{\frac{x}{y^3}}=\dfrac{\frac{x^2}{y}\bullet y^3}{\frac{x}{y^3}\bullet y^3}=\dfrac{x^2y^2}{x}=xy^2$

7. Simplifying the complex fraction: $\dfrac{\frac{4x^3}{y^6}}{\frac{8x^2}{y^7}}=\dfrac{\frac{4x^3}{y^6}\bullet y^7}{\frac{8x^2}{y^7}\bullet y^7}=\dfrac{4x^3y}{8x^2}=\dfrac{xy}{2}$

9. Simplifying the complex fraction: $\dfrac{y+\frac{1}{x}}{x+\frac{1}{y}}=\dfrac{\left(y+\frac{1}{x}\right)\bullet xy}{\left(x+\frac{1}{y}\right)\bullet xy}=\dfrac{xy^2+y}{x^2y+x}=\dfrac{y(xy+1)}{x(xy+1)}=\dfrac{y}{x}$

11. Simplifying the complex fraction: $\dfrac{1+\frac{1}{a}}{1-\frac{1}{a}}=\dfrac{\left(1+\frac{1}{a}\right)\bullet a}{\left(1-\frac{1}{a}\right)\bullet a}=\dfrac{a+1}{a-1}$

13. Simplifying the complex fraction: $\dfrac{\frac{x+1}{x^2-9}}{\frac{2}{x+3}}=\dfrac{\frac{x+1}{(x+3)(x-3)}\bullet(x+3)(x-3)}{\frac{2}{x+3}\bullet(x+3)(x-3)}=\dfrac{x+1}{2(x-3)}$

15. Simplifying the complex fraction: $\dfrac{\frac{1}{a+2}}{\frac{1}{a^2-a-6}}=\dfrac{\frac{1}{a+2}\bullet(a-3)(a+2)}{\frac{1}{(a-3)(a+2)}\bullet(a-3)(a+2)}=\dfrac{a-3}{1}=a-3$

17. Simplifying the complex fraction: $\dfrac{1-\frac{9}{y^2}}{1-\frac{1}{y}-\frac{6}{y^2}}=\dfrac{\left(1-\frac{9}{y^2}\right)\bullet y^2}{\left(1-\frac{1}{y}-\frac{6}{y^2}\right)\bullet y^2}=\dfrac{y^2-9}{y^2-y-6}=\dfrac{(y+3)(y-3)}{(y+2)(y-3)}=\dfrac{y+3}{y+2}$

19. Simplifying the complex fraction: $\dfrac{\frac{1}{y}+\frac{1}{x}}{\frac{1}{xy}}=\dfrac{\left(\frac{1}{y}+\frac{1}{x}\right)\bullet xy}{\left(\frac{1}{xy}\right)\bullet xy}=\dfrac{x+y}{1}=x+y$

21. Simplifying the complex fraction: $\dfrac{1-\frac{1}{a^2}}{1-\frac{1}{a}}=\dfrac{\left(1-\frac{1}{a^2}\right)\cdot a^2}{\left(1-\frac{1}{a}\right)\cdot a^2}=\dfrac{a^2-1}{a^2-a}=\dfrac{(a+1)(a-1)}{a(a-1)}=\dfrac{a+1}{a}$

23. Simplifying the complex fraction: $\dfrac{\frac{1}{10x}-\frac{y}{10x^2}}{\frac{1}{10}-\frac{y}{10x}}=\dfrac{\left(\frac{1}{10x}-\frac{y}{10x^2}\right)\cdot 10x^2}{\left(\frac{1}{10}-\frac{y}{10x}\right)\cdot 10x^2}=\dfrac{x-y}{x^2-xy}=\dfrac{1(x-y)}{x(x-y)}=\dfrac{1}{x}$

25. Simplifying the complex fraction: $\dfrac{\frac{1}{a+1}+2}{\frac{1}{a+1}+3}=\dfrac{\left(\frac{1}{a+1}+2\right)\cdot(a+1)}{\left(\frac{1}{a+1}+3\right)\cdot(a+1)}=\dfrac{1+2(a+1)}{1+3(a+1)}=\dfrac{1+2a+2}{1+3a+3}=\dfrac{2a+3}{3a+4}$

27. Simplifying each term in the sequence:

$$2+\frac{1}{2+1}=2+\frac{1}{3}=\frac{6}{3}+\frac{1}{3}=\frac{7}{3}$$

$$2+\frac{1}{2+\frac{1}{2+1}}=2+\frac{1}{\frac{7}{3}}=2+\frac{3}{7}=\frac{14}{7}+\frac{3}{7}=\frac{17}{7}$$

$$2+\frac{2}{2+\frac{1}{2+\frac{1}{2+1}}}=2+\frac{1}{\frac{17}{7}}=2+\frac{7}{17}=\frac{34}{17}+\frac{7}{17}=\frac{41}{17}$$

29. Completing the table:

Number x	Reciprocal $\frac{1}{x}$	Quotient $\frac{x}{1/x}$	Square x^2
1	1	1	1
2	$\frac{1}{2}$	4	4
3	$\frac{1}{3}$	9	9
4	$\frac{1}{4}$	16	16

31. Completing the table:

Number x	Reciprocal $\frac{1}{x}$	Sum $1+\frac{1}{x}$	Quotient $\frac{1+\frac{1}{x}}{\frac{1}{x}}$
1	1	2	2
2	$\frac{1}{2}$	$\frac{3}{2}$	3
3	$\frac{1}{3}$	$\frac{4}{3}$	4
4	$\frac{1}{4}$	$\frac{5}{4}$	5

33. Simplifying: $6\left(\frac{1}{2}\right)=3$

35. Simplifying: $\frac{0}{5}=0$

37. Simplifying: $\frac{5}{0}$ is undefined

39. Simplifying: $1-\frac{5}{2}=\frac{2}{2}-\frac{5}{2}=-\frac{3}{2}$

41. Simplifying: $6\left(\frac{x}{3}+\frac{5}{2}\right)=6\cdot\frac{x}{3}+6\cdot\frac{5}{2}=2x+15$

43. Simplifying: $x^2\left(1-\frac{5}{x}\right)=x^2\cdot 1-x^2\cdot\frac{5}{x}=x^2-5x$

45. Solving the equation:

$$\begin{aligned}2x+15&=3\\2x&=-12\\x&=-6\end{aligned}$$

47. Solving the equation:

$$\begin{aligned}-2x-9&=x-3\\-3x-9&=-3\\-3x&=6\\x&=-2\end{aligned}$$

49. Simplifying using the division method: $\dfrac{\frac{5x^2}{x^2-16}}{\frac{2x}{x+4}} = \dfrac{5x^2}{x^2-16} \div \dfrac{2x}{x+4} = \dfrac{5x^2}{x^2-16} \cdot \dfrac{x+4}{2x} = \dfrac{5x^2(x+4)}{2x(x+4)(x-4)} = \dfrac{5x}{2(x-4)}$

The correct answer is d.

7.5 Equations Involving Rational Expressions

1. Multiplying both sides of the equation by 6:

$$6\left(\frac{x}{3}+\frac{1}{2}\right)=6\left(-\frac{1}{2}\right)$$
$$2x+3=-3$$
$$2x=-6$$
$$x=-3$$

Since $x=-3$ checks in the original equation, the solution is $x=-3$.

3. Multiplying both sides of the equation by $5a$:

$$5a\left(\frac{4}{a}\right)=5a\left(\frac{1}{5}\right)$$
$$20=a$$

Since $a=20$ checks in the original equation, the solution is $a=20$.

5. Multiplying both sides of the equation by x:

$$x\left(\frac{3}{x}+1\right)=x\left(\frac{2}{x}\right)$$
$$3+x=2$$
$$x=-1$$

Since $x=-1$ checks in the original equation, the solution is $x=-1$.

7. Multiplying both sides of the equation by $5a$:

$$5a\left(\frac{3}{a}-\frac{2}{a}\right)=5a\left(\tfrac{1}{5}\right)$$
$$15-10=a$$
$$a=5$$

Since $a=5$ checks in the original equation, the solution is $a=5$.

9. Multiplying both sides of the equation by $2x$:

$$2x\left(\frac{3}{x}+2\right)=2x\left(\frac{1}{2}\right)$$
$$6+4x=x$$
$$6=-3x$$
$$x=-2$$

Since $x=-2$ checks in the original equation, the solution is $x=-2$.

11. Multiplying both sides of the equation by $4y$:

$$4y\left(\frac{1}{y}-\frac{1}{2}\right)=4y\left(-\frac{1}{4}\right)$$
$$4-2y=-y$$
$$4=y$$

Since $y=4$ checks in the original equation, the solution is $y=4$.

13. Multiplying both sides of the equation by x^2:

$$x^2\left(1-\frac{8}{x}\right)=x^2\left(-\frac{15}{x^2}\right)$$
$$x^2-8x=-15$$
$$x^2-8x+15=0$$
$$(x-3)(x-5)=0$$
$$x=3,5$$

Both $x=3$ and $x=5$ check in the original equation.

15. Multiplying both sides of the equation by $2x$:

$$2x\left(\frac{x}{2}-\frac{4}{x}\right)=2x\left(-\frac{7}{2}\right)$$
$$x^2-8=-7x$$
$$x^2+7x-8=0$$
$$(x+8)(x-1)=0$$
$$x=-8,1$$

Both $x=-8$ and $x=1$ check in the original equation.

17. Multiplying both sides of the equation by 6:

$$6\left(\frac{x-3}{2}+\frac{2x}{3}\right)=6\left(\frac{5}{6}\right)$$
$$3(x-3)+2(2x)=5$$
$$3x-9+4x=5$$
$$7x-9=5$$
$$7x=14$$
$$x=2$$

Since $x=2$ checks in the original equation, the solution is $x=2$.

19. Multiplying both sides of the equation by 12:

$$12\left(\frac{x+1}{3}+\frac{x-3}{4}\right)=12\left(\frac{1}{6}\right)$$
$$4(x+1)+3(x-3)=2$$
$$4x+4+3x-9=2$$
$$7x-5=2$$
$$7x=7$$
$$x=1$$

Since $x=1$ checks in the original equation, the solution is $x=1$.

21. Multiplying both sides of the equation by $5(x+2)$:

$$5(x+2)\bullet\frac{6}{x+2}=5(x+2)\bullet\frac{3}{5}$$
$$30=3x+6$$
$$24=3x$$
$$x=8$$

Since $x=8$ checks in the original equation, the solution is $x=8$.

23. Multiplying both sides of the equation by $(y-2)(y-3)$:

$$(y-2)(y-3)\bullet\frac{3}{y-2}=(y-2)(y-3)\bullet\frac{2}{y-3}$$
$$3(y-3)=2(y-2)$$
$$3y-9=2y-4$$
$$y=5$$

Since $y=5$ checks in the original equation, the solution is $y=5$.

25. Multiplying both sides of the equation by $3(x-2)$:

$$\begin{aligned}3(x-2)\left(\frac{x}{x-2}+\frac{2}{3}\right)&=3(x-2)\left(\frac{2}{x-2}\right)\\3x+2(x-2)&=6\\3x+2x-4&=6\\5x-4&=6\\5x&=10\\x&=2\end{aligned}$$

Since $x=2$ does not check in the original equation, there is no solution.

27. Multiplying both sides of the equation by $2(x-2)$:

$$\begin{aligned}2(x-2)\left(\frac{x}{x-2}+\frac{3}{2}\right)&=2(x-2)\bullet\frac{9}{2(x-2)}\\2x+3(x-2)&=9\\2x+3x-6&=9\\5x-6&=9\\5x&=15\\x&=3\end{aligned}$$

Since $x=3$ checks in the original equation, the solution is $x=3$.

29. Multiplying both sides of the equation by $x^2+5x+6=(x+2)(x+3)$:

$$\begin{aligned}(x+2)(x+3)\left(\frac{5}{x+2}+\frac{1}{x+3}\right)&=(x+2)(x+3)\bullet\frac{-1}{(x+2)(x+3)}\\5(x+3)+1(x+2)&=-1\\5x+15+x+2&=-1\\6x+17&=-1\\6x&=-18\\x&=-3\end{aligned}$$

Since $x=-3$ does not check in the original equation, there is no solution.

31. Multiplying both sides of the equation by $x^2-4=(x+2)(x-2)$:

$$\begin{aligned}(x+2)(x-2)\left(\frac{8}{x^2-4}+\frac{3}{x+2}\right)&=(x+2)(x-2)\bullet\frac{1}{x-2}\\8+3(x-2)&=1(x+2)\\8+3x-6&=x+2\\3x+2&=x+2\\2x&=0\\x&=0\end{aligned}$$

Since $x=0$ checks in the original equation, the solution is $x=0$.

33. Multiplying both sides of the equation by $2(a-3)$:

$$2(a-3)\left(\frac{a}{2}+\frac{3}{a-3}\right)=2(a-3)\cdot\frac{a}{a-3}$$
$$a(a-3)+6=2a$$
$$a^2-3a+6=2a$$
$$a^2-5a+6=0$$
$$(a-2)(a-3)=0$$
$$a=2,3$$

Since $a=3$ does not check in the original equation, the solution is $a=2$.

35. Since $y^2-4=(y+2)(y-2)$ and $y^2+2y=y(y+2)$, the LCD is $y(y+2)(y-2)$. Multiplying by the LCD:

$$y(y+2)(y-2)\cdot\frac{6}{(y+2)(y-2)}=y(y+2)(y-2)\cdot\frac{4}{y(y+2)}$$
$$6y=4(y-2)$$
$$6y=4y-8$$
$$2y=-8$$
$$y=-4$$

Since $y=-4$ checks in the original equation, the solution is $y=-4$.

37. Since $a^2-9=(a+3)(a-3)$ and $a^2+a-12=(a+4)(a-3)$, the LCD is $(a+3)(a-3)(a+4)$. Multiplying by the LCD:

$$(a+3)(a-3)(a+4)\cdot\frac{2}{(a+3)(a-3)}=(a+3)(a-3)(a+4)\cdot\frac{3}{(a+4)(a-3)}$$
$$2(a+4)=3(a+3)$$
$$2a+8=3a+9$$
$$-a+8=9$$
$$-a=1$$
$$a=-1$$

Since $a=-1$ checks in the original equation, the solution is $a=-1$.

39. Multiplying both sides of the equation by $x^2-4x-5=(x-5)(x+1)$:

$$(x-5)(x+1)\left(\frac{3x}{x-5}-\frac{2x}{x+1}\right)=(x-5)(x+1)\cdot\frac{-42}{(x-5)(x+1)}$$
$$3x(x+1)-2x(x-5)=-42$$
$$3x^2+3x-2x^2+10x=-42$$
$$x^2+13x+42=0$$
$$(x+7)(x+6)=0$$
$$x=-7,-6$$

Both $x=-7$ and $x=-6$ check in the original equation.

41. Multiplying both sides of the equation by $x^2+5x+6=(x+2)(x+3)$:

$$(x+2)(x+3)\left(\frac{2x}{x+2}\right)=(x+2)(x+3)\left(\frac{x}{x+3}-\frac{3}{x^2+5x+6}\right)$$
$$2x(x+3)=x(x+2)-3$$
$$2x^2+6x=x^2+2x-3$$
$$x^2+4x+3=0$$
$$(x+3)(x+1)=0$$
$$x=-3,-1$$

Since $x=-3$ does not check in the original equation, the solution is $x=-1$.

43. **a.** Solving the equation:

$$5x-1=0$$
$$5x=1$$
$$x=\frac{1}{5}$$

b. Solving the equation:

$$\frac{5}{x}-1=0$$
$$x\left(\frac{5}{x}-1\right)=x(0)$$
$$5-x=0$$
$$x=5$$

c. Solving the equation:

$$\frac{x}{5}-1=\frac{2}{3}$$
$$15\left(\frac{x}{5}-1\right)=15\left(\frac{2}{3}\right)$$
$$3x-15=10$$
$$3x=25$$
$$x=\frac{25}{3}$$

d. Solving the equation:

$$\frac{5}{x}-1=\frac{2}{3}$$
$$3x\left(\frac{5}{x}-1\right)=3x\left(\frac{2}{3}\right)$$
$$15-3x=2x$$
$$15=5x$$
$$x=3$$

e. Solving the equation:

$$\frac{5}{x^2}+5=\frac{26}{x}$$
$$x^2\left(\frac{5}{x^2}+5\right)=x^2\left(\frac{26}{x}\right)$$
$$5+5x^2=26x$$
$$5x^2-26x+5=0$$
$$(5x-1)(x-5)=0$$
$$x=\frac{1}{5},5$$

45. **a.** Dividing: $\frac{7}{a^2-5a-6} \div \frac{a+2}{a+1} = \frac{7}{(a-6)(a+1)} \bullet \frac{a+1}{a+2} = \frac{7}{(a-6)(a+2)}$

b. Adding:

$$\begin{aligned}\frac{7}{a^2-5a-6}+\frac{a+2}{a+1} &= \frac{7}{(a-6)(a+1)}+\frac{a+2}{a+1}\bullet\frac{a-6}{a-6}\\ &= \frac{7}{(a-6)(a+1)}+\frac{a^2-4a-12}{(a-6)(a+1)}\\ &= \frac{a^2-4a-5}{(a-6)(a+1)}\\ &= \frac{(a-5)(a+1)}{(a-6)(a+1)}\\ &= \frac{a-5}{a-6}\end{aligned}$$

c. Solving the equation:

$$\begin{aligned}\frac{7}{a^2-5a-6}+\frac{a+2}{a+1} &= 2\\ \frac{7}{(a-6)(a+1)}+\frac{a+2}{a+1} &= 2\\ (a-6)(a+1)\left(\frac{7}{(a-6)(a+1)}+\frac{a+2}{a+1}\right) &= 2(a-6)(a+1)\\ 7+(a-6)(a+2) &= 2(a-6)(a+1)\\ 7+a^2-4a-12 &= 2a^2-10a-12\\ 0 &= a^2-6a-7\\ 0 &= (a-7)(a+1)\\ a &= -1,7\end{aligned}$$

Upon checking, only $a=7$ checks in the original equation.

47. Solving the equation:

$$\begin{aligned}21 &= 6x\\ x &= \frac{21}{6}=\frac{7}{2}\end{aligned}$$

49. Solving the equation:

$$\begin{aligned}x^2+x &= 6\\ x^2+x-6 &= 0\\ (x+3)(x-2) &= 0\\ x &= -3,2\end{aligned}$$

51. Multiplying both sides of the equation by $2(x+3)$:

$$\begin{aligned}2(x+3)\left(\frac{2x}{x+3}-\frac{1}{2}\right) &= 2(x+3)\bullet\frac{6}{x+3}\\ 2(2x)-1(x+3) &= 2(6)\\ 4x-x-3 &= 12\\ 3x-3 &= 12\\ 3x &= 15\\ x &= 5\end{aligned}$$

Since $x=5$ checks in the original equation, the solution is $x=5$. The correct answer is a.

7.6 Proportions

1. Solving the proportion:

$$\frac{x}{2}=\frac{6}{12}$$
$$12x=12$$
$$x=1$$

3. Solving the proportion:

$$\frac{2}{5}=\frac{4}{x}$$
$$2x=20$$
$$x=10$$

5. Solving the proportion:

$$\frac{10}{20}=\frac{20}{x}$$
$$10x=400$$
$$x=40$$

7. Solving the proportion:

$$\frac{a}{3}=\frac{5}{12}$$
$$12a=15$$
$$a=\frac{15}{12}=\frac{5}{4}$$

9. Solving the proportion:

$$\frac{2}{x}=\frac{6}{7}$$
$$6x=14$$
$$x=\frac{14}{6}=\frac{7}{3}$$

11. Solving the proportion:

$$\frac{x+1}{3}=\frac{4}{x}$$
$$x^2+x=12$$
$$x^2+x-12=0$$
$$(x+4)(x-3)=0$$
$$x=-4,3$$

13. Solving the proportion:

$$\frac{x}{2}=\frac{8}{x}$$
$$x^2=16$$
$$x^2-16=0$$
$$(x+4)(x-4)=0$$
$$x=-4,4$$

15. Solving the proportion:

$$\frac{4}{a+2}=\frac{a}{2}$$
$$a^2+2a=8$$
$$a^2+2a-8=0$$
$$(a+4)(a-2)=0$$
$$a=-4,2$$

17. Solving the proportion:

$$\frac{1}{x}=\frac{x-5}{6}$$
$$x^2-5x=6$$
$$x^2-5x-6=0$$
$$(x-6)(x+1)=0$$
$$x=-1,6$$

19. Solving the proportion:

$$\frac{26}{x-3}=\frac{38}{x+3}$$
$$26(x+3)=38(x-3)$$
$$26x+78=38x-114$$
$$-12x=-192$$
$$x=16$$

21. Comparing hits to games, the proportion is:

$$\frac{6}{18}=\frac{x}{45}$$
$$18x=270$$
$$x=15$$

He will get 15 hits in 45 games.

23. Comparing ml alcohol to ml water, the proportion is:

$$\frac{12}{16}=\frac{x}{28}$$
$$16x=336$$
$$x=21$$

The solution will have 21 ml of alcohol.

25. Comparing grams of fat to total grams, the proportion is:

$$\frac{13}{100}=\frac{x}{350}$$
$$100x=4550$$
$$x=45.5$$

There are 45.5 grams of fat in 350 grams of ice cream.

27. Comparing inches on the map to actual miles, the proportion is:

$$\frac{3.5}{100}=\frac{x}{420}$$
$$100x=1470$$
$$x=14.7$$

They are 14.7 inches apart on the map.

29. Comparing miles to hours, the proportion is:

$$\frac{245}{5}=\frac{x}{7}$$
$$5x=1715$$
$$x=343$$

He will travel 343 miles.

31. Multiplying both sides of the equation by $2x$:

$$2x\left(\frac{1}{x}+\frac{1}{2x}\right)=2x\left(\frac{9}{2}\right)$$
$$2+1=9x$$
$$9x=3$$
$$x=\frac{1}{3}$$

33. Multiplying both sides of the equation by $30x$:

$$30x\left(\frac{1}{10}-\frac{1}{15}\right)=30x\left(\frac{1}{x}\right)$$
$$3x-2x=30$$
$$x=30$$

35. Evaluating when $x=-6$: $y=-\frac{6}{-6}=1$

37. Evaluating when $x=2$: $y=-\frac{6}{2}=-3$

39. Solving the proportion:

$$\frac{4}{x+3}=\frac{x}{10}$$
$$40=x^2+3x$$
$$x^2+3x-40=0$$
$$(x+8)(x-5)=0$$
$$x=-8,5$$

The correct answer is d.

7.7 Applications

1. Let x and $3x$ represent the two numbers. The equation is:

$$\frac{1}{x}+\frac{1}{3x}=\frac{16}{3}$$
$$3x\left(\frac{1}{x}+\frac{1}{3x}\right)=3x\left(\frac{16}{3}\right)$$
$$3+1=16x$$
$$16x=4$$
$$x=\frac{1}{4}$$
$$3x=\frac{3}{4}$$

The numbers are $\frac{1}{4}$ and $\frac{3}{4}$.

3. Let x represent the number. The equation is:

$$x+\frac{1}{x}=\frac{13}{6}$$
$$6x\left(x+\frac{1}{x}\right)=6x\left(\frac{13}{6}\right)$$
$$6x^2+6=13x$$
$$6x^2-13x+6=0$$
$$(3x-2)(2x-3)=0$$
$$x=\frac{2}{3},\frac{3}{2}$$

The number is either $\frac{2}{3}$ or $\frac{3}{2}$.

5. Let x represent the number. The equation is:

$$\frac{7+x}{9+x}=\frac{5}{7}$$
$$7(9+x)\bullet\frac{7+x}{9+x}=7(9+x)\bullet\frac{5}{7}$$
$$7(7+x)=5(9+x)$$
$$49+7x=45+5x$$
$$49+2x=45$$
$$2x=-4$$
$$x=-2$$

The number is –2.

7. Let x and $x + 2$ represent the two integers. The equation is:

$$\frac{1}{x}+\frac{1}{x+2}=\frac{5}{12}$$
$$12x(x+2)\left(\frac{1}{x}+\frac{1}{x+2}\right)=12x(x+2)\left(\frac{5}{12}\right)$$
$$12(x+2)+12x=5x(x+2)$$
$$12x+24+12x=5x^2+10x$$
$$0=5x^2-14x-24$$
$$(5x+6)(x-4)=0$$
$$x=4 \quad \left(x=-\frac{6}{5}\text{ is impossible}\right)$$
$$x+2=6$$

The integers are 4 and 6.

9. Let x represent the rate of the boat in still water:

	d	r	t
Upstream	26	$x-3$	$\frac{26}{x-3}$
Downstream	38	$x+3$	$\frac{38}{x+3}$

The equation is:

$$\frac{26}{x-3}=\frac{38}{x+3}$$
$$(x+3)(x-3)\bullet\frac{26}{x-3}=(x+3)(x-3)\bullet\frac{38}{x+3}$$
$$26(x+3)=38(x-3)$$
$$26x+78=38x-114$$
$$-12x+78=-114$$
$$-12x=-192$$
$$x=16$$

The speed of the boat in still water is 16 mph.

11. Let x represent the plane speed in still air:

	d	r	t
Against Wind	140	$x-20$	$\frac{140}{x-20}$
With Wind	160	$x+20$	$\frac{160}{x+20}$

The equation is:

$$\frac{140}{x-20}=\frac{160}{x+20}$$

$$(x+20)(x-20)\cdot\frac{140}{x-20}=(x+20)(x-20)\cdot\frac{160}{x+20}$$

$$140(x+20)=160(x-20)$$

$$140x+2800=160x-3200$$

$$-20x+2800=-3200$$

$$-20x=-6000$$

$$x=300$$

The plane speed in still air is 300 mph.

13. Let x and $x + 20$ represent the rates of each plane:

	d	r	t
Plane 1	285	$x+20$	$\frac{285}{x+20}$
Plane 2	255	x	$\frac{255}{x}$

The equation is:

$$\frac{285}{x+20}=\frac{255}{x}$$

$$x(x+20)\cdot\frac{285}{x+20}=x(x+20)\cdot\frac{255}{x}$$

$$285x=255(x+20)$$

$$285x=255x+5100$$

$$30x=5100$$

$$x=170$$

$$x+20=190$$

The plane speeds are 170 mph and 190 mph.

15. Let x represent her rate downhill:

	d	r	t
Level Ground	2	$x-3$	$\frac{2}{x-3}$
Downhill	6	x	$\frac{6}{x}$

The equation is:

$$\begin{aligned}\frac{2}{x-3}+\frac{6}{x}&=1\\ x(x-3)\left(\frac{2}{x-3}+\frac{6}{x}\right)&=x(x-3)\cdot 1\\ 2x+6(x-3)&=x(x-3)\\ 2x+6x-18&=x^2-3x\\ 8x-18&=x^2-3x\\ 0&=x^2-11x+18\\ 0&=(x-2)(x-9)\\ x&=9 \quad (x=2 \text{ is impossible})\end{aligned}$$

Tina runs 9 mph on the downhill part of the course.

17. Let x represent her rate on level ground:

	d	r	t
Level Ground	4	x	$\frac{4}{x}$
Downhill	5	$x+2$	$\frac{5}{x+2}$

The equation is:

$$\begin{aligned}\frac{4}{x}+\frac{5}{x+2}&=1\\ x(x+2)\left(\frac{4}{x}+\frac{5}{x+2}\right)&=x(x+2)\cdot 1\\ 4(x+2)+5x&=x(x+2)\\ 4x+8+5x&=x^2+2x\\ 9x+8&=x^2+2x\\ 0&=x^2-7x-8\\ 0&=(x-8)(x+1)\\ x&=8 \quad (x=-1 \text{ is impossible})\end{aligned}$$

Jerri jogs 8 mph on level ground.

19. Let t represent the time to wax the family car with both Jason and Kevin working together. The equation is:

$$\begin{aligned}\frac{1}{90}+\frac{1}{60}&=\frac{1}{t}\\ 180t\left(\frac{1}{90}+\frac{1}{60}\right)&=180t\cdot\frac{1}{t}\\ 2t+3t&=180\\ 5t&=180\\ t&=36\end{aligned}$$

It would take 36 minutes to wax the car with both Jason and Kevin working together.

21. Let t represent the time for Peggy to install the car stereo alone. The equation is:

$$\frac{1}{45}+\frac{1}{t}=\frac{1}{30}$$
$$90t\left(\frac{1}{45}+\frac{1}{t}\right)=90t\cdot\frac{1}{30}$$
$$2t+90=3t$$
$$90=t$$

It would take 90 minutes for Peggy to install the car stereo working alone.

23. Let t represent the time to fill the pool with both pipes left open. The equation is:

$$\frac{1}{12}-\frac{1}{15}=\frac{1}{t}$$
$$60t\left(\frac{1}{12}-\frac{1}{15}\right)=60t\cdot\frac{1}{t}$$
$$5t-4t=60$$
$$t=60$$

It will take 60 hours to fill the pool with both pipes left open.

25. Let t represent the time to fill the bathtub with both faucets open. The equation is:

$$\frac{1}{10}+\frac{1}{12}=\frac{1}{t}$$
$$60t\left(\frac{1}{10}+\frac{1}{12}\right)=60t\cdot\frac{1}{t}$$
$$6t+5t=60$$
$$11t=60$$
$$t=\frac{60}{11}=5\frac{5}{11}$$

It will take $5\frac{5}{11}$ minutes to fill the tub with both faucets open.

27. Let t represent the time to fill the sink with both the faucet and the drain left open. The equation is:

$$\frac{1}{3}-\frac{1}{4}=\frac{1}{t}$$
$$12t\left(\frac{1}{3}-\frac{1}{4}\right)=12t\cdot\frac{1}{t}$$
$$4t-3t=12$$
$$t=12$$

It will take 12 minutes for the sink to overflow with both the faucet and drain left open.

29. Substituting $x=4$: $y=5(4)=20$

31. Substituting $x=10$: $y=\frac{20}{10}=2$

33. Substituting $y=50$:

$$50=2x^2$$
$$2x^2-50=0$$
$$2\left(x^2-25\right)=0$$
$$2(x+5)(x-5)=0$$
$$x=-5,5$$

35. Substituting $y = 15$ and $x = 3$:

$$15 = K(3)$$
$$K = 5$$

37. Substituting $y = 32$ and $x = 4$:

$$32 = K(4)^2$$
$$32 = 16K$$
$$K = 2$$

39. Let x represent the number. The equation is: $x+\frac{2}{x}=\frac{41}{12}$. The correct answer is c.

41. Let x represent the time to fill the hot tub with both the hose and inlet pipe. The equation is: $\frac{1}{10}+\frac{1}{8}=\frac{1}{x}$

The correct answer is d.

7.8 Variation

1. This is a direct variation with constant $K = 10$.

3. This is an inverse variation with constant $K = 40$.

5. This is a direct variation with constant $K = \frac{4}{3}\pi$.

7. This is an inverse variation with constant $K = \frac{1}{9}$.

9. The variation equation is $C = Kr^2$.

11. The variation equation is $P = \frac{K}{V}$.

13. The variation equation is $R = K\sqrt{n}$.

15. The variation equation is $y = Kx$. Finding K:

$$10 = K \cdot 5$$
$$K = 2$$

So $y = 2x$. Substituting $x = 4$: $y = 2 \cdot 4 = 8$

17. The variation equation is $y = Kx$. Finding K:

$$30 = K \cdot (-15)$$
$$K = -2$$

So $y = -2x$. Substituting $y = 8$:

$$-2x = 8$$
$$x = -4$$

19. The variation equation is $y = \frac{K}{x}$. Finding K:

$$5 = \frac{K}{2}$$
$$K = 10$$

So $y = \frac{10}{x}$. Substituting $x = 5$: $y = \frac{10}{5} = 2$

21. The variation equation is $y = \frac{K}{x}$. Finding K:

$$5 = \frac{K}{3}$$
$$K = 15$$

So $y = \frac{15}{x}$. Substituting $y = 15$:

$$\frac{15}{x} = 15$$
$$15 = 15x$$
$$x = 1$$

23. The variation equation is $y = Kx^2$. Finding K:

$$75 = K \cdot 5^2$$
$$75 = 25K$$
$$K = 3$$

So $y = 3x^2$. Substituting $x = 1$: $y = 3 \cdot 1^2 = 3 \cdot 1 = 3$

25. The variation equation is $z = \frac{K}{w^2}$. Finding K:

$$5 = \frac{K}{2^2}$$
$$5 = \frac{K}{4}$$
$$K = 20$$

So $z = \frac{20}{w^2}$. Substituting $w = 6$: $z = \frac{20}{6^2} = \frac{20}{56} = \frac{5}{9}$

27. The variation equation is $F = K\sqrt{h}$. Finding K:

$$24 = K\sqrt{4}$$
$$24 = 2K$$
$$K = 12$$

So $F = 12\sqrt{h}$. Substituting $h = 25$: $F = 12\sqrt{25} = 12 \cdot 5 = 60$

29. The variation equation is $t = Kd$. Finding K:

$$42 = K \cdot 2$$
$$K = 21$$

So $t = 21d$. Substituting $d = 4$: $t = 21 \cdot 4 = 84$ pounds

31. The variation equation is $P = KI^2$. Finding K:

$$30 = K \cdot 2^2$$
$$30 = 4K$$
$$K = \frac{15}{2}$$

So $P = \frac{15}{2}I^2$. Substituting $I = 7$: $P = \frac{15}{2} \cdot 7^2 = \frac{15}{2} \cdot 49 = \frac{735}{2} = 367.5$

33. The variation equation is $M = Kh$. Finding K:

$$157 = K \cdot 20$$
$$K = 7.85$$

So $M = 7.85h$. Substituting $h = 30$: $M = 7.85 \cdot 30 = \$235.50$

35. The variation equation is $F = \frac{K}{d^2}$. Finding K:

$$150 = \frac{K}{4000^2}$$
$$150 = \frac{K}{1.6 \times 10^7}$$
$$K = 2.4 \times 10^9$$

So $F = \frac{2.4 \times 10^9}{d^2}$. Substituting $d = 5000$: $F = \frac{2.4 \times 10^9}{(5000)^2} = \frac{2.4 \times 10^9}{2.5 \times 10^7} = 96$ pounds

37. The variation equation is $I=\frac{K}{R}$. Finding K:

$$30=\frac{K}{2}$$
$$K=60$$

So $I=\frac{60}{R}$. Substituting $R=5$: $I=\frac{60}{5}=12$ amperes

39. Reducing the rational expression: $\frac{x^2-x-6}{x^2-9}=\frac{(x+2)(x-3)}{(x+3)(x-3)}=\frac{x+2}{x+3}$

41. Performing the operations: $\frac{x^2-25}{x+4}\cdot\frac{2x+8}{x^2-9x+20}=\frac{(x+5)(x-5)}{x+4}\cdot\frac{2(x+4)}{(x-4)(x-5)}=\frac{2(x+5)(x-5)(x+4)}{(x+4)(x-4)(x-5)}=\frac{2(x+5)}{x-4}$

43. Performing the operations: $\frac{x}{x^2-16}+\frac{4}{x^2-16}=\frac{x+4}{x^2-16}=\frac{x+4}{(x+4)(x-4)}=\frac{1}{x-4}$

45. Simplifying the complex fraction: $\frac{1-\frac{25}{x^2}}{1-\frac{8}{x}+\frac{15}{x^2}}=\frac{\left(1-\frac{25}{x^2}\right)\cdot x^2}{\left(1-\frac{8}{x}+\frac{15}{x^2}\right)\cdot x^2}=\frac{x^2-25}{x^2-8x+15}=\frac{(x+5)(x-5)}{(x-5)(x-3)}=\frac{x+5}{x-3}$

47. Multiplying each side of the equation by $x^2-9=(x+3)(x-3)$:

$$(x+3)(x-3)\left(\frac{x}{(x+3)(x-3)}-\frac{3}{x-3}\right)=(x+3)(x-3)\cdot\frac{1}{x+3}$$
$$x-3(x+3)=x-3$$
$$x-3x-9=x-3$$
$$-2x-9=x-3$$
$$-3x=6$$
$$x=-2$$

Since $x=-2$ checks in the original equation, the solution is $x=-2$.

49. Let t represent the time to fill the pool with both pipes open. The equation is:

$$\frac{1}{8}-\frac{1}{12}=\frac{1}{t}$$
$$24t\left(\frac{1}{8}-\frac{1}{12}\right)=24t\cdot\frac{1}{t}$$
$$3t-2t=24$$
$$t=24$$

It will take 24 hours to fill the pool with both pipes open.

51. The variation equation is $y=Kx$. Finding K:

$$8=K\cdot 12$$
$$K=\frac{2}{3}$$

So $y=\frac{2}{3}x$. Substituting $x=36$: $y=\frac{2}{3}(36)=24$

53. The variation equation is $f=\frac{K}{h^2}$. The correct answer is b.

55. The variation equation is $I=\frac{K}{R}$. Finding K:

$$36=\frac{K}{2}$$
$$K=72$$

So $I=\frac{72}{R}$. Substituting $R=9$: $I=\frac{72}{9}=8$ amperes. The correct answer is a.

Chapter 7 Test

1. Evaluating for $x=-2$: $\frac{2x+5}{3x^2-2x-1}=\frac{2(-2)+5}{3(-2)^2-2(-2)-1}=\frac{-4+5}{12+4-1}=\frac{1}{15}$

2. Evaluating for $x=-2$: $\frac{x^2-4}{x^2+4}=\frac{(-2)^2-4}{(-2)^2+4}=\frac{4-4}{4+4}=\frac{0}{8}=0$

3. The denominator is 0 when $x=5$.

4. Setting the denominator equal to 0:

$$x^2+x-12=0$$
$$(x+4)(x-3)=0$$
$$x=-4,3$$

5. Reducing the rational expression: $\frac{x^2-9}{x^2-6x+9}=\frac{(x+3)(x-3)}{(x-3)^2}=\frac{x+3}{x-3}$

6. Reducing the rational expression: $\frac{15a+30}{5a^2-10a-40}=\frac{15(a+2)}{5(a^2-2a-8)}=\frac{15(a+2)}{5(a+2)(a-4)}=\frac{3}{a-4}$

7. Performing the operations: $\frac{2x-6}{3}\bullet\frac{9}{4x-12}=\frac{2(x-3)}{3}\bullet\frac{9}{4(x-3)}=\frac{18(x-3)}{12(x-3)}=\frac{3}{2}$

8. Performing the operations:

$$\frac{x^2-9}{x-4}\div\frac{x+3}{x^2-16}=\frac{x^2-9}{x-4}\bullet\frac{x^2-16}{x+3}=\frac{(x+3)(x-3)}{x-4}\bullet\frac{(x+4)(x-4)}{x+3}=\frac{(x+3)(x-3)(x+4)(x-4)}{(x+3)(x-4)}=(x-3)(x+4)$$

9. Performing the operations:

$$\begin{aligned}\frac{x^2+x-6}{x^2+4x+3}\div\frac{x^2+2x-8}{2x^2-x-3}&=\frac{x^2+x-6}{x^2+4x+3}\bullet\frac{2x^2-x-3}{x^2+2x-8}\\&=\frac{(x-2)(x+3)}{(x+1)(x+3)}\bullet\frac{(2x-3)(x+1)}{(x+4)(x-2)}\\&=\frac{(x-2)(x+3)(2x-3)(x+1)}{(x+1)(x+3)(x+4)(x-2)}\\&=\frac{2x-3}{x+4}\end{aligned}$$

10. Performing the operations: $(x^2-16)\left(\frac{x-1}{x-4}\right)=\frac{(x+4)(x-4)}{1}\bullet\frac{x-1}{x-4}=\frac{(x+4)(x-4)(x-1)}{x-4}=(x+4)(x-1)$

11. Performing the operations: $\frac{7}{x-8}-\frac{9}{x-8}=\frac{7-9}{x-8}=\frac{-2}{x-8}$

12. Performing the operations:

$$\frac{x}{x^2-16}+\frac{3}{3x-12}=\frac{x}{(x+4)(x-4)}+\frac{3}{3(x-4)}$$
$$=\frac{x}{(x+4)(x-4)}+\frac{1\bullet(x+4)}{(x-4)(x+4)}$$
$$=\frac{x}{(x+4)(x-4)}+\frac{x+4}{(x+4)(x-4)}$$
$$=\frac{2x+4}{(x+4)(x-4)}$$
$$=\frac{2(x+2)}{(x+4)(x-4)}$$

13. Performing the operations:

$$\frac{3}{(x-3)(x+3)}-\frac{1}{(x-3)(x-1)}=\frac{3\bullet(x-1)}{(x-3)(x+3)(x-1)}-\frac{1\bullet(x+3)}{(x-3)(x-1)(x+3)}$$
$$=\frac{3x-3}{(x-3)(x+3)(x-1)}-\frac{x+3}{(x-3)(x+3)(x-1)}$$
$$=\frac{2x-6}{(x-3)(x+3)(x-1)}$$
$$=\frac{2(x-3)}{(x-3)(x+3)(x-1)}$$
$$=\frac{2}{(x+3)(x-1)}$$

14. Simplifying the complex fraction: $\dfrac{1+\frac{2}{x}}{1-\frac{2}{x}}=\dfrac{\left(1+\frac{2}{x}\right)\bullet x}{\left(1-\frac{2}{x}\right)\bullet x}=\dfrac{x+2}{x-2}$

15. Simplifying the complex fraction: $\dfrac{1-\frac{9}{x^2}}{1-\frac{1}{x}-\frac{6}{x^2}}=\dfrac{\left(1-\frac{9}{x^2}\right)\bullet x^2}{\left(1-\frac{1}{x}-\frac{6}{x^2}\right)\bullet x^2}=\dfrac{x^2-9}{x^2-x-6}=\dfrac{(x+3)(x-3)}{(x-3)(x+2)}=\dfrac{x+3}{x+2}$

16. Multiplying both sides of the equation by 35:

$$35\bullet\frac{3}{5}=35\bullet\frac{x+3}{7}$$
$$21=5(x+3)$$
$$21=5x+15$$
$$6=5x$$
$$x=\frac{6}{5}$$

Since $x=\frac{6}{5}$ checks in the original equation, the solution is $x=\frac{6}{5}$.

17. Multiplying both sides of the equation by $x(x-3)$:

$$x(x-3)\cdot\frac{25}{x-3}=x(x-3)\cdot\frac{7}{x}$$
$$25x=7(x-3)$$
$$25x=7x-21$$
$$18x=-21$$
$$x=-\frac{7}{6}$$

Since $x=-\frac{7}{6}$ checks in the original equation, the solution is $x=-\frac{7}{6}$.

18. Multiplying both sides of the equation by $x^2-2x-3=(x-3)(x+1)$:

$$(x-3)(x+1)\left(\frac{6}{x-3}-\frac{5}{x+1}\right)=(x-3)(x+1)\cdot\frac{7}{(x-3)(x+1)}$$
$$6(x+1)-5(x-3)=7$$
$$6x+6-5x+15=7$$
$$x+21=7$$
$$x=-14$$

Since $x=-14$ checks in the original equation, the solution is $x=-14$.

19. The ratio of alcohol to water is given by: $\frac{29\text{ ml}}{87\text{ ml}}=\frac{1}{3}$

The ratio of alcohol to total volume is given by: $\frac{29\text{ ml}}{116\text{ ml}}=\frac{1}{4}$

20. Comparing defective parts to total parts, the proportion is:

$$\frac{6}{150}=\frac{x}{2{,}550}$$
$$150x=15{,}300$$
$$x=102$$

The machine can be expected to produce 102 defective parts.

21. Let x represent the speed of the boat in still water. Completing the table:

	r	t	d
Upstream	$x-3$	3	$3(x-3)$
Downstream	$x+3$	2	$2(x+3)$

The equation is:

$$3(x-3)=2(x+3)$$
$$3x-9=2x+6$$
$$x=15$$

The speed of the boat in still water is 15 mph.

22. Let t represent the time to empty the pool with both pipes open. The equation is:

$$\frac{1}{8}-\frac{1}{12}=\frac{1}{t}$$
$$24t\left(\frac{1}{8}-\frac{1}{12}\right)=24t\cdot\frac{1}{t}$$
$$3t-2t=24$$
$$t=24$$

It will take 24 hours to empty the pool with both pipes open.

23. The variation equation is $y = Kx^3$. Finding *K*:

$$16 = K \cdot 2^3$$
$$16 = 8K$$
$$K = 2$$

So $y = 2x^3$. Substituting $x = 3$: $y = 2 \cdot 3^3 = 2 \cdot 27 = 54$

24. The variation equation is $y = \frac{K}{x^2}$. Finding *K*:

$$8 = \frac{K}{3^2}$$
$$K = 8 \cdot 9 = 72$$

So $y = \frac{72}{x^2}$. Substituting $x = 6$: $y = \frac{72}{6^2} = \frac{72}{36} = 2$

Chapter 8
Roots and Radical Expressions

8.1 Definitions and Common Roots

1. Finding the root: $\sqrt{4}=2$
3. Finding the root: $-\sqrt{4}=-2$
5. Finding the root: $\sqrt{-16}$ is not a real number
7. Finding the root: $-\sqrt{144}=-12$
9. Finding the root: $\sqrt{625}=25$
11. Finding the root: $\sqrt{-25}$ is not a real number
13. Finding the root: $-\sqrt{64}=-8$
15. Finding the root: $-\sqrt{100}=-10$
17. Finding the root: $\sqrt{\frac{1}{4}}=\frac{1}{2}$
19. Finding the root: $-\sqrt{\frac{1}{100}}=-\frac{1}{10}$
21. Finding the root: $\sqrt{\frac{9}{49}}=\frac{3}{7}$
23. Finding the root: $-\sqrt{\frac{4}{121}}=-\frac{2}{11}$
25. Finding the root: $-\sqrt{1,225}=-35$
27. Finding the root: $\sqrt[3]{1}=1$
29. Finding the root: $\sqrt[3]{-8}=-2$
31. Finding the root: $-\sqrt[3]{125}=-5$
33. Finding the root: $\sqrt[3]{-1}=-1$
35. Finding the root: $\sqrt[3]{-27}=-3$
37. Finding the root: $\sqrt[3]{\frac{1}{64}}=\frac{1}{4}$
39. Finding the root: $-\sqrt[4]{16}=-2$
41. Finding the root: $\sqrt[4]{81}=3$
43. Finding the root: $\sqrt[5]{32}=2$
45. Simplifying the expression: $\sqrt{y^2}=y$
47. Simplifying the expression: $\sqrt{25x^2}=5x$
49. Simplifying the expression: $\sqrt{a^2b^2}=ab$
51. Simplifying the expression: $\sqrt{(a+b)^2}=a+b$
53. Simplifying the expression: $\sqrt{81x^2y^2}=9xy$
55. Simplifying the expression: $\sqrt[3]{x^3}=x$
57. Simplifying the expression: $\sqrt[3]{8x^3}=2x$
59. Simplifying the expression: $\sqrt{x^4}=x^2$
61. Simplifying the expression: $\sqrt{36a^6}=6a^3$
63. Simplifying the expression: $\sqrt{25a^8b^4}=5a^4b^2$
65. Approximating the expression: $\sqrt{2}\approx 1.414$
67. Approximating the expression: $-\sqrt{17}\approx -4.123$
69. a. Approximating the expression: $\sqrt{18}\approx 4.243$
    b. Approximating the expression: $3\sqrt{2}\approx 4.243$
71. a. Approximating the expression: $\sqrt{50}\approx 7.071$
    b. Approximating the expression: $5\sqrt{2}\approx 7.071$
73. a. Factoring the expression: $8+6x=2(4+3x)$
    b. Factoring the expression: $8+6\sqrt{3}=2(4+3\sqrt{3})$
75. Simplifying the expression: $\sqrt{9}+\sqrt{16}=3+4=7$
77. Simplifying the expression: $\sqrt{9+16}=\sqrt{25}=5$
79. Simplifying the expression: $\sqrt{144}+\sqrt{25}=12+5=17$
81. Simplifying the expression: $\sqrt{144+25}=\sqrt{169}=13$

83. **a.** Approximating the expression: $\frac{1+\sqrt{5}}{2} \approx \frac{1+2.236}{2} = \frac{3.236}{2} = 1.618$

b. Approximating the expression: $\frac{1-\sqrt{5}}{2} \approx \frac{1-2.236}{2} = \frac{-1.236}{2} = -0.618$

c. Approximating the expression: $\frac{1+\sqrt{5}}{2} + \frac{1-\sqrt{5}}{2} \approx 1.618 - 0.618 = 1$

85. **a.** Evaluating the root: $\sqrt{9} = 3$

b. Evaluating the root: $\sqrt{900} = 30$

c. Evaluating the root: $\sqrt{0.09} = 0.3$

87. Simplifying each expression:

$$\frac{5+\sqrt{49}}{2} = \frac{5+7}{2} = \frac{12}{2} = 6 \qquad \frac{5-\sqrt{49}}{2} = \frac{5-7}{2} = \frac{-2}{2} = -1$$

89. **a.** Simplifying the expression: $\frac{2\sqrt{3}}{10} = \frac{\sqrt{3}}{5}$

b. Simplifying the expression: $\frac{5\sqrt{2}}{15} = \frac{\sqrt{2}}{3}$

c. Simplifying the expression: $\frac{15+6\sqrt{3}}{12} = \frac{3\left(5+2\sqrt{3}\right)}{12} = \frac{5+2\sqrt{3}}{4}$

d. Simplifying the expression: $\frac{5+10\sqrt{6}}{5} = \frac{5\left(1+2\sqrt{6}\right)}{5} = 1+2\sqrt{6}$

91. Finding the hypotenuse: $x = \sqrt{3^2+4^2} = \sqrt{9+16} = \sqrt{25} = 5$

93. Finding the hypotenuse: $x = \sqrt{5^2+10^2} = \sqrt{25+100} = \sqrt{125} \approx 11.2$

95. Let x represent the length of the hypotenuse, then: $x = \sqrt{18^2+24^2} = \sqrt{324+576} = \sqrt{900} = 30$ feet

97. Drawing the figure:

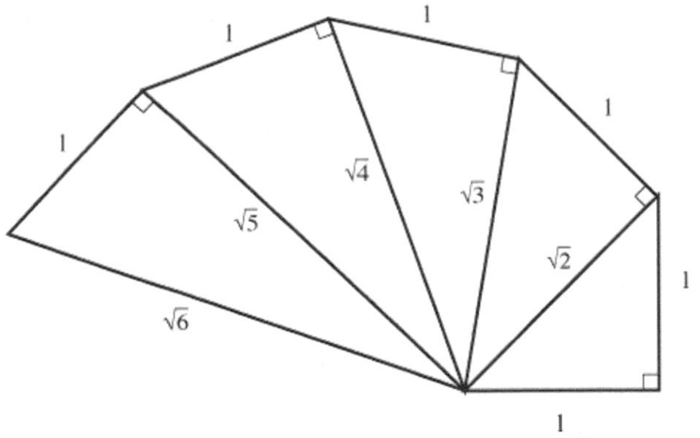

99. The sequence is: $\sqrt{1^2+1} = \sqrt{1+1} = \sqrt{2}$, $\sqrt{\left(\sqrt{2}\right)^2+1} = \sqrt{2+1} = \sqrt{3}$, $\sqrt{\left(\sqrt{3}\right)^2+1} = \sqrt{3+1} = \sqrt{4} = 2$, ...

101. Simplifying: $3 \cdot \sqrt{16} = 3 \cdot 4 = 12$

103. Factoring: $75 = 25 \cdot 3$

105. Factoring: $50 = 25 \cdot 2$

107. Factoring: $40 = 4 \cdot 10$

109. Factoring: $x^2 = x^2$

111. Factoring: $12x^2 = 4x^2 \cdot 3$

113. Factoring: $50x^3y^2 = 25x^2y^2 \cdot 2x$

115. Finding the root: $\sqrt{64} = 8$. The correct answer is b.

117. Simplifying the expression: $\sqrt{16x^2} = 4x$. The correct answer is a.

119. Let x represent the length of the hypotenuse, then: $x = \sqrt{1^2+3^2} = \sqrt{1+9} = \sqrt{10}$ meters. The correct answer is d.

8.2 Simplifying Form and Properties of Radicals

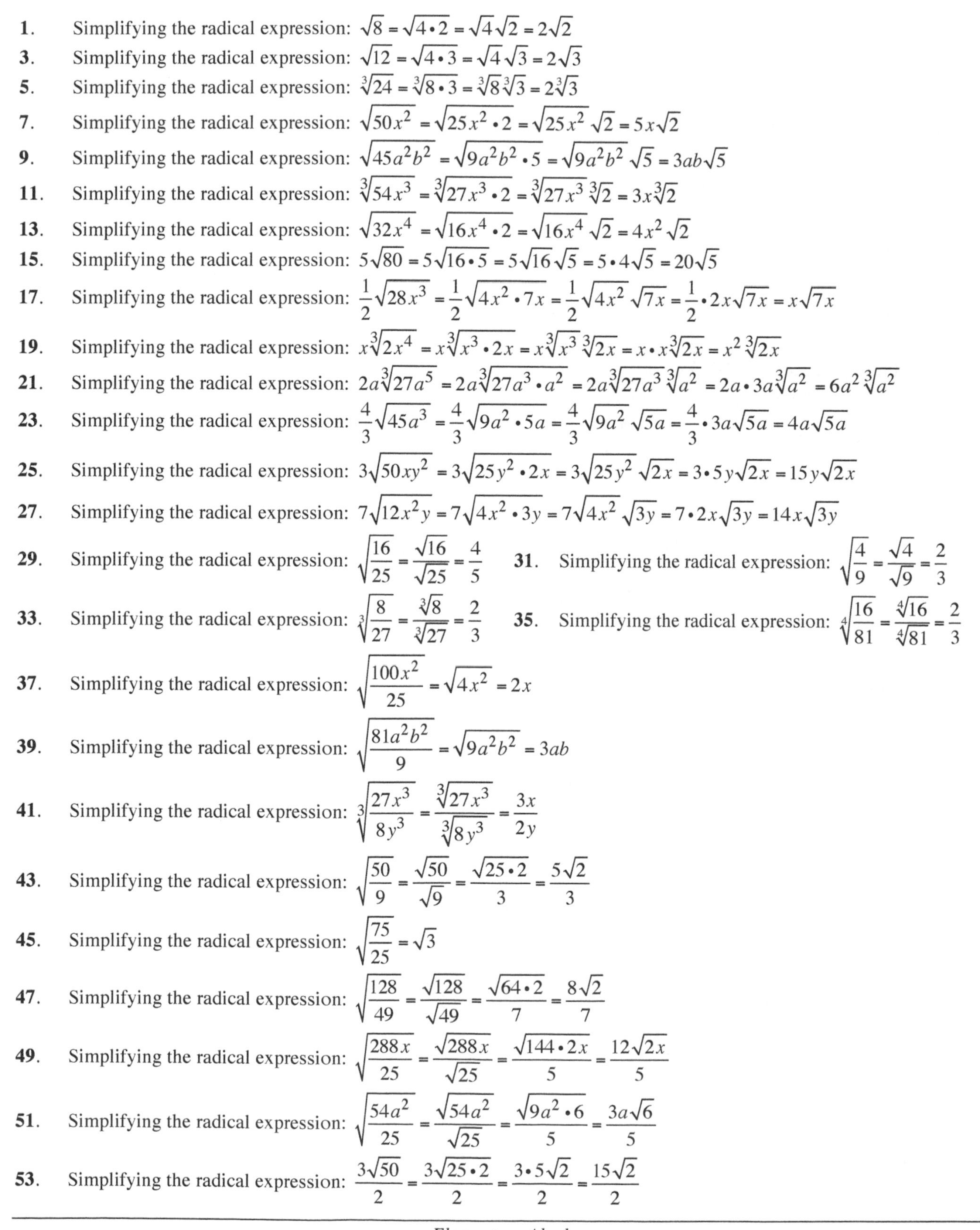

1. Simplifying the radical expression: $\sqrt{8}=\sqrt{4\cdot 2}=\sqrt{4}\sqrt{2}=2\sqrt{2}$
3. Simplifying the radical expression: $\sqrt{12}=\sqrt{4\cdot 3}=\sqrt{4}\sqrt{3}=2\sqrt{3}$
5. Simplifying the radical expression: $\sqrt[3]{24}=\sqrt[3]{8\cdot 3}=\sqrt[3]{8}\sqrt[3]{3}=2\sqrt[3]{3}$
7. Simplifying the radical expression: $\sqrt{50x^2}=\sqrt{25x^2\cdot 2}=\sqrt{25x^2}\sqrt{2}=5x\sqrt{2}$
9. Simplifying the radical expression: $\sqrt{45a^2b^2}=\sqrt{9a^2b^2\cdot 5}=\sqrt{9a^2b^2}\sqrt{5}=3ab\sqrt{5}$
11. Simplifying the radical expression: $\sqrt[3]{54x^3}=\sqrt[3]{27x^3\cdot 2}=\sqrt[3]{27x^3}\sqrt[3]{2}=3x\sqrt[3]{2}$
13. Simplifying the radical expression: $\sqrt{32x^4}=\sqrt{16x^4\cdot 2}=\sqrt{16x^4}\sqrt{2}=4x^2\sqrt{2}$
15. Simplifying the radical expression: $5\sqrt{80}=5\sqrt{16\cdot 5}=5\sqrt{16}\sqrt{5}=5\cdot 4\sqrt{5}=20\sqrt{5}$
17. Simplifying the radical expression: $\frac{1}{2}\sqrt{28x^3}=\frac{1}{2}\sqrt{4x^2\cdot 7x}=\frac{1}{2}\sqrt{4x^2}\sqrt{7x}=\frac{1}{2}\cdot 2x\sqrt{7x}=x\sqrt{7x}$
19. Simplifying the radical expression: $x\sqrt[3]{2x^4}=x\sqrt[3]{x^3\cdot 2x}=x\sqrt[3]{x^3}\sqrt[3]{2x}=x\cdot x\sqrt[3]{2x}=x^2\sqrt[3]{2x}$
21. Simplifying the radical expression: $2a\sqrt[3]{27a^5}=2a\sqrt[3]{27a^3\cdot a^2}=2a\sqrt[3]{27a^3}\sqrt[3]{a^2}=2a\cdot 3a\sqrt[3]{a^2}=6a^2\sqrt[3]{a^2}$
23. Simplifying the radical expression: $\frac{4}{3}\sqrt{45a^3}=\frac{4}{3}\sqrt{9a^2\cdot 5a}=\frac{4}{3}\sqrt{9a^2}\sqrt{5a}=\frac{4}{3}\cdot 3a\sqrt{5a}=4a\sqrt{5a}$
25. Simplifying the radical expression: $3\sqrt{50xy^2}=3\sqrt{25y^2\cdot 2x}=3\sqrt{25y^2}\sqrt{2x}=3\cdot 5y\sqrt{2x}=15y\sqrt{2x}$
27. Simplifying the radical expression: $7\sqrt{12x^2y}=7\sqrt{4x^2\cdot 3y}=7\sqrt{4x^2}\sqrt{3y}=7\cdot 2x\sqrt{3y}=14x\sqrt{3y}$
29. Simplifying the radical expression: $\sqrt{\frac{16}{25}}=\frac{\sqrt{16}}{\sqrt{25}}=\frac{4}{5}$
31. Simplifying the radical expression: $\sqrt{\frac{4}{9}}=\frac{\sqrt{4}}{\sqrt{9}}=\frac{2}{3}$
33. Simplifying the radical expression: $\sqrt[3]{\frac{8}{27}}=\frac{\sqrt[3]{8}}{\sqrt[3]{27}}=\frac{2}{3}$
35. Simplifying the radical expression: $\sqrt[4]{\frac{16}{81}}=\frac{\sqrt[4]{16}}{\sqrt[4]{81}}=\frac{2}{3}$
37. Simplifying the radical expression: $\sqrt{\frac{100x^2}{25}}=\sqrt{4x^2}=2x$
39. Simplifying the radical expression: $\sqrt{\frac{81a^2b^2}{9}}=\sqrt{9a^2b^2}=3ab$
41. Simplifying the radical expression: $\sqrt[3]{\frac{27x^3}{8y^3}}=\frac{\sqrt[3]{27x^3}}{\sqrt[3]{8y^3}}=\frac{3x}{2y}$
43. Simplifying the radical expression: $\sqrt{\frac{50}{9}}=\frac{\sqrt{50}}{\sqrt{9}}=\frac{\sqrt{25\cdot 2}}{3}=\frac{5\sqrt{2}}{3}$
45. Simplifying the radical expression: $\sqrt{\frac{75}{25}}=\sqrt{3}$
47. Simplifying the radical expression: $\sqrt{\frac{128}{49}}=\frac{\sqrt{128}}{\sqrt{49}}=\frac{\sqrt{64\cdot 2}}{7}=\frac{8\sqrt{2}}{7}$
49. Simplifying the radical expression: $\sqrt{\frac{288x}{25}}=\frac{\sqrt{288x}}{\sqrt{25}}=\frac{\sqrt{144\cdot 2x}}{5}=\frac{12\sqrt{2x}}{5}$
51. Simplifying the radical expression: $\sqrt{\frac{54a^2}{25}}=\frac{\sqrt{54a^2}}{\sqrt{25}}=\frac{\sqrt{9a^2\cdot 6}}{5}=\frac{3a\sqrt{6}}{5}$
53. Simplifying the radical expression: $\frac{3\sqrt{50}}{2}=\frac{3\sqrt{25\cdot 2}}{2}=\frac{3\cdot 5\sqrt{2}}{2}=\frac{15\sqrt{2}}{2}$

55. Simplifying the radical expression: $\frac{7\sqrt{28y^2}}{3}=\frac{7\sqrt{4y^2\cdot 7}}{3}=\frac{7\cdot 2y\sqrt{7}}{3}=\frac{14y\sqrt{7}}{3}$

57. Simplifying the radical expression: $\frac{2\sqrt{20x^2y^3}}{3}=\frac{2\sqrt{4x^2y^2\cdot 5y}}{3}=\frac{2\cdot 2xy\sqrt{5y}}{3}=\frac{4xy\sqrt{5y}}{3}$

59. Simplifying the radical expression: $\frac{6\sqrt{54a^2b^3}}{5}=\frac{6\sqrt{9a^2b^2\cdot 6b}}{5}=\frac{6\cdot 3ab\sqrt{6b}}{5}=\frac{18ab\sqrt{6b}}{5}$

61. Simplifying the radical expression: $\frac{5\sqrt{72a^2b^2}}{\sqrt{36}}=\frac{5\sqrt{36a^2b^2\cdot 2}}{6}=\frac{5\cdot 6ab\sqrt{2}}{6}=\frac{30ab\sqrt{2}}{6}=5ab\sqrt{2}$

63. Simplifying the expression: $\frac{\sqrt{20}}{4}=\frac{\sqrt{4\cdot 5}}{4}=\frac{2\sqrt{5}}{4}=\frac{\sqrt{5}}{2}$

65. Simplifying the expression: $\frac{4+\sqrt{12}}{2}=\frac{4+2\sqrt{3}}{2}=\frac{2(2+\sqrt{3})}{2}=2+\sqrt{3}$

67. Simplifying the expression: $\frac{\sqrt{12}}{4}=\frac{\sqrt{4\cdot 3}}{4}=\frac{2\sqrt{3}}{4}=\frac{\sqrt{3}}{2}$

69. Simplifying the expression: $\frac{9+\sqrt{27}}{3}=\frac{9+3\sqrt{3}}{3}=\frac{3(3+\sqrt{3})}{3}=3+\sqrt{3}$

71. Simplifying the expression: $\frac{8+\sqrt{28}}{6}=\frac{8+2\sqrt{7}}{6}=\frac{2(4+\sqrt{7})}{6}=\frac{4+\sqrt{7}}{3}$

73. Simplifying the expression: $\frac{10+\sqrt{75}}{5}=\frac{10+5\sqrt{3}}{5}=\frac{5(2+\sqrt{3})}{5}=2+\sqrt{3}$

75. Simplifying the expression: $\frac{-2-\sqrt{27}}{6}=\frac{-2-3\sqrt{3}}{6}$

77. Simplifying the expression: $\frac{-4-\sqrt{8}}{2}=\frac{-4-2\sqrt{2}}{2}=\frac{2(-2-\sqrt{2})}{2}=-2-\sqrt{2}$

79. **a.** Substituting the values: $\sqrt{b^2-4ac}=\sqrt{(4)^2-4(2)(-3)}=\sqrt{16+24}=\sqrt{40}=\sqrt{4\cdot 10}=2\sqrt{10}$

b. Substituting the values: $\sqrt{b^2-4ac}=\sqrt{(1)^2-4(1)(-6)}=\sqrt{1+24}=\sqrt{25}=5$

c. Substituting the values: $\sqrt{b^2-4ac}=\sqrt{(1)^2-4(1)(-11)}=\sqrt{1+44}=\sqrt{45}=\sqrt{9\cdot 5}=3\sqrt{5}$

d. Substituting the values: $\sqrt{b^2-4ac}=\sqrt{(6)^2-4(3)(2)}=\sqrt{36-24}=\sqrt{12}=\sqrt{4\cdot 3}=2\sqrt{3}$

81. **a.** Simplifying: $\sqrt{32x^{10}y^5}=\sqrt{16x^{10}y^4\cdot 2y}=\sqrt{16x^{10}y^4}\sqrt{2y}=4x^5y^2\sqrt{2y}$

b. Simplifying: $\sqrt[3]{32x^{10}y^5}=\sqrt[3]{8x^9y^3\cdot 4xy^2}=\sqrt[3]{8x^9y^3}\sqrt[3]{4xy^2}=2x^3y\sqrt[3]{4xy^2}$

c. Simplifying: $\sqrt[4]{32x^{10}y^5}=\sqrt[4]{16x^8y^4\cdot 2x^2y}=\sqrt[4]{16x^8y^4}\sqrt[4]{2x^2y}=2x^2y\sqrt[4]{2x^2y}$

d. Simplifying: $\sqrt[5]{32x^{10}y^5}=2x^2y$

83. **a.** Simplifying: $\sqrt{4}=2$

b. Simplifying: $\sqrt{0.04}=0.2$

c. Simplifying: $\sqrt{400}=20$

d. Simplifying: $\sqrt{0.0004}=0.02$

85. Completing the table:

x	$\sqrt{x}$	$2\sqrt{x}$	$\sqrt{4x}$
1	1	2	2
2	1.414	2.828	2.828
3	1.732	3.464	3.464
4	2	4	4

87. Completing the table:

x	$\sqrt{x}$	$3\sqrt{x}$	$\sqrt{9x}$
1	1	3	3
2	1.414	4.243	4.243
3	1.732	5.196	5.196
4	2	6	6

89. Substituting $h = 25$: $t = \sqrt{\frac{25}{16}} = \frac{5}{4}$ seconds

91. Approximating the expression: $\frac{1+\sqrt{5}}{2} \approx \frac{1+2.236}{2} = \frac{3.236}{2} = 1.618$

93. Simplifying the radical expression: $\sqrt{4x^3y^2} = \sqrt{4x^2y^2 \cdot x} = 2xy\sqrt{x}$

95. Simplifying the radical expression: $\sqrt{20} = \sqrt{4 \cdot 5} = \sqrt{4}\sqrt{5} = 2\sqrt{5}$

97. Simplifying the radical expression: $5\sqrt{12} = 5\sqrt{4 \cdot 3} = 5\sqrt{4}\sqrt{3} = 5 \cdot 2\sqrt{3} = 10\sqrt{3}$

99. Simplifying the radical expression: $\frac{3}{2}\sqrt{24} = \frac{3}{2}\sqrt{4 \cdot 6} = \frac{3}{2} \cdot 2\sqrt{6} = 3\sqrt{6}$

101. Combining like terms: $5x - 3x = 2x$

103. Combining like terms: $2x + 3y - 8x = 2x - 8x + 3y = -6x + 3y$

105. Simplifying the radical expression: $\sqrt{80} = \sqrt{16 \cdot 5} = \sqrt{16}\sqrt{5} = 4\sqrt{5}$. The correct answer is d.

107. Simplifying the radical expression: $\sqrt{\frac{60}{49}} = \frac{\sqrt{60}}{\sqrt{49}} = \frac{\sqrt{4}\sqrt{15}}{7} = \frac{2\sqrt{15}}{7}$. The correct answer is b.

8.3 Addition and Subtraction of Radical Expressions

1. Combining radicals: $3\sqrt{2} + 4\sqrt{2} = 7\sqrt{2}$

3. Combining radicals: $9\sqrt{5} - 7\sqrt{5} = 2\sqrt{5}$

5. Combining radicals: $\sqrt{3} + 6\sqrt{3} = 7\sqrt{3}$

7. Combining radicals: $\frac{5}{8}\sqrt{5} - \frac{3}{7}\sqrt{5} = \frac{35}{56}\sqrt{5} - \frac{24}{56}\sqrt{5} = \frac{11}{56}\sqrt{5}$

9. Combining radicals: $14\sqrt[3]{13} - \sqrt[3]{13} = 13\sqrt[3]{13}$

11. Combining radicals: $-3\sqrt[3]{10} + 9\sqrt[3]{10} = 6\sqrt[3]{10}$

13. Combining radicals: $5\sqrt{5} + \sqrt{5} = 6\sqrt{5}$

15. Combining radicals: $\sqrt{8} + 2\sqrt{2} = \sqrt{4 \cdot 2} + 2\sqrt{2} = 2\sqrt{2} + 2\sqrt{2} = 4\sqrt{2}$

17. Combining radicals: $3\sqrt{3} - \sqrt{27} = 3\sqrt{3} - \sqrt{9 \cdot 3} = 3\sqrt{3} - 3\sqrt{3} = 0$

19. Combining radicals: $5\sqrt{12} - 10\sqrt{48} = 5\sqrt{4 \cdot 3} - 10\sqrt{16 \cdot 3} = 5 \cdot 2\sqrt{3} - 10 \cdot 4\sqrt{3} = 10\sqrt{3} - 40\sqrt{3} = -30\sqrt{3}$

21. Combining radicals: $-\sqrt[3]{54} - \sqrt[3]{2} = -3\sqrt[3]{2} - \sqrt[3]{2} = -4\sqrt[3]{2}$

23. Combining radicals: $\frac{1}{5}\sqrt{75} - \frac{1}{2}\sqrt{12} = \frac{1}{5}\sqrt{25 \cdot 3} - \frac{1}{2}\sqrt{4 \cdot 3} = \frac{1}{5} \cdot 5\sqrt{3} - \frac{1}{2} \cdot 2\sqrt{3} = \sqrt{3} - \sqrt{3} = 0$

25. Combining radicals: $\frac{3}{4}\sqrt{8} + \frac{3}{10}\sqrt{75} = \frac{3}{4}\sqrt{4 \cdot 2} + \frac{3}{10}\sqrt{25 \cdot 3} = \frac{3}{4} \cdot 2\sqrt{2} + \frac{3}{10} \cdot 5\sqrt{3} = \frac{3}{2}\sqrt{2} + \frac{3}{2}\sqrt{3}$

27. Combining radicals: $\sqrt{27} - 2\sqrt{12} + \sqrt{3} = \sqrt{9 \cdot 3} - 2\sqrt{4 \cdot 3} + \sqrt{3} = 3\sqrt{3} - 4\sqrt{3} + \sqrt{3} = 0$

29. Combining radicals:

$$\frac{5}{6}\sqrt{72}-\frac{3}{8}\sqrt{8}+\frac{3}{10}\sqrt{50}=\frac{5}{6}\sqrt{36\cdot 2}-\frac{3}{8}\sqrt{4\cdot 2}+\frac{3}{10}\sqrt{25\cdot 2}$$
$$=\frac{5}{6}\cdot 6\sqrt{2}-\frac{3}{8}\cdot 2\sqrt{2}+\frac{3}{10}\cdot 5\sqrt{2}$$
$$=5\sqrt{2}-\frac{3}{4}\sqrt{2}+\frac{3}{2}\sqrt{2}$$
$$=\frac{20}{4}\sqrt{2}-\frac{3}{4}\sqrt{2}+\frac{6}{4}\sqrt{2}$$
$$=\frac{23}{4}\sqrt{2}$$

31. Combining radicals:

$$5\sqrt{7}+2\sqrt{28}-4\sqrt{63}=5\sqrt{7}+2\sqrt{4\cdot 7}-4\sqrt{9\cdot 7}=5\sqrt{7}+2\cdot 2\sqrt{7}-4\cdot 3\sqrt{7}=5\sqrt{7}+4\sqrt{7}-12\sqrt{7}=-3\sqrt{7}$$

33. Combining radicals:

$$5\sqrt[3]{16}-3\sqrt[3]{128}+\sqrt[3]{432}=5\sqrt[3]{8\cdot 2}-3\sqrt[3]{64\cdot 2}+\sqrt[3]{216\cdot 2}=5\cdot 2\sqrt[3]{2}-3\cdot 4\sqrt[3]{2}+6\sqrt[3]{2}=10\sqrt[3]{2}-12\sqrt[3]{2}+6\sqrt[3]{2}=4\sqrt[3]{2}$$

35. Combining radicals:

$$6\sqrt{24}-2\sqrt{12}+5\sqrt{27}=6\sqrt{4\cdot 6}-2\sqrt{4\cdot 3}+5\sqrt{9\cdot 3}$$
$$=6\cdot 2\sqrt{6}-2\cdot 2\sqrt{3}+5\cdot 3\sqrt{3}$$
$$=12\sqrt{6}-4\sqrt{3}+15\sqrt{3}$$
$$=12\sqrt{6}+11\sqrt{3}$$

37. Combining radicals:

$$6\sqrt{48}-\sqrt{72}-3\sqrt{300}=6\sqrt{16\cdot 3}-\sqrt{36\cdot 2}-3\sqrt{100\cdot 3}$$
$$=6\cdot 4\sqrt{3}-6\sqrt{2}-3\cdot 10\sqrt{3}$$
$$=24\sqrt{3}-6\sqrt{2}-30\sqrt{3}$$
$$=-6\sqrt{2}-6\sqrt{3}$$

39. Combining radicals: $\sqrt{x^3}+x\sqrt{x}=\sqrt{x^2\cdot x}+x\sqrt{x}=x\sqrt{x}+x\sqrt{x}=2x\sqrt{x}$

41. Combining radicals: $5\sqrt{3a^2}-a\sqrt{3}=5a\sqrt{3}-a\sqrt{3}=4a\sqrt{3}$

43. Combining radicals: $5\sqrt{8x^3}+x\sqrt{50x}=5\sqrt{4x^2\cdot 2x}+x\sqrt{25\cdot 2x}=5\cdot 2x\sqrt{2x}+x\cdot 5\sqrt{2x}=10x\sqrt{2x}+5x\sqrt{2x}=15x\sqrt{2x}$

45. Combining radicals:

$$3\sqrt{75x^3y}-2x\sqrt{3xy}=3\sqrt{25x^2\cdot 3xy}-2x\sqrt{3xy}=3\cdot 5x\sqrt{3xy}-2x\sqrt{3xy}=15x\sqrt{3xy}-2x\sqrt{3xy}=13x\sqrt{3xy}$$

47. Combining radicals: $\sqrt{20ab^2}-b\sqrt{45a}=\sqrt{4b^2\cdot 5a}-b\sqrt{9\cdot 5a}=2b\sqrt{5a}-3b\sqrt{5a}=-b\sqrt{5a}$

49. Combining radicals: $9\sqrt{18x^3}-2x\sqrt{48x}=9\sqrt{9x^2\cdot 2x}-2x\sqrt{16\cdot 3x}=9\cdot 3x\sqrt{2x}-2x\cdot 4\sqrt{3x}=27x\sqrt{2x}-8x\sqrt{3x}$

51. Combining radicals: $2\sqrt[3]{3x^3}+x\sqrt[3]{24}=2\sqrt[3]{x^3\cdot 3}+x\sqrt[3]{8\cdot 3}=2x\sqrt[3]{3}+2x\sqrt[3]{3}=4x\sqrt[3]{3}$

53. Combining radicals:

$$6\sqrt[3]{64a^5b^3}-5a\sqrt[3]{8a^2b^3}=6\sqrt[3]{64a^3b^3\cdot a^2}-5a\sqrt[3]{8b^3\cdot a^2}$$
$$=6\cdot 4ab\sqrt[3]{a^2}-5a\cdot 2b\sqrt[3]{a^2}$$
$$=24ab\sqrt[3]{a^2}-10ab\sqrt[3]{a^2}$$
$$=14ab\sqrt[3]{a^2}$$

55. Combining radicals:

$$7\sqrt{50x^2y}+8x\sqrt{8y}-7\sqrt{32x^2y}=7\sqrt{25x^2\cdot 2y}+8x\sqrt{4\cdot 2y}-7\sqrt{16x^2\cdot 2y}$$
$$=7\cdot 5x\sqrt{2y}+8x\cdot 2\sqrt{2y}-7\cdot 4x\sqrt{2y}$$
$$=35x\sqrt{2y}+16x\sqrt{2y}-28x\sqrt{2y}$$
$$=23x\sqrt{2y}$$

57. Simplifying the expression: $\frac{8-\sqrt{24}}{6}=\frac{8-\sqrt{4\cdot 6}}{6}=\frac{8-2\sqrt{6}}{6}=\frac{2(4-\sqrt{6})}{6}=\frac{4-\sqrt{6}}{3}$

59. Simplifying the expression: $\frac{6+\sqrt{8}}{2}=\frac{6+\sqrt{4\cdot 2}}{2}=\frac{6+2\sqrt{2}}{2}=\frac{2(3+\sqrt{2})}{2}=3+\sqrt{2}$

61. Simplifying the expression: $\frac{-10+\sqrt{50}}{10}=\frac{-10+\sqrt{25\cdot 2}}{10}=\frac{-10+5\sqrt{2}}{10}=\frac{5(-2+\sqrt{2})}{10}=\frac{-2+\sqrt{2}}{2}$

63. **a.** Simplifying by combining like terms: $3x+4x=7x$
b. Simplifying by combining like terms: $3y+4y=7y$
c. Simplifying by combining like terms: $3\sqrt{5}+4\sqrt{5}=7\sqrt{5}$

65. **a.** Simplifying by combining like terms: $x+6x=7x$
b. Simplifying by combining like terms: $t+6t=7t$
c. Simplifying by combining like terms: $\sqrt{x}+6\sqrt{x}=7\sqrt{x}$

67. Completing the table:

x	$\sqrt{x^2+9}$	$x+3$
1	3.162	4
2	3.606	5
3	4.243	6
4	5	7
5	5.831	8
6	6.708	9

69. Completing the table:

x	$\sqrt{x+3}$	$\sqrt{x}+\sqrt{3}$
1	2	2.732
2	2.236	3.146
3	2.449	3.464
4	2.646	3.732
5	2.828	3.968
6	3	4.182

71. The correct statement is: $4\sqrt{3}+5\sqrt{3}=9\sqrt{3}$

73. Simplifying the radical expression: $\sqrt{12}=\sqrt{4\cdot 3}=\sqrt{4}\sqrt{3}=2\sqrt{3}$

75. Simplifying the radical expression: $\sqrt{20}=\sqrt{4\cdot 5}=\sqrt{4}\sqrt{5}=2\sqrt{5}$

77. Simplifying the radical expression: $\sqrt[3]{16}=\sqrt[3]{8\cdot 2}=\sqrt[3]{8}\sqrt[3]{2}=2\sqrt[3]{2}$

79. Multiplying the expressions: $x(2x+5)=2x^2+5x$

81. Multiplying the expressions: $(x+1)^2=(x+1)(x+1)=x^2+2x+1$

83. Multiplying the expressions: $(x+8)(x-2)=x^2-2x+8x-16=x^2+6x-16$

85. Multiplying the expressions: $(x+6)(x-6)=x^2-(6)^2=x^2-36$

87. Multiplying the expressions: $(2x+3)(3x-4)=6x^2-8x+9x-12=6x^2+x-12$

89. Combining like terms: $5+7\sqrt{5}+2\sqrt{5}+14=(5+14)+(7\sqrt{5}+2\sqrt{5})=19+9\sqrt{5}$

91. Combining like terms: $x-7\sqrt{x}+3\sqrt{x}-21=x-21+(-7\sqrt{x}+3\sqrt{x})=x-4\sqrt{x}-21$

93. The correct answer is c.

8.4 Multiplication of Radicals

1. Simplifying: $\left(\sqrt{2}\right)^2=\left(\sqrt{2}\right)\left(\sqrt{2}\right)=2$

3. Simplifying: $\left(-\sqrt{7}\right)^2=\left(-\sqrt{7}\right)\left(-\sqrt{7}\right)=7$

5. Simplifying: $\left(\sqrt{x}\right)^2=\left(\sqrt{x}\right)\left(\sqrt{x}\right)=x$

7. Simplifying: $\left(\sqrt[3]{4}\right)^3=\left(\sqrt[3]{4}\right)\left(\sqrt[3]{4}\right)\left(\sqrt[3]{4}\right)=4$

9. Simplifying: $\left(\sqrt[3]{-10}\right)^3=\left(\sqrt[3]{-10}\right)\left(\sqrt[3]{-10}\right)\left(\sqrt[3]{-10}\right)=-10$

11. Simplifying: $\left(\sqrt[3]{a}\right)^3=\left(\sqrt[3]{a}\right)\left(\sqrt[3]{a}\right)\left(\sqrt[3]{a}\right)=a$

13. Multiplying the radicals: $\sqrt{3}\sqrt{2}=\sqrt{6}$

15. Multiplying the radicals: $\sqrt{6}\sqrt{2}=\sqrt{12}=\sqrt{4\bullet 3}=2\sqrt{3}$

17. Multiplying the radicals: $\left(2\sqrt{3}\right)\left(5\sqrt{7}\right)=10\sqrt{21}$

19. Multiplying the radicals: $\left(4\sqrt{3}\right)\left(2\sqrt{6}\right)=8\sqrt{18}=8\sqrt{9\bullet 2}=8\bullet 3\sqrt{2}=24\sqrt{2}$

21. Multiplying the radicals: $\left(\sqrt[3]{3}\right)\left(\sqrt[3]{4}\right)=\sqrt[3]{12}$

23. Multiplying the radicals: $\left(\sqrt[3]{6}\right)\left(\sqrt[3]{9}\right)=\sqrt[3]{54}=\sqrt[3]{27\bullet 2}=3\sqrt[3]{2}$

25. Multiplying the radicals: $\left(9\sqrt[3]{12}\right)\left(2\sqrt[3]{2}\right)=18\sqrt[3]{24}=18\sqrt[3]{8\bullet 3}=18\bullet 2\sqrt[3]{3}=36\sqrt[3]{3}$

27. Multiplying the radicals: $\sqrt{2}\left(\sqrt{3}-1\right)=\sqrt{6}-\sqrt{2}$

29. Multiplying the radicals: $\sqrt{2}\left(\sqrt{3}+\sqrt{2}\right)=\sqrt{6}+\sqrt{4}=\sqrt{6}+2$

31. Multiplying the radicals: $\sqrt{3}\left(2\sqrt{2}+\sqrt{3}\right)=2\sqrt{6}+\sqrt{9}=2\sqrt{6}+3$

33. Multiplying the radicals: $\sqrt{3}\left(2\sqrt{3}-\sqrt{5}\right)=2\sqrt{9}-\sqrt{15}=2\bullet 3-\sqrt{15}=6-\sqrt{15}$

35. Multiplying the radicals: $2\sqrt{3}\left(\sqrt{2}+\sqrt{5}\right)=2\sqrt{6}+2\sqrt{15}$

37. Multiplying the radicals: $5\sqrt[3]{4}\left(\sqrt[3]{2}+3\sqrt[3]{5}\right)=5\sqrt[3]{8}+15\sqrt[3]{20}=10+15\sqrt[3]{20}$

39. Multiplying the radicals: $\left(\sqrt{2}+1\right)^2=\left(\sqrt{2}\right)^2+2\left(\sqrt{2}\right)(1)+(1)^2=2+2\sqrt{2}+1=3+2\sqrt{2}$

41. Multiplying the radicals: $\left(\sqrt{x}+3\right)^2=\left(\sqrt{x}\right)^2+2\left(\sqrt{x}\right)(3)+(3)^2=x+6\sqrt{x}+9$

43. Multiplying the radicals: $\left(5-\sqrt{2}\right)^2=(5)^2-2(5)\left(\sqrt{2}\right)+\left(\sqrt{2}\right)^2=25-10\sqrt{2}+2=27-10\sqrt{2}$

45. Multiplying the radicals: $\left(\sqrt{a}-\frac{1}{2}\right)^2=\left(\sqrt{a}\right)^2-2\left(\sqrt{a}\right)\left(\frac{1}{2}\right)+\left(\frac{1}{2}\right)^2=a-\sqrt{a}+\frac{1}{4}$

47. Multiplying the radicals: $\left(\sqrt{3}+\sqrt{7}\right)^2=\left(\sqrt{3}+\sqrt{7}\right)\left(\sqrt{3}+\sqrt{7}\right)=3+\sqrt{21}+\sqrt{21}+7=10+2\sqrt{21}$

49. Multiplying the radicals:
$$\left(\sqrt{2}-\sqrt{10}\right)^2=\left(\sqrt{2}-\sqrt{10}\right)\left(\sqrt{2}-\sqrt{10}\right)=2-\sqrt{20}-\sqrt{20}+10=12-2\sqrt{20}=12-2\sqrt{4\bullet 5}=12-4\sqrt{5}$$

51. Multiplying the radicals: $\left(\sqrt{5}+3\right)\left(\sqrt{5}+2\right)=5+3\sqrt{5}+2\sqrt{5}+6=11+5\sqrt{5}$

53. Multiplying the radicals: $\left(\sqrt{2}-5\right)\left(\sqrt{2}+6\right)=2-5\sqrt{2}+6\sqrt{2}-30=-28+\sqrt{2}$

55. Multiplying the radicals: $\left(2\sqrt{7}+3\right)\left(3\sqrt{7}-4\right)=6\sqrt{49}+9\sqrt{7}-8\sqrt{7}-12=42+\sqrt{7}-12=30+\sqrt{7}$

57. Multiplying the radicals: $\left(2\sqrt{x}+4\right)\left(3\sqrt{x}+2\right)=6x+12\sqrt{x}+4\sqrt{x}+8=6x+16\sqrt{x}+8$

59. Multiplying the radicals: $\left(\sqrt{3}+\frac{1}{2}\right)\left(\sqrt{2}+\frac{1}{3}\right)=\sqrt{6}+\frac{1}{3}\sqrt{3}+\frac{1}{2}\sqrt{2}+\frac{1}{6}$

61. Multiplying the radicals: $\left(\sqrt{a}+\frac{1}{3}\right)\left(\sqrt{a}+\frac{2}{3}\right)=a+\frac{2}{3}\sqrt{a}+\frac{1}{3}\sqrt{a}+\frac{2}{9}=a+\sqrt{a}+\frac{2}{9}$

63. Multiplying the radicals: $(\sqrt{5}-2)(\sqrt{5}+2)=(\sqrt{5})^2-(2)^2=5-4=1$

65. Multiplying the radicals: $(5+\sqrt{2})(5-\sqrt{2})=(5)^2-(\sqrt{2})^2=25-2=23$

67. Multiplying the radicals: $(\sqrt{7}-\sqrt{3})(\sqrt{7}+\sqrt{3})=(\sqrt{7})^2-(\sqrt{3})^2=7-3=4$

69. Multiplying the radicals: $(\sqrt{x}+6)(\sqrt{x}-6)=(\sqrt{x})^2-(6)^2=x-36$

71. Multiplying the radicals: $(7\sqrt{a}+2\sqrt{b})(7\sqrt{a}-2\sqrt{b})=(7\sqrt{a})^2-(2\sqrt{b})^2=49a-4b$

73. **a**. Subtracting the radicals: $(\sqrt{7}+\sqrt{3})-(\sqrt{7}-\sqrt{3})=\sqrt{7}+\sqrt{3}-\sqrt{7}+\sqrt{3}=2\sqrt{3}$

b. Multiplying the radicals: $(\sqrt{7}+\sqrt{3})(\sqrt{7}-\sqrt{3})=(\sqrt{7})^2-(\sqrt{3})^2=7-3=4$

c. Squaring the radicals: $(\sqrt{7}-\sqrt{3})^2=(\sqrt{7}-\sqrt{3})(\sqrt{7}-\sqrt{3})=7-\sqrt{21}-\sqrt{21}+3=10-2\sqrt{21}$

d. Simplifying the radicals: $(\sqrt{7})^2+(\sqrt{3})^2=7+3=10$

75. **a**. Adding the radicals: $(\sqrt{x}+2)+(\sqrt{x}-2)=\sqrt{x}+2+\sqrt{x}-2=2\sqrt{x}$

b. Multiplying the radicals: $(\sqrt{x}+2)(\sqrt{x}-2)=(\sqrt{x})^2-(2)^2=x-4$

c. Squaring the radicals: $(\sqrt{x}+2)^2=(\sqrt{x}+2)(\sqrt{x}+2)=x+2\sqrt{x}+2\sqrt{x}+4=x+4\sqrt{x}+4$

d. Simplifying the radicals: $(\sqrt{x})^2+(2)^2=x+4$

77. The correct statement is: $2(3\sqrt{5})=6\sqrt{5}$

79. The correct statement is: $(\sqrt{3}+7)^2=(\sqrt{3})^2+2(\sqrt{3})(7)+(7)^2=3+14\sqrt{3}+49=52+14\sqrt{3}$

81. Simplifying the radical: $\sqrt{12}=\sqrt{4\cdot 3}=\sqrt{4}\sqrt{3}=2\sqrt{3}$

83. Simplifying the radical: $\sqrt{4x^2y^2}=2xy$

85. Simplifying the radical: $\sqrt{20x^2y^3}=\sqrt{4x^2y^2\cdot 5y}=2xy\sqrt{5y}$

87. Simplifying the radical: $(\sqrt{6})^2=(\sqrt{6})(\sqrt{6})=6$

89. Multiplying the radicals: $(\sqrt{5})(\sqrt{3})=\sqrt{15}$

91. Multiplying the radicals: $(\sqrt[3]{4})(\sqrt[3]{2})=\sqrt[3]{8}=2$

93. Multiplying the radicals: $\left(\sqrt[3]{5xy^2}\right)\left(\sqrt[3]{25x^2y}\right)=\sqrt[3]{125x^3y^3}=5xy$

95. Multiplying the radicals: $(3-\sqrt{5})(3+\sqrt{5})=(3)^2-(\sqrt{5})^2=9-5=4$

97. Multiplying the radicals: $(\sqrt{7}-\sqrt{2})(\sqrt{7}+\sqrt{2})=(\sqrt{7})^2-(\sqrt{2})^2=7-2=5$

99. Multiplying the radicals: $(2\sqrt{6})(5\sqrt{10})=10\sqrt{60}=10\sqrt{4\cdot 15}=20\sqrt{15}$. The correct answer is b.

101. Multiplying the radicals: $(4+\sqrt{3})(4-\sqrt{3})=(4)^2-(\sqrt{3})^2=16-3=13$. The correct answer is c.

8.5 Division of Radicals

1. Performing the division: $\dfrac{\sqrt{21}}{\sqrt{3}}=\sqrt{\dfrac{21}{3}}=\sqrt{7}$

3. Performing the division: $\dfrac{\sqrt{35}}{\sqrt{7}}=\sqrt{\dfrac{35}{7}}=\sqrt{5}$

5. Performing the division: $\dfrac{10\sqrt{15}}{5\sqrt{3}}=\dfrac{10}{5}\bullet\sqrt{\dfrac{15}{3}}=2\sqrt{5}$

7. Performing the division: $\dfrac{6\sqrt{21}}{3\sqrt{7}}=\dfrac{6}{3}\bullet\sqrt{\dfrac{21}{7}}=2\sqrt{3}$

9. Performing the division: $\dfrac{6\sqrt{35}}{12\sqrt{5}}=\dfrac{6}{12}\bullet\sqrt{\dfrac{35}{5}}=\dfrac{1}{2}\sqrt{7}=\dfrac{\sqrt{7}}{2}$

11. Performing the division: $\dfrac{\sqrt[3]{36}}{\sqrt[3]{3}}=\sqrt[3]{\dfrac{36}{3}}=\sqrt[3]{12}$

13. Performing the division: $\dfrac{\sqrt[3]{72}}{6\sqrt[3]{3}}=\dfrac{1}{6}\sqrt[3]{\dfrac{72}{3}}=\dfrac{1}{6}\sqrt[3]{24}=\dfrac{1}{6}\sqrt[3]{8\bullet 3}=\dfrac{1}{6}\bullet 2\sqrt[3]{3}=\dfrac{1}{3}\sqrt[3]{3}=\dfrac{\sqrt[3]{3}}{3}$

15. Simplifying the radical expression: $\sqrt{\dfrac{20}{5}}=\sqrt{4}=2$

17. Simplifying the radical expression: $\sqrt{\dfrac{20}{2}}=\sqrt{10}$

19. Simplifying the radical expression: $\sqrt{\dfrac{1}{2}}=\dfrac{\sqrt{1}}{\sqrt{2}}\bullet\dfrac{\sqrt{2}}{\sqrt{2}}=\dfrac{\sqrt{2}}{\sqrt{4}}=\dfrac{\sqrt{2}}{2}$

21. Simplifying the radical expression: $\sqrt{\dfrac{1}{3}}=\dfrac{\sqrt{1}}{\sqrt{3}}\bullet\dfrac{\sqrt{3}}{\sqrt{3}}=\dfrac{\sqrt{3}}{\sqrt{9}}=\dfrac{\sqrt{3}}{3}$

23. Simplifying the radical expression: $\sqrt{\dfrac{2}{5}}=\dfrac{\sqrt{2}}{\sqrt{5}}\bullet\dfrac{\sqrt{5}}{\sqrt{5}}=\dfrac{\sqrt{10}}{\sqrt{25}}=\dfrac{\sqrt{10}}{5}$

25. Simplifying the radical expression: $\sqrt{\dfrac{3}{2}}=\dfrac{\sqrt{3}}{\sqrt{2}}\bullet\dfrac{\sqrt{2}}{\sqrt{2}}=\dfrac{\sqrt{6}}{\sqrt{4}}=\dfrac{\sqrt{6}}{2}$

27. Simplifying the radical expression: $\sqrt{\dfrac{45}{6}}=\sqrt{\dfrac{15}{2}}=\dfrac{\sqrt{15}}{\sqrt{2}}\bullet\dfrac{\sqrt{2}}{\sqrt{2}}=\dfrac{\sqrt{30}}{\sqrt{4}}=\dfrac{\sqrt{30}}{2}$

29. Simplifying the radical expression: $\sqrt[3]{\dfrac{3}{5}}=\dfrac{\sqrt[3]{3}}{\sqrt[3]{5}}\bullet\dfrac{\sqrt[3]{25}}{\sqrt[3]{25}}=\dfrac{\sqrt[3]{75}}{\sqrt[3]{125}}=\dfrac{\sqrt[3]{75}}{5}$

31. Simplifying the radical expression: $\sqrt[3]{\dfrac{15}{4}}=\dfrac{\sqrt[3]{15}}{\sqrt[3]{4}}\bullet\dfrac{\sqrt[3]{2}}{\sqrt[3]{2}}=\dfrac{\sqrt[3]{30}}{\sqrt[3]{8}}=\dfrac{\sqrt[3]{30}}{2}$

33. Simplifying the radical expression: $\sqrt[3]{\dfrac{21}{6}}=\sqrt[3]{\dfrac{7}{2}}=\dfrac{\sqrt[3]{7}}{\sqrt[3]{2}}\bullet\dfrac{\sqrt[3]{4}}{\sqrt[3]{4}}=\dfrac{\sqrt[3]{28}}{\sqrt[3]{8}}=\dfrac{\sqrt[3]{28}}{2}$

35. Simplifying the radical expression: $\sqrt{\dfrac{4x^2y^2}{2}}=\sqrt{2x^2y^2}=xy\sqrt{2}$

37. Simplifying the radical expression: $\sqrt{\dfrac{5x^2y}{3}}=\dfrac{\sqrt{5x^2y}}{\sqrt{3}}\bullet\dfrac{\sqrt{3}}{\sqrt{3}}=\dfrac{\sqrt{15x^2y}}{\sqrt{9}}=\dfrac{x\sqrt{15y}}{3}$

39. Simplifying the radical expression: $\sqrt{\frac{16a^4}{5}}=\frac{\sqrt{16a^4}}{\sqrt{5}}\cdot\frac{\sqrt{5}}{\sqrt{5}}=\frac{4a^2\sqrt{5}}{\sqrt{25}}=\frac{4a^2\sqrt{5}}{5}$

41. Simplifying the radical expression: $\sqrt{\frac{72a^5}{5b}}=\frac{\sqrt{72a^5}}{\sqrt{5b}}\cdot\frac{\sqrt{5b}}{\sqrt{5b}}=\frac{\sqrt{360a^5b}}{\sqrt{25b^2}}=\frac{\sqrt{36a^4\cdot 10ab}}{5b}=\frac{6a^2\sqrt{10ab}}{5b}$

43. Simplifying the radical expression: $\sqrt{\frac{20x^2y^3}{3z}}=\frac{\sqrt{20x^2y^3}}{\sqrt{3z}}\cdot\frac{\sqrt{3z}}{\sqrt{3z}}=\frac{\sqrt{60x^2y^3z}}{\sqrt{9z^2}}=\frac{\sqrt{4x^2y^2\cdot 15yz}}{3z}=\frac{2xy\sqrt{15yz}}{3z}$

45. Simplifying the radical expression: $\sqrt[3]{\frac{12x^3y^4}{3}}=\sqrt[3]{4x^3y^4}=\sqrt[3]{x^3y^3\cdot 4y}=xy\sqrt[3]{4y}$

47. Simplifying the radical expression: $\sqrt[3]{\frac{3y^3}{4x}}=\frac{\sqrt[3]{3y^3}}{\sqrt[3]{4x}}\cdot\frac{\sqrt[3]{2x^2}}{\sqrt[3]{2x^2}}=\frac{\sqrt[3]{6x^2y^3}}{\sqrt[3]{8x^3}}=\frac{y\sqrt[3]{6x^2}}{2x}$

49. Rationalizing the denominator: $\frac{8}{3-\sqrt{5}}\cdot\frac{3+\sqrt{5}}{3+\sqrt{5}}=\frac{8(3+\sqrt{5})}{9-5}=\frac{8(3+\sqrt{5})}{4}=2(3+\sqrt{5})=6+2\sqrt{5}$

51. Rationalizing the denominator: $\frac{\sqrt{3}}{\sqrt{2}+4}=\frac{\sqrt{3}}{\sqrt{2}+4}\cdot\frac{\sqrt{2}-4}{\sqrt{2}-4}=\frac{\sqrt{6}-4\sqrt{3}}{2-16}=\frac{\sqrt{6}-4\sqrt{3}}{-14}=\frac{4\sqrt{3}-\sqrt{6}}{14}$

53. Rationalizing the denominator: $\frac{\sqrt{x}+2}{\sqrt{x}-2}\cdot\frac{\sqrt{x}+2}{\sqrt{x}+2}=\frac{x+2\sqrt{x}+2\sqrt{x}+4}{x-4}=\frac{x+4\sqrt{x}+4}{x-4}$

55. Rationalizing the denominator: $\frac{\sqrt{3}}{\sqrt{5}-\sqrt{2}}\cdot\frac{\sqrt{5}+\sqrt{2}}{\sqrt{5}+\sqrt{2}}=\frac{\sqrt{15}+\sqrt{6}}{5-2}=\frac{\sqrt{15}+\sqrt{6}}{3}$

57. Rationalizing the denominator: $\frac{\sqrt{5}}{\sqrt{5}+\sqrt{2}}\cdot\frac{\sqrt{5}-\sqrt{2}}{\sqrt{5}-\sqrt{2}}=\frac{\sqrt{25}-\sqrt{10}}{5-2}=\frac{5-\sqrt{10}}{3}$

59. Rationalizing the denominator: $\frac{\sqrt{3}+\sqrt{2}}{\sqrt{3}-\sqrt{2}}\cdot\frac{\sqrt{3}+\sqrt{2}}{\sqrt{3}+\sqrt{2}}=\frac{3+\sqrt{6}+\sqrt{6}+2}{3-2}=\frac{5+2\sqrt{6}}{1}=5+2\sqrt{6}$

61. Rationalizing the denominator: $\frac{\sqrt{7}-\sqrt{3}}{\sqrt{7}+\sqrt{3}}\cdot\frac{\sqrt{7}-\sqrt{3}}{\sqrt{7}-\sqrt{3}}=\frac{7-\sqrt{21}-\sqrt{21}+3}{7-3}=\frac{10-2\sqrt{21}}{4}=\frac{2(5-\sqrt{21})}{4}=\frac{5-\sqrt{21}}{2}$

63. Rationalizing the denominator: $\frac{\sqrt{5}-\sqrt{2}}{\sqrt{5}+\sqrt{3}}\cdot\frac{\sqrt{5}-\sqrt{3}}{\sqrt{5}-\sqrt{3}}=\frac{5-\sqrt{15}-\sqrt{10}+\sqrt{6}}{5-3}=\frac{5-\sqrt{15}-\sqrt{10}+\sqrt{6}}{2}$

65. Completing the table:

x	$\sqrt{x}$	$\frac{1}{\sqrt{x}}$	$\frac{\sqrt{x}}{x}$
1	1	1	1
2	1.414	0.707	0.707
3	1.732	0.577	0.577
4	2	0.5	0.5
5	2.236	0.447	0.447
6	2.449	0.408	0.408

67. Completing the table:

x	$\sqrt{x^2}$	$\sqrt{x^3}$	$x\sqrt{x}$
1	1	1	1
2	2	2.828	2.828
3	3	5.196	5.196
4	4	8	8
5	5	11.180	11.180
6	6	14.697	14.697

69. **a.** Adding the radicals: $(5+\sqrt{2})+(5-\sqrt{2})=5+\sqrt{2}+5-\sqrt{2}=10$

b. Multiplying the radicals: $(5+\sqrt{2})(5-\sqrt{2})=(5)^2-(\sqrt{2})^2=25-2=23$

c. Squaring the radicals: $(5+\sqrt{2})^2=(5+\sqrt{2})(5+\sqrt{2})=25+5\sqrt{2}+5\sqrt{2}+2=27+10\sqrt{2}$

d. Dividing the radicals: $\frac{5+\sqrt{2}}{5-\sqrt{2}}=\frac{5+\sqrt{2}}{5-\sqrt{2}}\cdot\frac{5+\sqrt{2}}{5+\sqrt{2}}=\frac{25+10\sqrt{2}+2}{25-2}=\frac{27+10\sqrt{2}}{23}$

71. Substituting $h = 24$: $d = \sqrt{\frac{3(24)}{2}} = \sqrt{\frac{72}{2}} = \sqrt{36} = 6$ miles

73. Simplifying: $7^2 = 7 \cdot 7 = 49$

75. Simplifying: $(-9)^2 = (-9)(-9) = 81$

77. Simplifying: $\left(\sqrt{x+1}\right)^2 = x+1$

79. Simplifying: $\left(\sqrt{2x-3}\right)^2 = 2x-3$

81. Simplifying: $(x+3)^2 = x^2 + 2(3x) + 3^2 = x^2 + 6x + 9$

83. Solving the equation:

$$\begin{aligned} 3a-2 &= 4 \\ 3a &= 6 \\ a &= 2 \end{aligned}$$

85. Solving the equation:

$$\begin{aligned} x+15 &= x^2+6x+9 \\ 0 &= x^2+5x-6 \\ (x+6)(x-1) &= 0 \\ x &= -6,1 \end{aligned}$$

87. **a.** Substituting $x = -6$: $\sqrt{-6+15} = \sqrt{9} = 3 \neq -6+3$. No, $x = -6$ is not a solution to the equation.
b. Substituting $x = 1$: $\sqrt{1+15} = \sqrt{16} = 4 = 1+3$. Yes, $x = 1$ is a solution to the equation.

89. Evaluating the expression for the given values of x:

$x = -4$: $y = 3\sqrt{-4}$, which is undefined

$x = -1$: $y = 3\sqrt{-1}$, which is undefined

$x = 0$: $y = 3\sqrt{0} = 0$

$x = 1$: $y = 3\sqrt{1} = 3$

$x = 4$: $y = 3\sqrt{4} = 3 \cdot 2 = 6$

$x = 9$: $y = 3\sqrt{9} = 3 \cdot 3 = 9$

$x = 16$: $y = 3\sqrt{16} = 3 \cdot 4 = 12$

91. Performing the division: $\frac{12\sqrt{10}}{3\sqrt{2}} = \frac{12}{3} \cdot \sqrt{\frac{10}{2}} = 4\sqrt{5}$. The correct answer is a.

93. Simplifying the radical expression: $\sqrt[3]{\frac{2x}{9y}} = \frac{\sqrt[3]{2x}}{\sqrt[3]{9y}} \cdot \frac{\sqrt[3]{3y^2}}{\sqrt[3]{3y^2}} = \frac{\sqrt[3]{6xy^2}}{\sqrt[3]{27y^3}} = \frac{\sqrt[3]{6xy^2}}{3y}$. The correct answer is a.

8.6 Equations Involving Radicals

1. Solving the equation:

$$\begin{aligned} \sqrt{x+1} &= 2 \\ \left(\sqrt{x+1}\right)^2 &= (2)^2 \\ x+1 &= 4 \\ x &= 3 \end{aligned}$$

This value checks in the original equation.

3. Solving the equation:

$$\begin{aligned} \sqrt{x+5} &= 7 \\ \left(\sqrt{x+5}\right)^2 &= (7)^2 \\ x+5 &= 49 \\ x &= 44 \end{aligned}$$

This value checks in the original equation.

5. Solving the equation:

$$\begin{aligned} \sqrt{x-9} &= -6 \\ \left(\sqrt{x-9}\right)^2 &= (-6)^2 \\ x-9 &= 36 \\ x &= 45 \end{aligned}$$

Since this value does not check, the solution set is $\varnothing$.

7. Solving the equation:

$$\begin{aligned} \sqrt{x-5} &= -4 \\ \left(\sqrt{x-5}\right)^2 &= (-4)^2 \\ x-5 &= 16 \\ x &= 21 \end{aligned}$$

Since this value does not check, the solution set is $\varnothing$.

9. Solving the equation:

$$\begin{aligned} \sqrt{x-8} &= 0 \\ \left(\sqrt{x-8}\right)^2 &= (0)^2 \\ x-8 &= 0 \\ x &= 8 \end{aligned}$$

This value checks in the original equation.

11. Solving the equation:

$$\begin{aligned} \sqrt{2x+1} &= 3 \\ \left(\sqrt{2x+1}\right)^2 &= (3)^2 \\ 2x+1 &= 9 \\ 2x &= 8 \\ x &= 4 \end{aligned}$$

This value checks in the original equation.

13. Solving the equation:

$$\begin{aligned} \sqrt{2x-3} &= -5 \\ \left(\sqrt{2x-3}\right)^2 &= (-5)^2 \\ 2x-3 &= 25 \\ 2x &= 28 \\ x &= 14 \end{aligned}$$

Since this value does not check, the solution set is $\varnothing$.

15. Solving the equation:

$$\begin{aligned} \sqrt{3x+6} &= 2 \\ \left(\sqrt{3x+6}\right)^2 &= (2)^2 \\ 3x+6 &= 4 \\ 3x &= -2 \\ x &= -\frac{2}{3} \end{aligned}$$

This value checks in the original equation.

17. Solving the equation:

$$\begin{aligned} 2\sqrt{x} &= 10 \\ \sqrt{x} &= 5 \\ \left(\sqrt{x}\right)^2 &= (5)^2 \\ x &= 25 \end{aligned}$$

This value checks in the original equation.

19. Solving the equation:

$$\begin{aligned} 3\sqrt{a} &= 6 \\ \sqrt{a} &= 2 \\ \left(\sqrt{a}\right)^2 &= (2)^2 \\ a &= 4 \end{aligned}$$

This value checks in the original equation.

21. Solving the equation:

$$\begin{aligned} \sqrt{3x+4} - 3 &= 2 \\ \sqrt{3x+4} &= 5 \\ \left(\sqrt{3x+4}\right)^2 &= (5)^2 \\ 3x+4 &= 25 \\ 3x &= 21 \\ x &= 7 \end{aligned}$$

This value checks in the original equation.

23. Solving the equation:

$$\begin{aligned} \sqrt{5y-4} - 2 &= 4 \\ \sqrt{5y-4} &= 6 \\ \left(\sqrt{5y-4}\right)^2 &= (6)^2 \\ 5y-4 &= 36 \\ 5y &= 40 \\ y &= 8 \end{aligned}$$

This value checks in the original equation.

25. Solving the equation:

$$\begin{aligned} \sqrt{2x+1} + 5 &= 2 \\ \sqrt{2x+1} &= -3 \\ \left(\sqrt{2x+1}\right)^2 &= (-3)^2 \\ 2x+1 &= 9 \\ 2x &= 8 \\ x &= 4 \end{aligned}$$

Since this value does not check, the solution set is $\varnothing$.

27. Solving the equation:

$$\begin{aligned} \sqrt{x+3} &= x-3 \\ \left(\sqrt{x+3}\right)^2 &= (x-3)^2 \\ x+3 &= x^2-6x+9 \\ 0 &= x^2-7x+6 \\ 0 &= (x-6)(x-1) \\ x &= 6,1 \end{aligned}$$

Since $x=1$ does not check in the original equation, the solution is $x=6$.

29. Solving the equation:

$$\begin{aligned} \sqrt{a+2} &= a+2 \\ \left(\sqrt{a+2}\right)^2 &= (a+2)^2 \\ a+2 &= a^2+4a+4 \\ 0 &= a^2+3a+2 \\ 0 &= (a+2)(a+1) \\ a &= -2,-1 \end{aligned}$$

Both values check in the original equation.

31. Solving the equation:

$$\begin{aligned} \sqrt{2x+9} &= x+5 \\ \left(\sqrt{2x+9}\right)^2 &= (x+5)^2 \\ 2x+9 &= x^2+10x+25 \\ 0 &= x^2+8x+16 \\ 0 &= (x+4)^2 \\ x &= -4 \end{aligned}$$

This value checks in the original equation.

33. Solving the equation:

$$\begin{aligned} \sqrt{y-4} &= y-6 \\ \left(\sqrt{y-4}\right)^2 &= (y-6)^2 \\ y-4 &= y^2-12y+36 \\ 0 &= y^2-13y+40 \\ 0 &= (y-8)(y-5) \\ y &= 5,8 \end{aligned}$$

Since $y=5$ does not check in the original equation, the solution is $y=8$.

35. Solving the equation:

$$\begin{aligned} \sqrt{3x-5} &= \sqrt{x+3} \\ \left(\sqrt{3x-5}\right)^2 &= \left(\sqrt{x+3}\right)^2 \\ 3x-5 &= x+3 \\ 2x &= 8 \\ x &= 4 \end{aligned}$$

This value checks in the original equation.

37. Solving the equation:

$$\begin{aligned} \sqrt{a+3} &= 4-\sqrt{a-1} \\ \left(\sqrt{a+3}\right)^2 &= \left(4-\sqrt{a-1}\right)^2 \\ a+3 &= 16-8\sqrt{a-1}+a-1 \\ -12 &= -8\sqrt{a-1} \\ 2\sqrt{a-1} &= 3 \\ \left(2\sqrt{a-1}\right)^2 &= (3)^2 \\ 4a-4 &= 9 \\ 4a &= 13 \\ a &= \frac{13}{4} \end{aligned}$$

This value checks in the original equation.

39. Solving the equation:

$$\begin{aligned}
\sqrt{3y-2} &= \sqrt{y+7}-1 \\
\left(\sqrt{3y-2}\right)^2 &= \left(\sqrt{y+7}-1\right)^2 \\
3y-2 &= y+7-2\sqrt{y+7}+1 \\
2y-10 &= -2\sqrt{y+7} \\
\sqrt{y+7} &= 5-y \\
\left(\sqrt{y+7}\right)^2 &= (5-y)^2 \\
y+7 &= 25-10y+y^2 \\
y^2-11y+18 &= 0 \\
(y-2)(y-9) &= 0 \\
y &= 2,9
\end{aligned}$$

Since $y=9$ does not check in the original equation, the solution is $y=2$.

41. Solving the equation:

$$\begin{aligned}
\sqrt[3]{x+4} &= -2 \\
\left(\sqrt[3]{x+4}\right)^3 &= (-2)^3 \\
x+4 &= -8 \\
x &= -12
\end{aligned}$$

43. Solving the equation:

$$\begin{aligned}
\sqrt[3]{x-7}-4 &= 0 \\
\sqrt[3]{x-7} &= 4 \\
\left(\sqrt[3]{x-7}\right)^3 &= (4)^3 \\
x-7 &= 64 \\
x &= 71
\end{aligned}$$

45. Solving the equation:

$$\begin{aligned}
\sqrt[3]{2x-3}+7 &= 2 \\
\sqrt[3]{2x-3} &= -5 \\
\left(\sqrt[3]{2x-3}\right)^3 &= (-5)^3 \\
2x-3 &= -125 \\
2x &= -122 \\
x &= -61
\end{aligned}$$

47. **a.** Solving the equation:

$$\begin{aligned}
\sqrt{y}-4 &= 6 \\
\sqrt{y} &= 10 \\
\left(\sqrt{y}\right)^2 &= (10)^2 \\
y &= 100
\end{aligned}$$

This value checks in the original equation.

b. Solving the equation:

$$\begin{aligned}
\sqrt{y-4} &= 6 \\
\left(\sqrt{y-4}\right)^2 &= (6)^2 \\
y-4 &= 36 \\
y &= 40
\end{aligned}$$

This value checks in the original equation.

c. Solving the equation:

$$\begin{aligned}
\sqrt{y-4} &= -6 \\
\left(\sqrt{y-4}\right)^2 &= (-6)^2 \\
y-4 &= 36 \\
y &= 40
\end{aligned}$$

Since this value does not check, the solution set is $\varnothing$.

d. Solving the equation:

$$\sqrt{y-4} = y-6$$
$$\left(\sqrt{y-4}\right)^2 = (y-6)^2$$
$$y-4 = y^2 - 12y + 36$$
$$y^2 - 13y + 40 = 0$$
$$(y-5)(y-8) = 0$$
$$y = 5, 8$$

Since $y = 5$ does not check in the original equation, the solution is $y = 8$.

49. **a.** Solving the equation:

$$x - 3 = 0$$
$$x = 3$$

b. Solving the equation:

$$\sqrt{x} - 3 = 0$$
$$\sqrt{x} = 3$$
$$\left(\sqrt{x}\right)^2 = (3)^2$$
$$x = 9$$

This value checks in the original equation.

c. Solving the equation:

$$\sqrt{x-3} = 0$$
$$\left(\sqrt{x-3}\right)^2 = (0)^2$$
$$x - 3 = 0$$
$$x = 3$$

This value checks in the original equation.

d. Solving the equation:

$$\sqrt{x} + 3 = 0$$
$$\sqrt{x} = -3$$
$$\left(\sqrt{x}\right)^2 = (-3)^2$$
$$x = 9$$

Since this value does not check, the solution set is $\varnothing$.

e. Solving the equation:

$$\sqrt{x} + 3 = 5$$
$$\sqrt{x} = 2$$
$$\left(\sqrt{x}\right)^2 = (2)^2$$
$$x = 4$$

This value checks in the original equation.

f. Solving the equation:

$$\sqrt{x} + 3 = -5$$
$$\sqrt{x} = -8$$
$$\left(\sqrt{x}\right)^2 = (-8)^2$$
$$x = 64$$

Since this value does not check, the solution set is $\varnothing$.

g. Solving the equation:

$$x - 3 = \sqrt{5-x}$$
$$(x-3)^2 = \left(\sqrt{5-x}\right)^2$$
$$x^2 - 6x + 9 = 5 - x$$
$$x^2 - 5x + 4 = 0$$
$$(x-1)(x-4) = 0$$
$$x = 1, 4$$

Since $x = 1$ does not check in the original equation, the solution is $x = 4$.

51. Completing the table:

Length L (feet)	1	2	3	4	5	6
Time T (seconds)	1.11	1.57	1.92	2.22	2.48	2.72

Drawing a line graph:

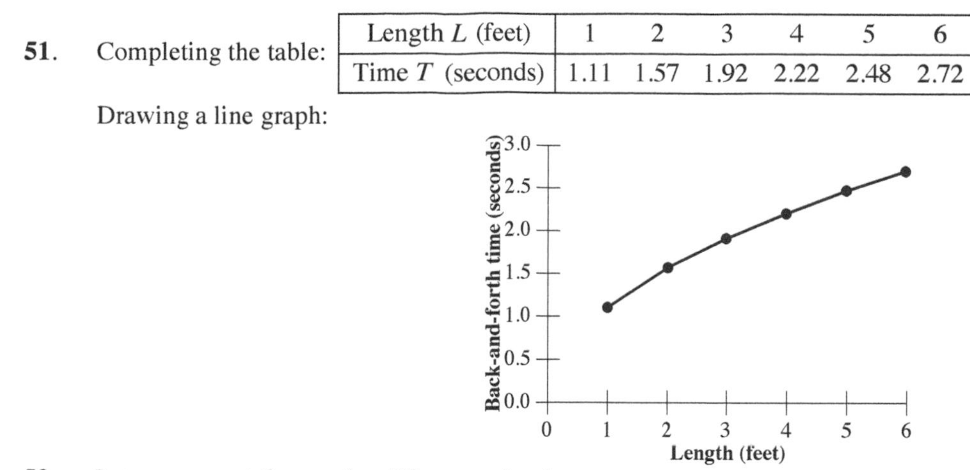

53. Let x represent the number. The equation is:

$$x+2=\sqrt{8x}$$
$$(x+2)^2=\left(\sqrt{8x}\right)^2$$
$$x^2+4x+4=8x$$
$$x^2-4x+4=0$$
$$(x-2)^2=0$$
$$x-2=0$$
$$x=2$$

The number is 2.

55. Let x represent the number. The equation is:

$$x-3=2\sqrt{x}$$
$$(x-3)^2=\left(2\sqrt{x}\right)^2$$
$$x^2-6x+9=4x$$
$$x^2-10x+9=0$$
$$(x-9)(x-1)=0$$
$$x=9 \qquad (x=1 \text{ does not check in the original equation})$$

The number is 9.

57. Substituting $T = 2$:

$$2 = \frac{11}{7}\sqrt{\frac{L}{2}}$$
$$14 = 11\sqrt{\frac{L}{2}}$$
$$\frac{14}{11} = \sqrt{\frac{L}{2}}$$
$$\left(\frac{14}{11}\right)^2 = \left(\sqrt{\frac{L}{2}}\right)^2$$
$$\frac{196}{121} = \frac{L}{2}$$
$$L = \frac{392}{121} \approx 3.2 \text{ feet}$$

59. Reducing the rational expression: $\frac{x^2-x-6}{x^2-9} = \frac{(x-3)(x+2)}{(x+3)(x-3)} = \frac{x+2}{x+3}$

61. Performing the operations: $\frac{x^2-25}{x+4} \bullet \frac{2x+8}{x^2-9x+20} = \frac{(x+5)(x-5)}{x+4} \bullet \frac{2(x+4)}{(x-4)(x-5)} = \frac{2(x+5)(x-5)(x+4)}{(x+4)(x-4)(x-5)} = \frac{2(x+5)}{x-4}$

63. Performing the operations: $\frac{x}{x^2-16} + \frac{4}{x^2-16} = \frac{x+4}{x^2-16} = \frac{x+4}{(x+4)(x-4)} = \frac{1}{x-4}$

65. Performing the operations: $\frac{1-\frac{25}{x^2}}{1-\frac{8}{x}+\frac{15}{x^2}} \bullet \frac{x^2}{x^2} = \frac{x^2-25}{x^2-8x+15} = \frac{(x+5)(x-5)}{(x-5)(x-3)} = \frac{x+5}{x-3}$

67. Multiplying each side of the equation by $x^2 - 9 = (x+3)(x-3)$:

$$(x+3)(x-3)\left(\frac{x}{x^2-9} - \frac{3}{x-3}\right) = (x+3)(x-3) \bullet \frac{1}{x+3}$$
$$x - 3(x+3) = x - 3$$
$$x - 3x - 9 = x - 3$$
$$-2x - 9 = x - 3$$
$$-3x = 6$$
$$x = -2$$

This value checks in the original equation.

69. Let t represent the time to fill the pool with both pipes open. The equation is:

$$\frac{1}{8} - \frac{1}{12} = \frac{1}{t}$$
$$24t\left(\frac{1}{8} - \frac{1}{12}\right) = 24t \bullet \frac{1}{t}$$
$$3t - 2t = 24$$
$$t = 24$$

It will take 24 hours to fill the pool with both pipes left open.

71. The variation equation is $y = Kx$. Finding K:

$$8 = K \cdot 12$$
$$K = \frac{2}{3}$$

So $y = \frac{2}{3}x$. Substituting $x = 36$: $y = \frac{2}{3}(36) = 24$

73. Solving the equation:

$$\sqrt{x+4} + 2 = 2x$$
$$\sqrt{x+4} = 2x - 2$$
$$\left(\sqrt{x+4}\right)^2 = (2x-2)^2$$
$$x + 4 = 4x^2 - 8x + 4$$
$$4x^2 - 9x = 0$$
$$x(4x-9) = 0$$
$$x = 0, \frac{9}{4}$$

Since $x = 0$ does not check in the original equation, the solution is $x = \frac{9}{4}$. The correct answer is c.

Chapter 8 Test

1. Finding the root: $\sqrt{36} = 6$

2. Finding the root: $-\sqrt{81} = -9$

3. Finding the root: $\sqrt{-121}$ is not a real number

4. Finding the root: $\sqrt[3]{64} = 4$

5. Finding the root: $\sqrt[3]{-27} = -3$

6. Finding the root: $-\sqrt[4]{625} = -5$

7. Simplifying the expression: $\sqrt{72} = \sqrt{36 \cdot 2} = 6\sqrt{2}$

8. Simplifying the expression: $\sqrt{96} = \sqrt{16 \cdot 6} = 4\sqrt{6}$

9. Simplifying the expression: $4\sqrt{12y^2} = 4\sqrt{4y^2 \cdot 3} = 4 \cdot 2y\sqrt{3} = 8y\sqrt{3}$

10. Simplifying the expression: $\sqrt{\frac{18x^3y^5}{3}} = \sqrt{6x^3y^5} = \sqrt{x^2y^4 \cdot 6xy} = xy^2\sqrt{6xy}$

11. Simplifying the radical expression: $\sqrt{\frac{128}{25}} = \frac{\sqrt{128}}{\sqrt{25}} = \frac{\sqrt{64 \cdot 2}}{5} = \frac{8\sqrt{2}}{5}$

12. Simplifying the radical expression: $\sqrt[3]{40x^7y^8} = \sqrt[3]{8x^6y^6 \cdot 5xy^2} = \sqrt[3]{8x^6y^6}\sqrt[3]{5xy^2} = 2x^2y^2\sqrt[3]{5xy^2}$

13. Simplifying the expression: $\frac{8+2\sqrt{5}}{4} = \frac{2\left(4+\sqrt{5}\right)}{4} = \frac{4+\sqrt{5}}{2}$

14. Simplifying the expression: $\frac{-2+\sqrt{24}}{6} = \frac{-2+2\sqrt{6}}{6} = \frac{2\left(-1+\sqrt{6}\right)}{6} = \frac{-1+\sqrt{6}}{3}$

15. Combining the expressions: $3\sqrt{20} - \sqrt{45} = 3\sqrt{4 \cdot 5} - \sqrt{9 \cdot 5} = 3 \cdot 2\sqrt{5} - 3\sqrt{5} = 6\sqrt{5} - 3\sqrt{5} = 3\sqrt{5}$

16. Combining the expressions: $3m\sqrt{12} + 2\sqrt{3m^2} = 3m\sqrt{4 \cdot 3} + 2\sqrt{m^2 \cdot 3} = 3m \cdot 2\sqrt{3} + 2m\sqrt{3} = 6m\sqrt{3} + 2m\sqrt{3} = 8m\sqrt{3}$

17. Combining radicals: $\sqrt[3]{24a^3b^3} - 5a\sqrt[3]{3b^3} = \sqrt[3]{8a^3b^3 \cdot 3} - 5a\sqrt[3]{b^3 \cdot 3} = 2ab\sqrt[3]{3} - 5ab\sqrt[3]{3} = -3ab\sqrt[3]{3}$

18. Multiplying the expressions: $\sqrt{2}\left(\sqrt{3} - 3\right) = \sqrt{6} - 3\sqrt{2}$

19. Multiplying the expressions: $\left(\sqrt{3} - 5\right)\left(\sqrt{3} + 2\right) = \sqrt{9} + 2\sqrt{3} - 5\sqrt{3} - 10 = 3 - 3\sqrt{3} - 10 = -7 - 3\sqrt{3}$

20. Multiplying the expressions: $\left(\sqrt{x} - 9\right)\left(\sqrt{x} + 9\right) = \left(\sqrt{x}\right)^2 - (9)^2 = x - 81$

21. Multiplying the expressions: $\left(\sqrt{7}-\sqrt{2}\right)^2=\left(\sqrt{7}\right)^2-2\left(\sqrt{7}\right)\left(\sqrt{2}\right)+\left(\sqrt{2}\right)^2=7-2\sqrt{14}+2=9-2\sqrt{14}$

22. Performing the division: $\frac{\sqrt{42}}{\sqrt{7}}=\sqrt{\frac{42}{7}}=\sqrt{6}$

23. Performing the division: $\frac{10\sqrt[3]{48}}{5\sqrt[3]{3}}=\frac{10}{5}\sqrt[3]{\frac{48}{3}}=2\sqrt[3]{16}=2\sqrt[3]{8\cdot 2}=2\cdot 2\sqrt[3]{2}=4\sqrt[3]{2}$

24. Simplifying the expression: $\sqrt{\frac{2}{7}}=\frac{\sqrt{2}}{\sqrt{7}}\cdot\frac{\sqrt{7}}{\sqrt{7}}=\frac{\sqrt{14}}{\sqrt{49}}=\frac{\sqrt{14}}{7}$

25. Simplifying the radical expression: $\sqrt[3]{\frac{3}{2x^2}}=\frac{\sqrt[3]{3}}{\sqrt[3]{2x^2}}\cdot\frac{\sqrt[3]{4x}}{\sqrt[3]{4x}}=\frac{\sqrt[3]{12x}}{\sqrt[3]{8x^3}}=\frac{\sqrt[3]{12x}}{2x}$

26. Rationalizing the denominator: $\frac{5}{\sqrt{3}-1}\cdot\frac{\sqrt{3}+1}{\sqrt{3}+1}=\frac{5\sqrt{3}+5}{3-1}=\frac{5\sqrt{3}+5}{2}$

27. Rationalizing the denominator: $\frac{\sqrt{5}-\sqrt{7}}{\sqrt{5}+\sqrt{7}}\cdot\frac{\sqrt{5}-\sqrt{7}}{\sqrt{5}-\sqrt{7}}=\frac{5-\sqrt{35}-\sqrt{35}+7}{5-7}=\frac{12-2\sqrt{35}}{-2}=-6+\sqrt{35}$

28. Rationalizing the denominator: $\frac{\sqrt{x}}{\sqrt{x}-3}\cdot\frac{\sqrt{x}+3}{\sqrt{x}+3}=\frac{x+3\sqrt{x}}{x-9}$

29. Solving the equation:

$$\begin{aligned}\sqrt{4x-3}-3&=4\\ \sqrt{4x-3}&=7\\ \left(\sqrt{4x-3}\right)^2&=(7)^2\\ 4x-3&=49\\ 4x&=52\\ x&=13\end{aligned}$$

This value checks in the original equation.

30. Solving the equation:

$$\begin{aligned}\sqrt{2x+8}&=x+4\\ \left(\sqrt{2x+8}\right)^2&=(x+4)^2\\ 2x+8&=x^2+8x+16\\ 0&=x^2+6x+8\\ 0&=(x+4)(x+2)\\ x&=-4,-2\end{aligned}$$

Both values check in the original equation.

31. Solving the equation:

$$\begin{aligned}\sqrt{x+3}&=\sqrt{x+4}-1\\ \left(\sqrt{x+3}\right)^2&=\left(\sqrt{x+4}-1\right)^2\\ x+3&=x+4-2\sqrt{x+4}+1\\ -2&=-2\sqrt{x+4}\\ \sqrt{x+4}&=1\\ \left(\sqrt{x+4}\right)^2&=(1)^2\\ x+4&=1\\ x&=-3\end{aligned}$$

This value checks in the original equation.

32. Solving the equation:

$$\begin{aligned}\sqrt[3]{2x+7}&=-1\\ \left(\sqrt[3]{2x+7}\right)^3&=(-1)^3\\ 2x+7&=-1\\ 2x&=-8\\ x&=-4\end{aligned}$$

33. Let x represent the number. The equation is:

$$x-3=2\sqrt{x}$$
$$(x-3)^2=(2\sqrt{x})^2$$
$$x^2-6x+9=4x$$
$$x^2-10x+9=0$$
$$(x-1)(x-9)=0$$
$$x=1,9$$

Since $x = 1$ does not check in the original equation, the solution is $x = 9$. The number is 9.

34. Using the Pythagorean theorem:

$$x^2=(\sqrt{2})^2+3^2=2+9=11$$
$$x=\sqrt{11}$$

Chapter 9
Quadratic Equations

9.1 Square Root Property

1. Solving the equation by factoring:

$$\begin{aligned} x^2-25&=0\\ (x+5)(x-5)&=0\\ x&=-5,5 \end{aligned}$$

3. Solving the equation by factoring:

$$\begin{aligned} 4x^2-9&=0\\ (2x+3)(2x-3)&=0\\ x&=-\frac{3}{2},\frac{3}{2} \end{aligned}$$

5. Solving the equation by factoring:

$$\begin{aligned} 2x^2+3x-5&=0\\ (2x+5)(x-1)&=0\\ x&=-\frac{5}{2},1 \end{aligned}$$

7. Solving the equation by factoring:

$$\begin{aligned} a^2-a-6&=0\\ (a+2)(a-3)&=0\\ a&=-2,3 \end{aligned}$$

9. Solving the equation:

$$\begin{aligned} x^2&=25\\ x&=\pm\sqrt{25}\\ x&=\pm 5 \end{aligned}$$

11. Solving the equation:

$$\begin{aligned} 4x^2&=9\\ x^2&=\frac{9}{4}\\ x&=\pm\sqrt{\frac{9}{4}}\\ x&=\pm\frac{3}{2} \end{aligned}$$

13. Solving the equation:

$$\begin{aligned} a^2&=8\\ a&=\pm\sqrt{8}\\ a&=\pm 2\sqrt{2} \end{aligned}$$

15. Solving the equation:

$$\begin{aligned} 2x^2&=24\\ x^2&=12\\ x&=\pm\sqrt{12}\\ x&=\pm 2\sqrt{3} \end{aligned}$$

17. Solving the equation:

$$\begin{aligned} (x+2)^2&=4\\ x+2&=\pm\sqrt{4}\\ x+2&=-2,2\\ x&=-4,0 \end{aligned}$$

19. Solving the equation:

$$\begin{aligned} (x+1)^2&=16\\ x+1&=\pm\sqrt{16}\\ x+1&=-4,4\\ x&=-5,3 \end{aligned}$$

21. Solving the equation:

$$\begin{aligned}(a-5)^2 &= 75\\ a-5 &= \pm\sqrt{75}\\ a-5 &= \pm 5\sqrt{3}\\ a &= 5\pm 5\sqrt{3}\end{aligned}$$

23. Solving the equation:

$$\begin{aligned}(y+1)^2 &= 12\\ y+1 &= \pm\sqrt{12}\\ y+1 &= \pm 2\sqrt{3}\\ y &= -1\pm 2\sqrt{3}\end{aligned}$$

25. Solving the equation:

$$\begin{aligned}(2x+1)^2 &= 25\\ 2x+1 &= \pm\sqrt{25}\\ 2x+1 &= -5,5\\ 2x &= -6,4\\ x &= -3,2\end{aligned}$$

27. Solving the equation:

$$\begin{aligned}(4a-5)^2 &= 36\\ 4a-5 &= \pm\sqrt{36}\\ 4a-5 &= -6,6\\ 4a &= -1,11\\ a &= -\frac{1}{4},\frac{11}{4}\end{aligned}$$

29. Solving the equation:

$$\begin{aligned}(3y-1)^2 &= 20\\ 3y-1 &= \pm\sqrt{20}\\ 3y-1 &= \pm 2\sqrt{5}\\ 3y &= 1\pm 2\sqrt{5}\\ y &= \frac{1\pm 2\sqrt{5}}{3}\end{aligned}$$

31. Solving the equation:

$$\begin{aligned}(3x+6)^2 &= 27\\ 3x+6 &= \pm\sqrt{27}\\ 3x+6 &= \pm 3\sqrt{3}\\ 3x &= -6\pm 3\sqrt{3}\\ x &= \frac{-6\pm 3\sqrt{3}}{3}\\ x &= -2\pm\sqrt{3}\end{aligned}$$

33. Solving the equation:

$$\begin{aligned}(3x-9)^2 &= 27\\ 3x-9 &= \pm\sqrt{27}\\ 3x-9 &= \pm 3\sqrt{3}\\ 3x &= 9\pm 3\sqrt{3}\\ x &= \frac{9\pm 3\sqrt{3}}{3}\\ x &= 3\pm\sqrt{3}\end{aligned}$$

35. Solving the equation:

$$\begin{aligned}\left(x-\frac{1}{2}\right)^2 &= \frac{7}{4}\\ x-\frac{1}{2} &= \pm\sqrt{\frac{7}{4}}\\ x-\frac{1}{2} &= \frac{\pm\sqrt{7}}{2}\\ x &= \frac{1\pm\sqrt{7}}{2}\end{aligned}$$

37. Solving the equation:

$$\begin{aligned}\left(a+\frac{4}{5}\right)^2 &= \frac{12}{25}\\ a+\frac{4}{5} &= \pm\sqrt{\frac{12}{25}}\\ a+\frac{4}{5} &= \frac{\pm 2\sqrt{3}}{5}\\ a &= \frac{-4\pm 2\sqrt{3}}{5}\end{aligned}$$

39. Solving the equation:

$$\begin{aligned}x^2+10x+25 &= 7\\ (x+5)^2 &= 7\\ x+5 &= \pm\sqrt{7}\\ x &= -5\pm\sqrt{7}\end{aligned}$$

41. Solving the equation:

$$x^2-2x+1=9$$
$$(x-1)^2=9$$
$$x-1=\pm\sqrt{9}$$
$$x-1=-3,3$$
$$x=-2,4$$

43. Solving the equation:

$$x^2+12x+36=8$$
$$(x+6)^2=8$$
$$x+6=\pm\sqrt{8}$$
$$x+6=\pm 2\sqrt{2}$$
$$x=-6\pm 2\sqrt{2}$$

45. **a.** Solving the equation:

$$2x-1=0$$
$$2x=1$$
$$x=\frac{1}{2}$$

b. Solving the equation:

$$2x-1=4$$
$$2x=5$$
$$x=\frac{5}{2}$$

c. Solving the equation:

$$(2x-1)^2=4$$
$$2x-1=\pm\sqrt{4}$$
$$2x-1=-2,2$$
$$2x=-1,3$$
$$x=-\frac{1}{2},\frac{3}{2}$$

d. Solving the equation:

$$\frac{1}{2x}-1=\frac{1}{4}$$
$$4x\left(\frac{1}{2x}-1\right)=4x\left(\frac{1}{4}\right)$$
$$2-4x=x$$
$$2=5x$$
$$x=\frac{2}{5}$$

47. Checking the solution: $(x+1)^2=\left(-1+5\sqrt{2}+1\right)^2=\left(5\sqrt{2}\right)^2=25\cdot 2=50$

49. Let x represent the number. The equation is:

$$(x+3)^2=16$$
$$x+3=\pm\sqrt{16}$$
$$x+3=-4,4$$
$$x=-7,1$$

The number is either –7 or 1.

51. Let x represent the horizontal distance. Using the Pythagorean theorem:

$$x^2+472^2=4{,}097^2$$
$$x^2+222{,}784=16{,}785{,}409$$
$$x^2=16{,}562{,}625$$
$$x=\sqrt{16{,}562{,}625}\approx 4{,}070 \text{ feet}$$

53. Let x represent the height of the triangle. Draw the figure:

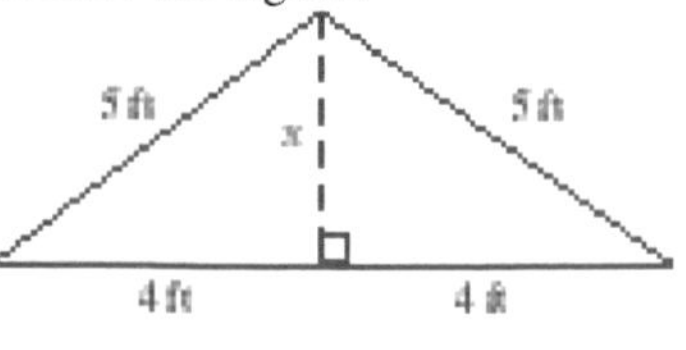

Using the Pythagorean theorem:

$$4^2+x^2=5^2$$
$$16+x^2=25$$
$$x^2=9$$
$$x=\sqrt{9}=3$$

The height is 3 feet.

55. Simplifying: $\left(\frac{1}{2}\cdot 18\right)^2=(9)^2=81$

57. Simplifying: $\left[\frac{1}{2}\cdot(-2)\right]^2=(-1)^2=1$

59. Simplifying: $\left(\frac{1}{2}\cdot 3\right)^2=\left(\frac{3}{2}\right)^2=\frac{9}{4}$

61. Simplifying: $\frac{2x^2+16}{2}=\frac{2(x^2+8)}{2}=x^2+8$

63. Factoring: $x^2+6x+9=(x+3)^2$

65. Factoring: $y^2-3y+\frac{9}{4}=\left(y-\frac{3}{2}\right)^2$

67. Solving the equation:

$$(x+2)^2=18$$
$$x+2=\pm\sqrt{18}$$
$$x+2=\pm 3\sqrt{2}$$
$$x=-2\pm 3\sqrt{2}$$

The correct answer is c.

9.2 Completing the Square

1. Completing the square: $x^2+12x+36=(x+6)^2$

3. Completing the square: $x^2-12x+36=(x-6)^2$

5. Completing the square: $x^2+4x+4=(x+2)^2$

7. Completing the square: $x^2-4x+4=(x-2)^2$

9. Completing the square: $a^2-10a+25=(a-5)^2$

11. Completing the square: $x^2+5x+\frac{25}{4}=\left(x+\frac{5}{2}\right)^2$

13. Completing the square: $y^2+y+\frac{1}{4}=\left(y+\frac{1}{2}\right)^2$

15. Completing the square: $x^2+\frac{1}{2}x+\frac{1}{16}=\left(x+\frac{1}{4}\right)^2$

17. Completing the square: $x^2+\frac{2}{3}x+\frac{1}{9}=\left(x+\frac{1}{3}\right)^2$

19. Solve by completing the square:

$$x^2+4x=12$$
$$x^2+4x+4=12+4$$
$$(x+2)^2=16$$
$$x+2=\pm 4$$
$$x=-2\pm 4$$
$$x=-6,2$$

21. Solve by completing the square:

$$x^2+12x=-27$$
$$x^2+12x+36=-27+36$$
$$(x+6)^2=9$$
$$x+6=\pm\sqrt{9}=\pm 3$$
$$x+6=-3,3$$
$$x=-9,-3$$

23. Solving the equation:

$$a^2-2a-5=0$$
$$a^2-2a+1=5+1$$
$$(a-1)^2=6$$
$$a-1=\pm\sqrt{6}$$
$$a=1\pm\sqrt{6}$$

25. Solving the equation:

$$y^2-8y+1=0$$
$$y^2-8y+16=-1+16$$
$$(y-4)^2=15$$
$$y-4=\pm\sqrt{15}$$
$$y=4\pm\sqrt{15}$$

27. Solving the equation:

$$x^2-5x-3=0$$
$$x^2-5x+\frac{25}{4}=3+\frac{25}{4}$$
$$\left(x-\frac{5}{2}\right)^2=\frac{37}{4}$$
$$x-\frac{5}{2}=\pm\frac{\sqrt{37}}{2}$$
$$x=\frac{5\pm\sqrt{37}}{2}$$

29. Solving the equation:

$$2x^2-4x-8=0$$
$$x^2-2x-4=0$$
$$x^2-2x+1=4+1$$
$$(x-1)^2=5$$
$$x-1=\pm\sqrt{5}$$
$$x=1\pm\sqrt{5}$$

31. Solving the equation:

$$x^2-10x=0$$
$$x^2-10x+25=0+25$$
$$(x-5)^2=25$$
$$x-5=\pm 5$$
$$x=5\pm 5$$
$$x=0,10$$

33. Solving the equation:

$$y^2+2y-15=0$$
$$y^2+2y=15$$
$$y^2+2y+1=15+1$$
$$(y+1)^2=16$$
$$y+1=\pm 4$$
$$y=-1\pm 4$$
$$y=-5,3$$

35. Solving the equation:

$$x^2+6x=-5$$
$$x^2+6x+9=-5+9$$
$$(x+3)^2=4$$
$$x+3=\pm 2$$
$$x=-3\pm 2$$
$$x=-5,-1$$

37. Solving the equation:

$$x^2-3x=-2$$
$$x^2-3x+\frac{9}{4}=-2+\frac{9}{4}$$
$$\left(x-\frac{3}{2}\right)^2=\frac{1}{4}$$
$$x-\frac{3}{2}=\pm\sqrt{\frac{1}{4}}$$
$$x-\frac{3}{2}=\pm\frac{1}{2}$$
$$x=\frac{3}{2}\pm\frac{1}{2}$$
$$x=1,2$$

39. Solve by completing the square:

$$\begin{aligned} 4x^2+8x-4&=0\\ x^2+2x-1&=0\\ x^2+2x&=1\\ x^2+2x+1&=1+1\\ (x+1)^2&=2\\ x+1&=\pm\sqrt{2}\\ x&=-1\pm\sqrt{2} \end{aligned}$$

41. Solve by completing the square:

$$\begin{aligned} 2x^2+2x-4&=0\\ x^2+x-2&=0\\ x^2+x&=2\\ x^2+x+\frac{1}{4}&=2+\frac{1}{4}\\ \left(x+\frac{1}{2}\right)^2&=\frac{9}{4}\\ x+\frac{1}{2}&=\pm\sqrt{\frac{9}{4}}\\ x+\frac{1}{2}&=\pm\frac{3}{2}\\ x&=-\frac{1}{2}\pm\frac{3}{2}\\ x&=-2,1 \end{aligned}$$

43. **a.** Solving by factoring:

$$\begin{aligned} x^2-6x&=0\\ x(x-6)&=0\\ x&=0,6 \end{aligned}$$

b. Solving by completing the square:

$$\begin{aligned} x^2-6x&=0\\ x^2-6x+9&=0+9\\ (x-3)^2&=9\\ x-3&=\pm\sqrt{9}\\ x-3&=-3,3\\ x&=0,6 \end{aligned}$$

45. **a.** Solving by factoring:

$$\begin{aligned} x^2+2x&=35\\ x^2+2x-35&=0\\ (x+7)(x-5)&=0\\ x&=-7,5 \end{aligned}$$

b. Solving by completing the square:

$$\begin{aligned} x^2+2x&=35\\ x^2+2x+1&=35+1\\ (x+1)^2&=36\\ x+1&=\pm\sqrt{36}\\ x+1&=-6,6\\ x&=-7,5 \end{aligned}$$

47. Using the Pythagorean theorem:

$$\begin{aligned} (3x)^2+(4x)^2&=14^2\\ 9x^2+16x^2&=196\\ 25x^2&=196\\ x^2&=\frac{196}{25}\\ x&=\sqrt{\frac{196}{25}}=2.8\\ 3x&=8.4\\ 4x&=11.2 \end{aligned}$$

The height of the screen is 8.4 inches and the width of the screen is 11.2 inches.

49. Solving by completing the square:

$$\begin{aligned} x^2-2x-1&=0 \\ x^2-2x&=1 \\ x^2-2x+1&=1+1 \\ (x-1)^2&=2 \\ x-1&=\pm\sqrt{2} \\ x&=1\pm\sqrt{2} \\ x&\approx -0.4,2.4 \end{aligned}$$

51. **a.** Adding the two solutions: $\left(-2+\sqrt{7}\right)+\left(-2-\sqrt{7}\right)=-4$

b. Multiplying the two solutions: $\left(-2+\sqrt{7}\right)\left(-2-\sqrt{7}\right)=(-2)^2-\left(\sqrt{7}\right)^2=4-7=-3$

53. Drawing the diagram:

55. Drawing the diagram:

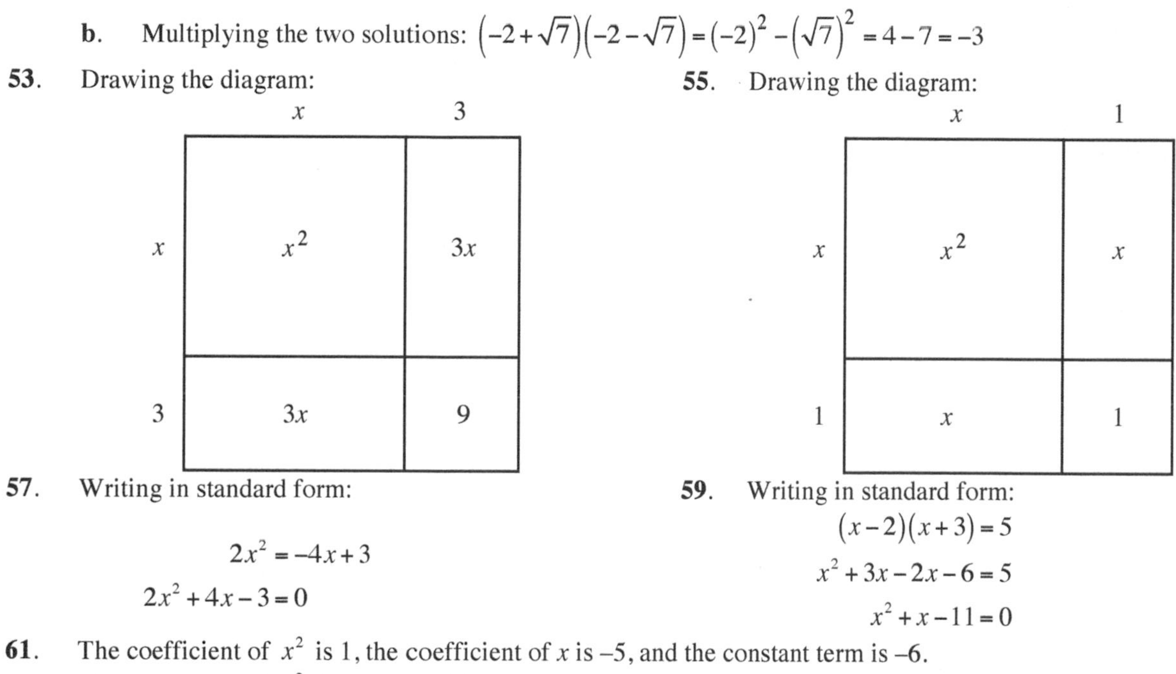

57. Writing in standard form:

$$\begin{aligned} 2x^2&=-4x+3 \\ 2x^2+4x-3&=0 \end{aligned}$$

59. Writing in standard form:

$$\begin{aligned} (x-2)(x+3)&=5 \\ x^2+3x-2x-6&=5 \\ x^2+x-11&=0 \end{aligned}$$

61. The coefficient of x^2 is 1, the coefficient of x is –5, and the constant term is –6.

63. The coefficient of x^2 is 2, the coefficient of x is 4, and the constant term is –3.

65. Evaluating when $a = 1$, $b = -5$, and $c = -6$: $b^2-4ac=(-5)^2-4(1)(-6)=25+24=49$

67. Evaluating when $a = 2$, $b = 4$, and $c = -3$: $b^2-4ac=(4)^2-4(2)(-3)=16+24=40$

69. Simplifying: $\dfrac{5+\sqrt{49}}{2}=\dfrac{5+7}{2}=\dfrac{12}{2}=6$

71. Simplifying: $\dfrac{-4-\sqrt{40}}{4}=\dfrac{-4-2\sqrt{10}}{4}=\dfrac{2\left(-2-\sqrt{10}\right)}{4}=\dfrac{-2-\sqrt{10}}{2}$

73. Since $x^2+8x+16=(x+4)^2$, the value is 16. The correct answer is c.

9.3 The Quadratic Formula

1. Evaluating when $a = 1, b = 2$, and $c = -3$:

$$\frac{-b \pm \sqrt{b^2 - 4ac}}{2a} = \frac{-2 \pm \sqrt{(2)^2 - 4(1)(-3)}}{2(1)} = \frac{-2 \pm \sqrt{4 + 12}}{2} = \frac{-2 \pm \sqrt{16}}{2} = \frac{-2 \pm 4}{2} = -3, 1$$

3. Evaluating when $a = 6, b = 1$, and $c = -2$:

$$\frac{-b \pm \sqrt{b^2 - 4ac}}{2a} = \frac{-1 \pm \sqrt{(1)^2 - 4(6)(-2)}}{2(6)} = \frac{-1 \pm \sqrt{1 + 48}}{12} = \frac{-1 \pm \sqrt{49}}{12} = \frac{-1 \pm 7}{12} = -\frac{2}{3}, \frac{1}{2}$$

5. Evaluating when $a = 1, b = 3$, and $c = -2$: $\frac{-b \pm \sqrt{b^2 - 4ac}}{2a} = \frac{-3 \pm \sqrt{(3)^2 - 4(1)(-2)}}{2(1)} = \frac{-3 \pm \sqrt{9 + 8}}{2} = \frac{-3 \pm \sqrt{17}}{2}$

7. Evaluating when $a = 2, b = -6$, and $c = 1$:

$$\frac{-b \pm \sqrt{b^2 - 4ac}}{2a} = \frac{6 \pm \sqrt{(-6)^2 - 4(2)(1)}}{2(2)} = \frac{6 \pm \sqrt{36 - 8}}{4} = \frac{6 \pm \sqrt{28}}{4} = \frac{6 \pm 2\sqrt{7}}{4} = \frac{3 \pm \sqrt{7}}{2}$$

9. Evaluating when $a = 1, b = -4$, and $c = -2$:

$$\frac{-b \pm \sqrt{b^2 - 4ac}}{2a} = \frac{4 \pm \sqrt{(-4)^2 - 4(1)(-2)}}{2(1)} = \frac{4 \pm \sqrt{16 + 8}}{2} = \frac{4 \pm \sqrt{24}}{2} = \frac{4 \pm 2\sqrt{6}}{2} = 2 \pm \sqrt{6}$$

11. Evaluating when $a = 3, b = -8$, and $c = 2$:

$$\frac{-b \pm \sqrt{b^2 - 4ac}}{2a} = \frac{8 \pm \sqrt{(-8)^2 - 4(3)(2)}}{2(3)} = \frac{8 \pm \sqrt{64 - 24}}{6} = \frac{8 \pm \sqrt{40}}{6} = \frac{8 \pm 2\sqrt{10}}{6} = \frac{4 \pm \sqrt{10}}{3}$$

13. Using the quadratic formula with $a = 1, b = 5$, and $c = 6$:

$$x = \frac{-b \pm \sqrt{b^2 - 4ac}}{2a} = \frac{-5 \pm \sqrt{(5)^2 - 4(1)(6)}}{2(1)} = \frac{-5 \pm \sqrt{25 - 24}}{2} = \frac{-5 \pm \sqrt{1}}{2} = \frac{-5 \pm 1}{2} = -3, -2$$

15. Using the quadratic formula with $a = 1, b = -4$, and $c = 1$:

$$x = \frac{-b \pm \sqrt{b^2 - 4ac}}{2a} = \frac{4 \pm \sqrt{(-4)^2 - 4(1)(1)}}{2(1)} = \frac{4 \pm \sqrt{16 - 4}}{2} = \frac{4 \pm \sqrt{12}}{2} = \frac{4 \pm 2\sqrt{3}}{2} = 2 \pm \sqrt{3}$$

17. Using the quadratic formula with $a = 1, b = 6$, and $c = 7$:

$$x = \frac{-b \pm \sqrt{b^2 - 4ac}}{2a} = \frac{-6 \pm \sqrt{(6)^2 - 4(1)(7)}}{2(1)} = \frac{-6 \pm \sqrt{36 - 28}}{2} = \frac{-6 \pm \sqrt{8}}{2} = \frac{-6 \pm 2\sqrt{2}}{2} = -3 \pm \sqrt{2}$$

19. Using the quadratic formula with $a = 4, b = 8$, and $c = 1$:

$$x = \frac{-b \pm \sqrt{b^2 - 4ac}}{2a} = \frac{-8 \pm \sqrt{(8)^2 - 4(4)(1)}}{2(4)} = \frac{-8 \pm \sqrt{64 - 16}}{8} = \frac{-8 \pm \sqrt{48}}{8} = \frac{-8 \pm 4\sqrt{3}}{8} = \frac{-2 \pm \sqrt{3}}{2}$$

21. First write the equation as $2x^2 - 3x - 5 = 0$. Using the quadratic formula with $a = 2, b = -3$, and $c = -5$:

$$x = \frac{-b \pm \sqrt{b^2 - 4ac}}{2a} = \frac{3 \pm \sqrt{(-3)^2 - 4(2)(-5)}}{2(2)} = \frac{3 \pm \sqrt{9 + 40}}{4} = \frac{3 \pm \sqrt{49}}{4} = \frac{3 \pm 7}{4} = -1, \frac{5}{2}$$

23. First write the equation as $2x^2 + 6x - 7 = 0$. Using the quadratic formula with $a = 2, b = 6$, and $c = -7$:

$$x = \frac{-b \pm \sqrt{b^2 - 4ac}}{2a} = \frac{-6 \pm \sqrt{(6)^2 - 4(2)(-7)}}{2(2)} = \frac{-6 \pm \sqrt{36 + 56}}{4} = \frac{-6 \pm \sqrt{92}}{4} = \frac{-6 \pm 2\sqrt{23}}{4} = \frac{-3 \pm \sqrt{23}}{2}$$

25. Using the quadratic formula with $a = 1, b = 3$, and $c = 2$:

$$x = \frac{-b \pm \sqrt{b^2 - 4ac}}{2a} = \frac{-3 \pm \sqrt{(3)^2 - 4(1)(2)}}{2(1)} = \frac{-3 \pm \sqrt{9-8}}{2} = \frac{-3 \pm \sqrt{1}}{2} = \frac{-3 \pm 1}{2} = -2, -1$$

27. Using the quadratic formula with $a = 1, b = 6$, and $c = 9$:

$$x = \frac{-b \pm \sqrt{b^2 - 4ac}}{2a} = \frac{-6 \pm \sqrt{(6)^2 - 4(1)(9)}}{2(1)} = \frac{-6 \pm \sqrt{36-36}}{2} = \frac{-6 \pm \sqrt{0}}{2} = \frac{-6 \pm 0}{2} = -3$$

29. Using the quadratic formula with $a = 1, b = -2$, and $c = 1$:

$$x = \frac{-b \pm \sqrt{b^2 - 4ac}}{2a} = \frac{2 \pm \sqrt{(-2)^2 - 4(1)(1)}}{2(1)} = \frac{2 \pm \sqrt{4-4}}{2} = \frac{2 \pm \sqrt{0}}{2} = \frac{2 \pm 0}{2} = 1$$

31. Using the quadratic formula with $a = 1, b = -5$, and $c = -7$:

$$x = \frac{-b \pm \sqrt{b^2 - 4ac}}{2a} = \frac{5 \pm \sqrt{(-5)^2 - 4(1)(-7)}}{2(1)} = \frac{5 \pm \sqrt{25+28}}{2} = \frac{5 \pm \sqrt{53}}{2}$$

33. First write the equation as $6x^2 - x - 2 = 0$. Using the quadratic formula with $a = 6, b = -1$, and $c = -2$:

$$x = \frac{-b \pm \sqrt{b^2 - 4ac}}{2a} = \frac{1 \pm \sqrt{(-1)^2 - 4(6)(-2)}}{2(6)} = \frac{1 \pm \sqrt{1+48}}{12} = \frac{1 \pm \sqrt{49}}{12} = \frac{1 \pm 7}{12} = -\frac{1}{2}, \frac{2}{3}$$

35. First simplify the equation:

$$(x-2)(x+1) = 3$$
$$x^2 - 2x + x - 2 = 3$$
$$x^2 - x - 5 = 0$$

Using the quadratic formula with $a = 1, b = -1$, and $c = -5$:

$$x = \frac{-b \pm \sqrt{b^2 - 4ac}}{2a} = \frac{1 \pm \sqrt{(-1)^2 - 4(1)(-5)}}{2(1)} = \frac{1 \pm \sqrt{1+20}}{2} = \frac{1 \pm \sqrt{21}}{2}$$

37. First write the equation as $3x^2 + 4x - 2 = 0$. Using the quadratic formula with $a = 3, b = 4$, and $c = -2$:

$$x = \frac{-b \pm \sqrt{b^2 - 4ac}}{2a} = \frac{-4 \pm \sqrt{(4)^2 - 4(3)(-2)}}{2(3)} = \frac{-4 \pm \sqrt{16+24}}{6} = \frac{-4 \pm \sqrt{40}}{6} = \frac{-4 \pm 2\sqrt{10}}{6} = \frac{-2 \pm \sqrt{10}}{3}$$

39. First write the equation as $2x^2 - 2x - 5 = 0$. Using the quadratic formula with $a = 2, b = -2$, and $c = -5$:

$$x = \frac{-b \pm \sqrt{b^2 - 4ac}}{2a} = \frac{1 \pm \sqrt{(-2)^2 - 4(2)(-5)}}{2(2)} = \frac{2 \pm \sqrt{4+40}}{4} = \frac{2 \pm \sqrt{44}}{4} = \frac{2 \pm 2\sqrt{11}}{4} = \frac{1 \pm \sqrt{11}}{2}$$

41. **a.** Solving by factoring:

$$3x^2 - 5x = 0$$
$$x(3x-5) = 0$$
$$x = 0, \frac{5}{3}$$

b. Using the quadratic formula: $x = \frac{-b \pm \sqrt{b^2 - 4ac}}{2a} = \frac{5 \pm \sqrt{(-5)^2 - 4(3)(0)}}{2(3)} = \frac{5 \pm \sqrt{25-0}}{6} = \frac{5 \pm 5}{6} = 0, \frac{5}{3}$

43. Solving the equation:

$$2x^3+3x^2-4x=0$$
$$x(2x^2+3x-4)=0$$
$$x=0, x=\frac{-3\pm\sqrt{(3)^2-4(2)(-4)}}{2(2)}=\frac{-3\pm\sqrt{9+32}}{4}=\frac{-3\pm\sqrt{41}}{4}$$

45. The expressions from **a** and **b** are equivalent, since: $\frac{6+2\sqrt{3}}{4}=\frac{2(3+\sqrt{3})}{4}=\frac{3+\sqrt{3}}{2}$

47. Solving the equation:

$$\frac{1}{2}x^2-\frac{1}{2}x-\frac{1}{6}=0$$
$$6\left(\frac{1}{2}x^2-\frac{1}{2}x-\frac{1}{6}\right)=6(0)$$
$$3x^2-3x-1=0$$
$$x=\frac{3\pm\sqrt{(-3)^2-4(3)(-1)}}{2(3)}=\frac{3\pm\sqrt{9+12}}{6}=\frac{3\pm\sqrt{21}}{6}$$

49. Solving the equation:

$$56=8+64t-16t^2$$
$$16t^2-64t+48=0$$
$$t^2-4t+3=0$$
$$(t-1)(t-3)=0$$
$$t=1,3$$

The arrow is 56 feet above the ground after 1 second and after 3 seconds.

51. Simplifying: $7x+5+8x-6=(7x+8x)+(5-6)=15x-1$

53. Simplifying: $(2-5x)-(3+7x)=2-5x-3-7x=(-5x-7x)+(2-3)=-12x-1$

55. Multiplying: $2x(8x-7)=16x^2-14x$

57. Multiplying: $(3x+6)(3x+4)=9x^2+12x+18x+24=9x^2+30x+24$

59. Rationalizing the denominator: $\frac{4}{3+\sqrt{2}}\cdot\frac{3-\sqrt{2}}{3-\sqrt{2}}=\frac{4(3-\sqrt{2})}{9-2}=\frac{4(3-\sqrt{2})}{7}=\frac{12-4\sqrt{2}}{7}$

61. Evaluating when $a=2$, $b=-5$, and $c=2$:

$$\frac{-b\pm\sqrt{b^2-4ac}}{2a}=\frac{5\pm\sqrt{(-5)^2-4(2)(2)}}{2(2)}=\frac{5\pm\sqrt{25-16}}{4}=\frac{5\pm\sqrt{9}}{4}=\frac{5\pm 3}{4}=\frac{1}{2},2$$

The correct answer is c.

9.4 Complex Numbers

1. Combining the complex numbers: $(3-2i)+3i=3+(-2i+3i)=3+i$

3. Combining the complex numbers: $(6+2i)-10i=6+(2i-10i)=6-8i$

5. Combining the complex numbers: $(11+9i)+(4+i)=(11+4)+(9i+i)=15+10i$

7. Combining the complex numbers: $(3+2i)+(6-i)=(3+6)+(2i-i)=9+i$

9. Combining the complex numbers: $(5+7i)-(6+8i)=5+7i-6-8i=-1-i$

11. Combining the complex numbers: $(9-i)+(2-i)=9-i+2-i=11-2i$

13. Combining the complex numbers: $(6+i)-4i-(2-i)=6+i-4i-2+i=4-2i$

15. Combining the complex numbers: $(6-11i)+3i+(2+i)=6-11i+3i+2+i=8-7i$

17. Combining the complex numbers: $(2+3i)-(6-2i)+(3-i)=2+3i-6+2i+3-i=-1+4i$

19. Multiplying the complex numbers: $3(2-i)=6-3i$

21. Multiplying the complex numbers: $2i(8-7i)=16i-14i^2=16i-14(-1)=14+16i$

23. Multiplying the complex numbers: $(2+i)(4-i)=8+4i-2i-i^2=8+2i-(-1)=8+2i+1=9+2i$

25. Multiplying the complex numbers: $(2+i)(3-5i)=6+3i-10i-5i^2=6-7i-5(-1)=6-7i+5=11-7i$

27. Multiplying the complex numbers: $(3+5i)(3-5i)=(3)^2-(5i)^2=9-25i^2=9-25(-1)=9+25=34$

29. Multiplying the complex numbers: $(2+i)(2-i)=(2)^2-(i)^2=4-i^2=4-(-1)=4+1=5$

31. Dividing the complex numbers: $\frac{2}{3-2i}\bullet\frac{3+2i}{3+2i}=\frac{2(3+2i)}{9-4i^2}=\frac{2(3+2i)}{9+4}=\frac{6+4i}{13}$

33. Dividing the complex numbers: $\frac{-3i}{2+3i}\bullet\frac{2-3i}{2-3i}=\frac{-6i+9i^2}{4-9i^2}=\frac{-6i-9}{4+9}=\frac{-9-6i}{13}$

35. Dividing the complex numbers: $\frac{i}{3-i}\bullet\frac{3+i}{3+i}=\frac{3i+i^2}{9-i^2}=\frac{3i-1}{9+1}=\frac{-1+3i}{10}$

37. Dividing the complex numbers: $\frac{2+i}{2-i}\bullet\frac{2+i}{2+i}=\frac{4+2i+2i+i^2}{4-i^2}=\frac{4+4i-1}{4+1}=\frac{3+4i}{5}$

39. Dividing the complex numbers:

$$\frac{4+5i}{3-6i}\bullet\frac{3+6i}{3+6i}=\frac{12+15i+24i+30i^2}{9-36i^2}=\frac{12+39i-30}{9+36}=\frac{-18+39i}{45}=\frac{3(-6+13i)}{45}=\frac{-6+13i}{15}$$

41. Multiplying using FOIL: $(x+3i)(x-3i)=x^2-(3i)^2=x^2-9i^2=x^2-9(-1)=x^2+9$

43. Simplifying: $\frac{1}{i}\bullet\frac{i}{i}=\frac{i}{i^2}=\frac{i}{-1}=-i$

45. Simplifying the radical: $\sqrt{36}=\sqrt{6^2}=6$

47. Simplifying the radical: $-\sqrt{75}=-\sqrt{25\bullet 3}=-5\sqrt{3}$

49. Solving the equation:

$$(x+2)^2=9$$
$$x+2=\pm\sqrt{9}$$
$$x+2=\pm 3$$
$$x=-2\pm 3$$
$$x=-5,1$$

51. Solving the equation:

$$\frac{1}{10}x^2-\frac{1}{5}x=\frac{1}{2}$$
$$10\left(\frac{1}{10}x^2-\frac{1}{5}x\right)=10\left(\frac{1}{2}\right)$$
$$x^2-2x=5$$
$$x^2-2x-5=0$$
$$x=\frac{2\pm\sqrt{(-2)^2-4(1)(-5)}}{2(1)}=\frac{2\pm\sqrt{4+20}}{2}=\frac{2\pm\sqrt{24}}{2}=\frac{2\pm2\sqrt{6}}{2}=1\pm\sqrt{6}$$

53. Solving the equation:

$$(2x-3)(2x-1)=4$$
$$4x^2-8x+3=4$$
$$4x^2-8x-1=0$$
$$x=\frac{8\pm\sqrt{(-8)^2-4(4)(-1)}}{2(4)}=\frac{8\pm\sqrt{64+16}}{8}=\frac{8\pm\sqrt{80}}{8}=\frac{8\pm4\sqrt{5}}{8}=\frac{2\pm\sqrt{5}}{2}$$

55. Adding the complex numbers: $(3-5i)+(4+2i)=(3+4)+(-5i+2i)=7-3i$

The correct answer is d.

57. Multiplying the complex numbers: $(3-5i)(4+2i)=12+6i-20i-10i^2=12-14i-10(-1)=12-14i+10=22-14i$

The correct answer is d.

9.5 Complex Solutions to Quadratic Equations

1. Writing as a complex number: $\sqrt{-16}=\sqrt{16(-1)}=\sqrt{16}\sqrt{-1}=4i$

3. Writing as a complex number: $-\sqrt{-49}=-\sqrt{49(-1)}=-\sqrt{49}\sqrt{-1}=-7i$

5. Writing as a complex number: $\sqrt{-6}=\sqrt{6(-1)}=\sqrt{6}\sqrt{-1}=i\sqrt{6}$

7. Writing as a complex number: $-\sqrt{-11}=-\sqrt{11(-1)}=-\sqrt{11}\sqrt{-1}=-i\sqrt{11}$

9. Writing as a complex number: $\sqrt{-32}=\sqrt{-16\cdot2}=\sqrt{-16}\sqrt{2}=4i\sqrt{2}$

11. Writing as a complex number: $\sqrt{-50}=\sqrt{-25\cdot2}=\sqrt{-25}\sqrt{2}=5i\sqrt{2}$

13. Writing as a complex number: $-\sqrt{-8}=-\sqrt{-4\cdot2}=-\sqrt{-4}\sqrt{2}=-2i\sqrt{2}$

15. Writing as a complex number: $\sqrt{-48}=\sqrt{-16\cdot3}=\sqrt{-16}\sqrt{3}=4i\sqrt{3}$

17. First write the equation as $x^2-2x+2=0$. Using $a=1$, $b=-2$, and $c=2$ in the quadratic formula:

$$x=\frac{-b\pm\sqrt{b^2-4ac}}{2a}=\frac{2\pm\sqrt{(-2)^2-4(1)(2)}}{2(1)}=\frac{2\pm\sqrt{4-8}}{2}=\frac{2\pm\sqrt{-4}}{2}=\frac{2\pm2i}{2}=1\pm i$$

19. Solving the equation:

$$x^2-4x=-4$$
$$x^2-4x+4=0$$
$$(x-2)^2=0$$
$$x-2=0$$
$$x=2$$

21. Solving the equation:

$$2x^2+5x=12$$
$$2x^2+5x-12=0$$
$$(2x-3)(x+4)=0$$
$$x=\frac{3}{2},-4$$

23. Solving the equation:

$$(x-2)^2=-4$$
$$x-2=\pm\sqrt{-4}$$
$$x-2=\pm 2i$$
$$x=2\pm 2i$$

25. Solving the equation:

$$\left(x+\frac{1}{2}\right)^2=-\frac{9}{4}$$
$$x+\frac{1}{2}=\pm\sqrt{-\frac{9}{4}}$$
$$x+\frac{1}{2}=\pm\frac{3}{2}i$$
$$x=-\frac{1}{2}\pm\frac{3}{2}i$$

27. Solving the equation:

$$\left(x-\frac{1}{2}\right)^2=-\frac{27}{36}$$
$$x-\frac{1}{2}=\pm\sqrt{-\frac{27}{36}}$$
$$x-\frac{1}{2}=\pm\frac{3i\sqrt{3}}{6}$$
$$x-\frac{1}{2}=\pm\frac{\sqrt{3}}{2}i$$
$$x=\frac{1}{2}\pm\frac{\sqrt{3}}{2}i$$

29. Using $a=1$, $b=1$, and $c=1$ in the quadratic formula:

$$x=\frac{-b\pm\sqrt{b^2-4ac}}{2a}=\frac{-1\pm\sqrt{(1)^2-4(1)(1)}}{2(1)}=\frac{-1\pm\sqrt{1-4}}{2}=\frac{-1\pm\sqrt{-3}}{2}=-\frac{1}{2}\pm\frac{\sqrt{3}}{2}i$$

31. Solving the equation:

$$x^2-5x+6=0$$
$$(x-2)(x-3)=0$$
$$x=2,3$$

33. First multiply by 6 to clear the equation of fractions:

$$6\left(\frac{1}{2}x^2+\frac{1}{3}x+\frac{1}{6}\right)=6(0)$$
$$3x^2+2x+1=0$$

Using $a=3$, $b=2$, and $c=1$ in the quadratic formula:

$$x=\frac{-b\pm\sqrt{b^2-4ac}}{2a}=\frac{-2\pm\sqrt{(2)^2-4(3)(1)}}{2(3)}=\frac{-2\pm\sqrt{4-12}}{6}=\frac{-2\pm\sqrt{-8}}{6}=\frac{-2\pm 2i\sqrt{2}}{6}=\frac{-1\pm i\sqrt{2}}{3}=-\frac{1}{3}\pm\frac{\sqrt{2}}{3}i$$

35. First multiply by 6 to clear the equation of fractions:

$$6\left(\frac{1}{3}x^2\right)=6\left(-\frac{1}{2}x+\frac{1}{3}\right)$$
$$2x^2=-3x+2$$
$$2x^2+3x-2=0$$
$$(2x-1)(x+2)=0$$
$$x=\frac{1}{2},-2$$

37. Simplifying the equation:

$$(x+2)(x-3)=5$$
$$x^2-x-6=5$$
$$x^2-x-11=0$$

Using $a = 1$, $b = -1$, and $c = -11$ in the quadratic formula:

$$x=\frac{-b\pm\sqrt{b^2-4ac}}{2a}=\frac{1\pm\sqrt{(-1)^2-4(1)(-11)}}{2(1)}=\frac{1\pm\sqrt{1+44}}{2}=\frac{1\pm\sqrt{45}}{2}=\frac{1\pm3\sqrt{5}}{2}$$

39. Simplifying the equation:

$$(x-5)(x-3)=-10$$
$$x^2-8x+15=-10$$
$$x^2-8x+25=0$$

Using $a = 1$, $b = -8$, and $c = 25$ in the quadratic formula:

$$x=\frac{-b\pm\sqrt{b^2-4ac}}{2a}=\frac{8\pm\sqrt{(-8)^2-4(1)(25)}}{2(1)}=\frac{8\pm\sqrt{64-100}}{2}=\frac{8\pm\sqrt{-36}}{2}=\frac{8\pm6i}{2}=4\pm3i$$

41. Simplifying the equation:

$$(2x-2)(x-3)=9$$
$$2x^2-8x+6=9$$
$$2x^2-8x-3=0$$

Using $a = 2$, $b = -8$, and $c = -3$ in the quadratic formula:

$$x=\frac{-b\pm\sqrt{b^2-4ac}}{2a}=\frac{8\pm\sqrt{(-8)^2-4(2)(-3)}}{2(2)}=\frac{8\pm\sqrt{64+24}}{4}=\frac{8\pm\sqrt{88}}{4}=\frac{8\pm2\sqrt{22}}{4}=\frac{4\pm\sqrt{22}}{2}$$

43. Substituting $x=2+2i$ into the equation:

$$x^2-4x+8=(2+2i)^2-4(2+2i)+8=4+8i+4i^2-8-8i+8=4+8i-4-8-8i+8=0$$

Yes, $x=2+2i$ is a solution to the equation.

45. The other solution is $3-7i$.

47. Completing the square: $x^2-6x+9=(x-3)^2$

49. Evaluating when $x = -4$: $y=(-4+1)^2-3=(-3)^2-3=9-3=6$

51. Evaluating when $x = -2$: $y=(-2+1)^2-3=(-1)^2-3=1-3=-2$

53. Evaluating when $x = 1$: $y=(1+1)^2-3=(2)^2-3=4-3=1$

55. Graphing the ordered pair:

57. Graphing the ordered pair:

59. Graphing the ordered pair:

61. Graphing the ordered pair:

63. Graphing the ordered pair:

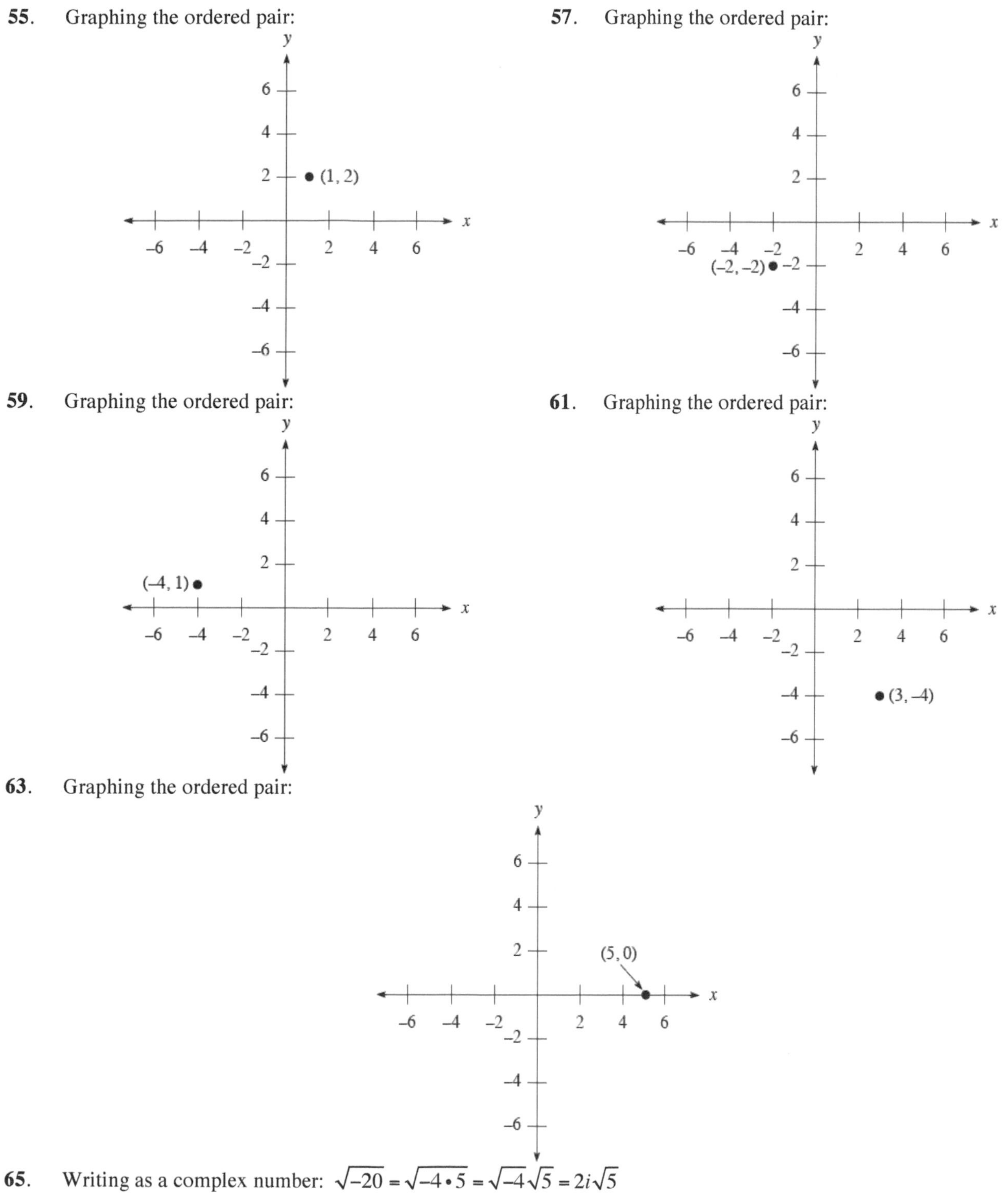

65. Writing as a complex number: $\sqrt{-20}=\sqrt{-4\cdot 5}=\sqrt{-4}\sqrt{5}=2i\sqrt{5}$
The correct answer is d.

9.6 Graphing Parabolas

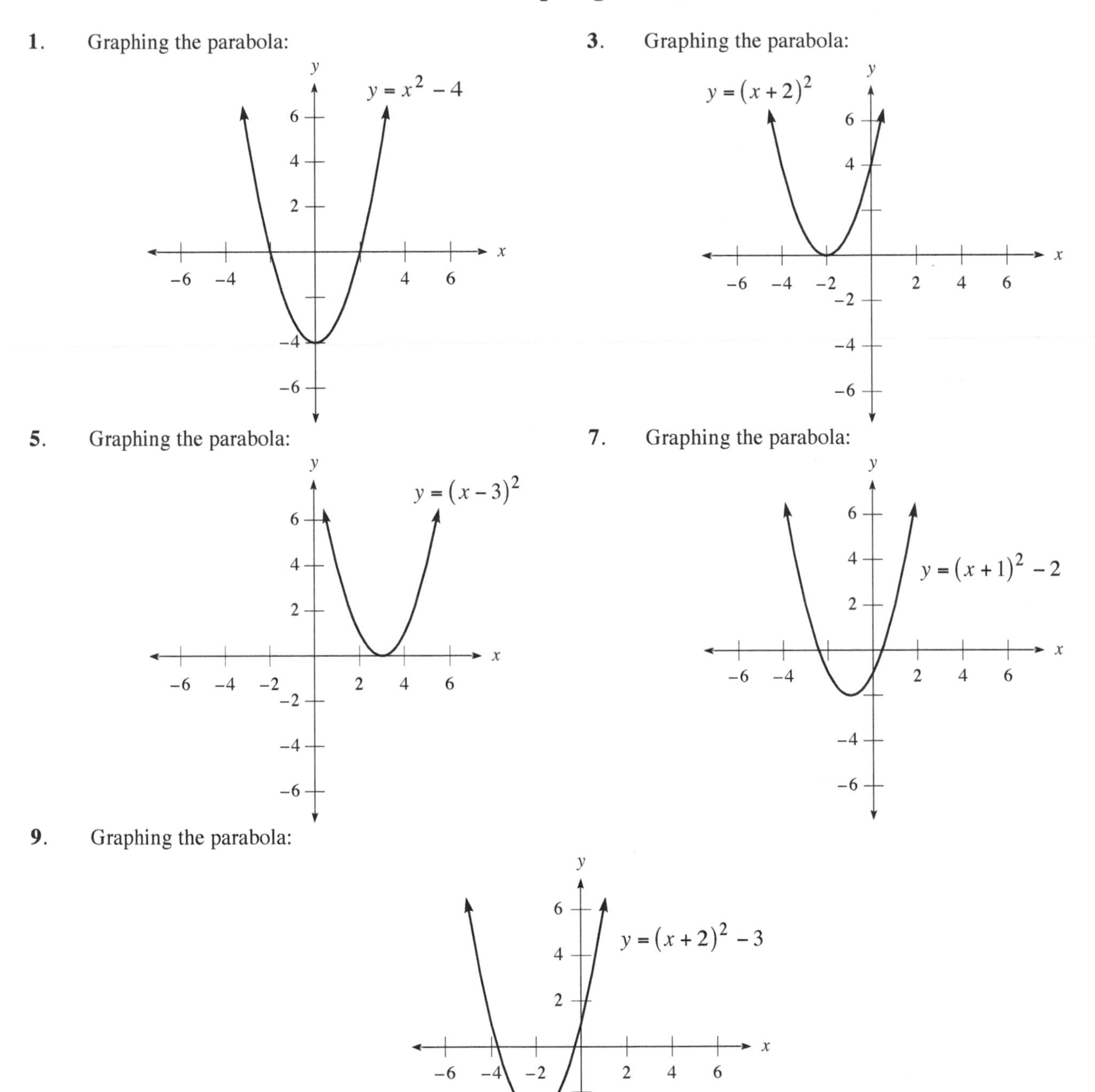

11. Begin by completing the square: $y=x^2+6x+5=\left(x^2+6x+9\right)+5-9=(x+3)^2-4$. Graphing the parabola:

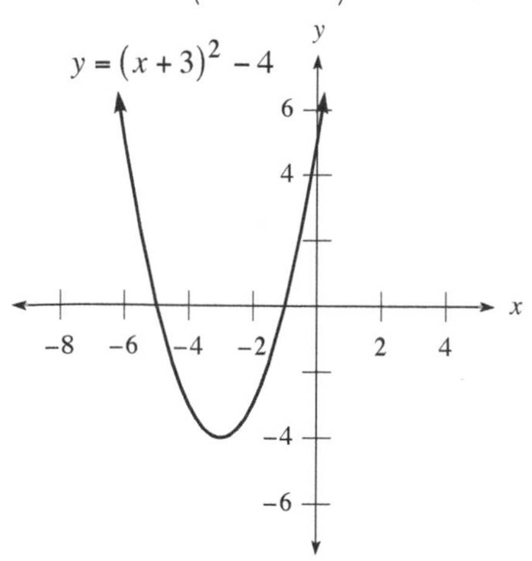

13. The x-intercepts are $\left(\pm\sqrt{5},0\right)$ and the y-intercept is $(0,-5)$. Graphing the parabola:

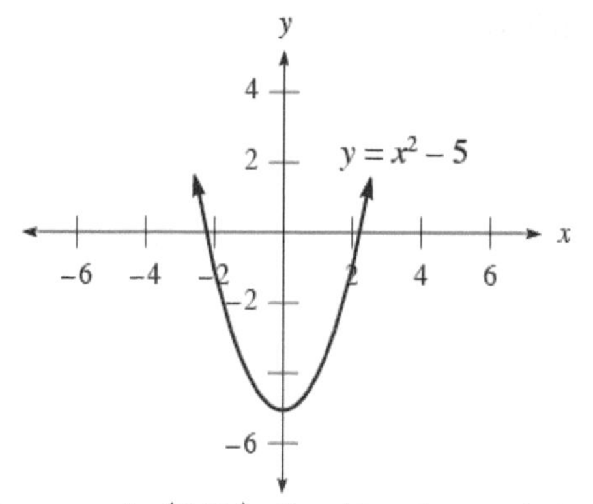

15. The x-intercept is $(5,0)$ and the y-intercept is $(0,25)$. Graphing the parabola:

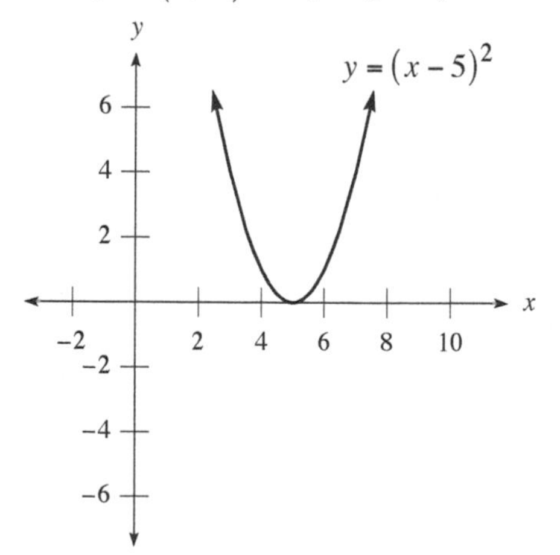

17. The x-intercepts are $\left(3\pm\sqrt{2},0\right)$ and the y-intercept is $(0,7)$. Graphing the parabola:

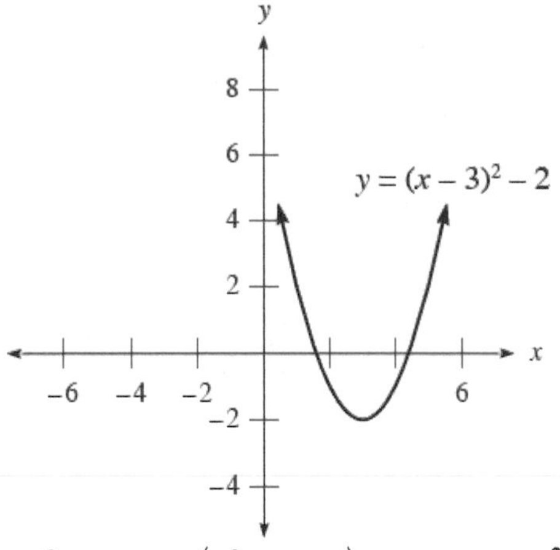

19. Begin by completing the square: $y=x^2-2x-3=\left(x^2-2x+1\right)-3-1=(x-1)^2-4$

The x-intercepts are $(-1,0)$ and $(3,0)$, and the y-intercept is $(0,-3)$. Graphing the parabola:

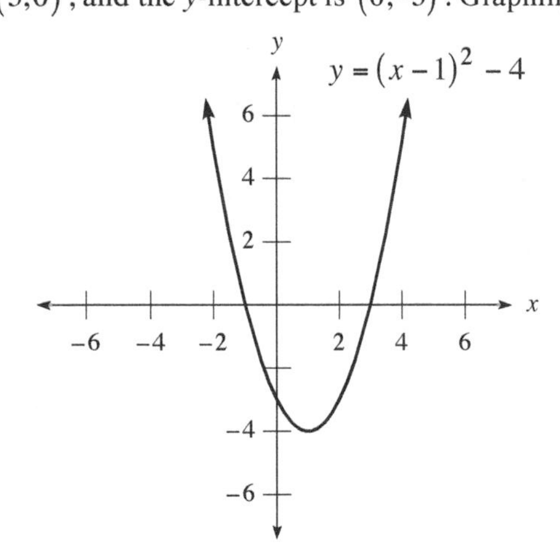

21. Begin by completing the square: $y=x^2-4x+2=\left(x^2-4x+4\right)-4+2=(x-2)^2-2$

The x-intercepts are $\left(2\pm\sqrt{2},0\right)$ and the y-intercept is $(0,2)$. Graphing the parabola:

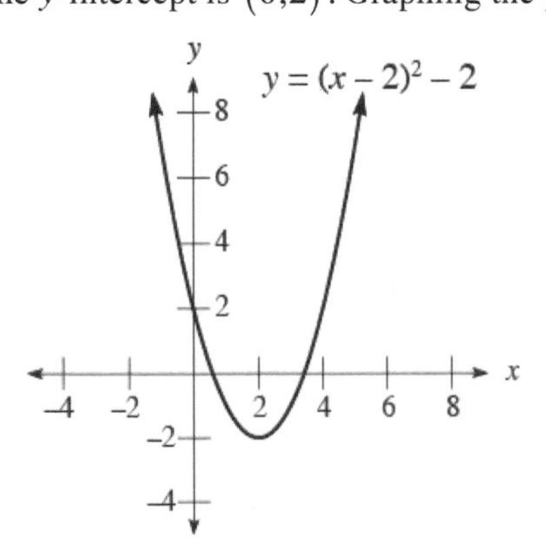

23. Begin by making a table of values:

x	y
−3	−5
−2	0
−1	3
0	4
1	3
2	0
3	−5

Graphing the parabola:

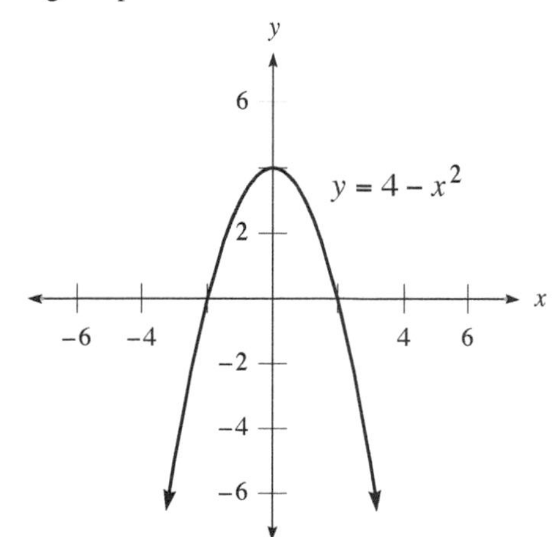

25. Begin by making a table of values:

x	y
−2	−5
−1	−2
0	−1
1	−2
2	−5

Graphing the parabola:

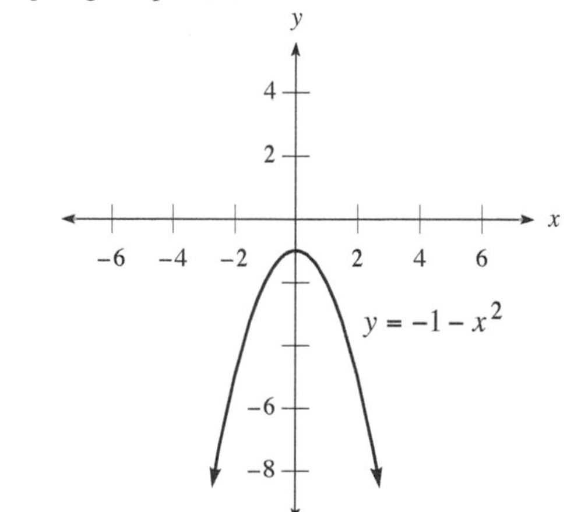

27. Graphing the line and the parabola:

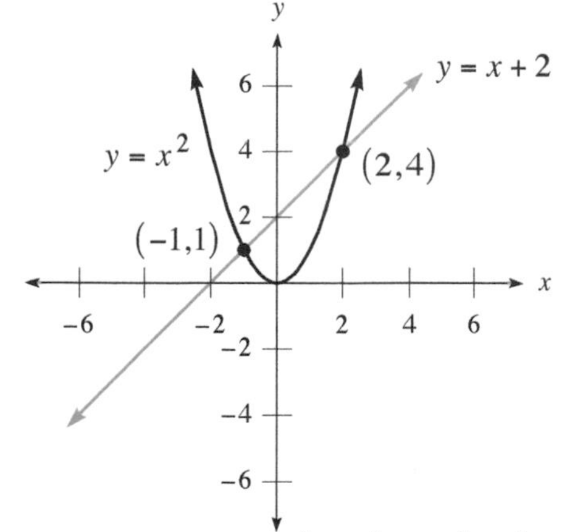

The intersection points are $(-1,1)$ and $(2,4)$.

29. Graphing the two parabolas:

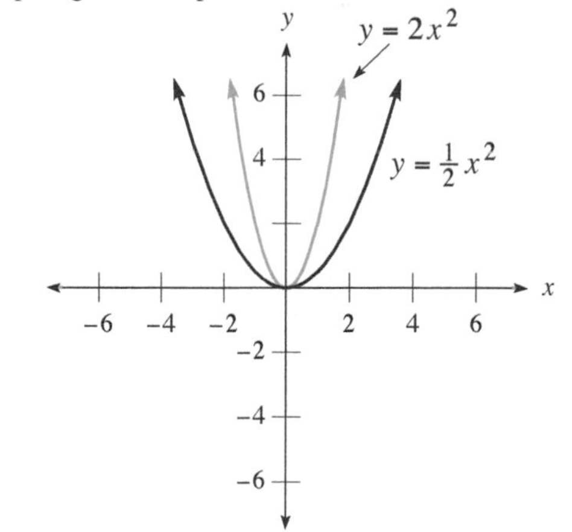

31. Finding the root: $\sqrt{49}=7$

33. Simplifying the radical: $\sqrt{50}=\sqrt{25\cdot 2}=5\sqrt{2}$

35. Simplifying the radical: $\sqrt{\frac{2}{5}}=\frac{\sqrt{2}}{\sqrt{5}}\cdot\frac{\sqrt{5}}{\sqrt{5}}=\frac{\sqrt{10}}{5}$

37. Performing the operations: $3\sqrt{12}+5\sqrt{27}=3\sqrt{4\cdot 3}+5\sqrt{9\cdot 3}=3\cdot 2\sqrt{3}+5\cdot 3\sqrt{3}=6\sqrt{3}+15\sqrt{3}=21\sqrt{3}$

39. Performing the operations: $\left(\sqrt{6}+2\right)\left(\sqrt{6}-5\right)=\sqrt{36}+2\sqrt{6}-5\sqrt{6}-10=6-3\sqrt{6}-10=-4-3\sqrt{6}$

41. Rationalizing the denominator: $\frac{8}{\sqrt{5}-\sqrt{3}}\cdot\frac{\sqrt{5}+\sqrt{3}}{\sqrt{5}+\sqrt{3}}=\frac{8\left(\sqrt{5}+\sqrt{3}\right)}{5-3}=\frac{8\left(\sqrt{5}+\sqrt{3}\right)}{2}=4\left(\sqrt{5}+\sqrt{3}\right)=4\sqrt{5}+4\sqrt{3}$

43. Solving the equation:

$$\begin{aligned}\sqrt{2x-5}&=3\\ \left(\sqrt{2x-5}\right)^2&=3^2\\ 2x-5&=9\\ 2x&=14\\ x&=7\end{aligned}$$

This solution checks in the original equation.

45. The ordered pair $(-6,5)$ satisfies the equation. The correct answer is b.

47. Setting $y = 0$ and using the quadratic formula: $x=\dfrac{-b\pm\sqrt{b^2-4ac}}{2a}=\dfrac{-3\pm\sqrt{(3)^2-4(1)(-1)}}{2(1)}=\dfrac{-3\pm\sqrt{9+4}}{2}=\dfrac{-3\pm\sqrt{13}}{2}$

The correct answer is a.

Chapter 9 Test

1. Solving the equation:

$$\begin{aligned}x^2-8x-9&=0\\ (x+1)(x-9)&=0\\ x&=-1,9\end{aligned}$$

2. Solving the equation:

$$\begin{aligned}(x-2)^2&=18\\ x-2&=\pm\sqrt{18}\\ x-2&=\pm3\sqrt{2}\\ x&=2\pm3\sqrt{2}\end{aligned}$$

3. Solving the equation:

$$\begin{aligned}\left(x-\frac{7}{3}\right)^2&=-\frac{50}{9}\\ x-\frac{7}{3}&=\pm\sqrt{-\frac{50}{9}}\\ x-\frac{7}{3}&=\pm\frac{\sqrt{50}}{3}i\\ x-\frac{7}{3}&=\pm\frac{5\sqrt{2}}{3}i\\ x&=\frac{7}{3}\pm\frac{5\sqrt{2}}{3}i\end{aligned}$$

4. Solving the equation:

$$\begin{aligned}\frac{1}{3}x^2&=\frac{5}{6}x-\frac{1}{2}\\ 6\left(\frac{1}{3}x^2\right)&=6\left(\frac{5}{6}x-\frac{1}{2}\right)\\ 2x^2&=5x-3\\ 2x^2-5x+3&=0\\ (2x-3)(x-1)&=0\\ x&=1,\frac{3}{2}\end{aligned}$$

5. Solving the equation:

$$\begin{aligned}4x^2+7x-2&=0\\ (4x-1)(x+2)&=0\\ x&=-2,\frac{1}{4}\end{aligned}$$

6. Solving the equation:

$$\begin{aligned}(x-4)(x+1)&=6\\ x^2-3x-4&=6\\ x^2-3x-10&=0\\ (x+2)(x-5)&=0\\ x&=-2,5\end{aligned}$$

7. Solving the equation:

$$\begin{aligned} 4x^2-20x+25&=0 \\ (2x-5)^2&=0 \\ 2x-5&=0 \\ x&=\frac{5}{2} \end{aligned}$$

8. Solving by completing the square:

$$\begin{aligned} x^2-4x-9&=0 \\ x^2-4x&=9 \\ x^2-4x+4&=9+4 \\ (x-2)^2&=13 \\ x-2&=\pm\sqrt{13} \\ x&=2\pm\sqrt{13} \end{aligned}$$

9. Writing as a complex number: $\sqrt{-36}=\sqrt{36(-1)}=\sqrt{36}\sqrt{-1}=6i$

10. Writing as a complex number: $\sqrt{-169}=\sqrt{169(-1)}=\sqrt{169}\sqrt{-1}=13i$

11. Writing as a complex number: $\sqrt{-45}=\sqrt{-9\bullet 5}=\sqrt{-9}\sqrt{5}=3i\sqrt{5}$

12. Writing as a complex number: $\sqrt{-12}=\sqrt{-4\bullet 3}=\sqrt{-4}\sqrt{3}=2i\sqrt{3}$

13. Combining the complex numbers: $(4i+3)+(4+6i)=(3+4)+(4i+6i)=7+10i$

14. Combining the complex numbers: $(5-3i)-(2-7i)=5-3i-2+7i=(5-2)+(-3i+7i)=3+4i$

15. Multiplying the complex numbers: $(5-i)(5+i)=(5)^2-(i)^2=25+1=26$

16. Multiplying the complex numbers: $(2+4i)(3-i)=6-2i+12i-4i^2=6+10i-4(-1)=10+10i$

17. Dividing the complex numbers: $\frac{i}{3+i}\bullet\frac{3-i}{3-i}=\frac{3i-i^2}{9-i^2}=\frac{3i+1}{9+1}=\frac{3i+1}{10}=\frac{1}{10}+\frac{3}{10}i$

18. Dividing the complex numbers: $\frac{4-i}{4+i}\bullet\frac{4-i}{4-i}=\frac{16-4i-4i+i^2}{16-i^2}=\frac{16-8i-1}{16+1}=\frac{15-8i}{17}=\frac{15}{17}-\frac{8}{17}i$

19. The vertex is (0,3). Sketching the graph:

20. The vertex is (–2,0). Sketching the graph:

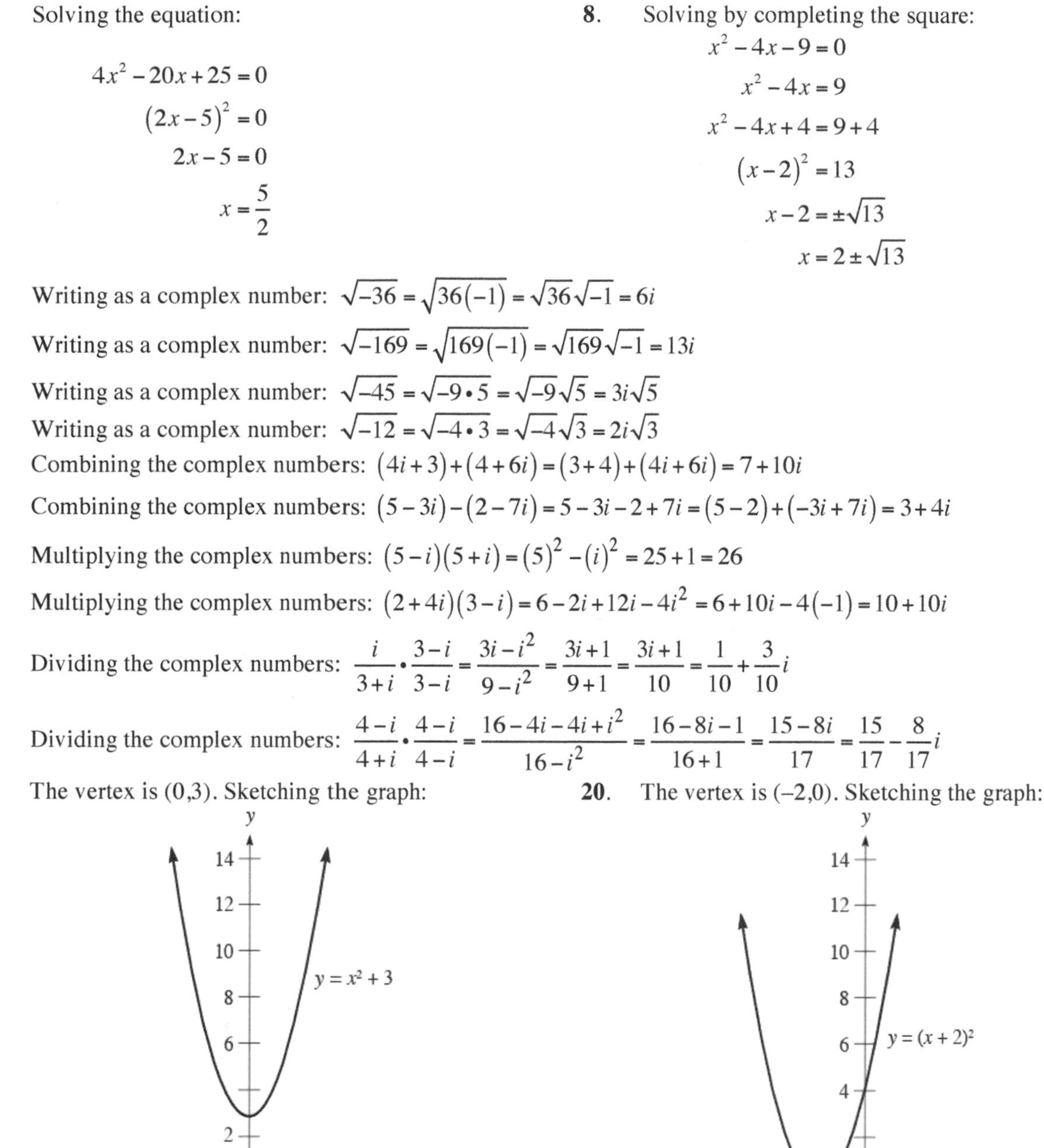

21. The vertex is (4,2). Sketching the graph:

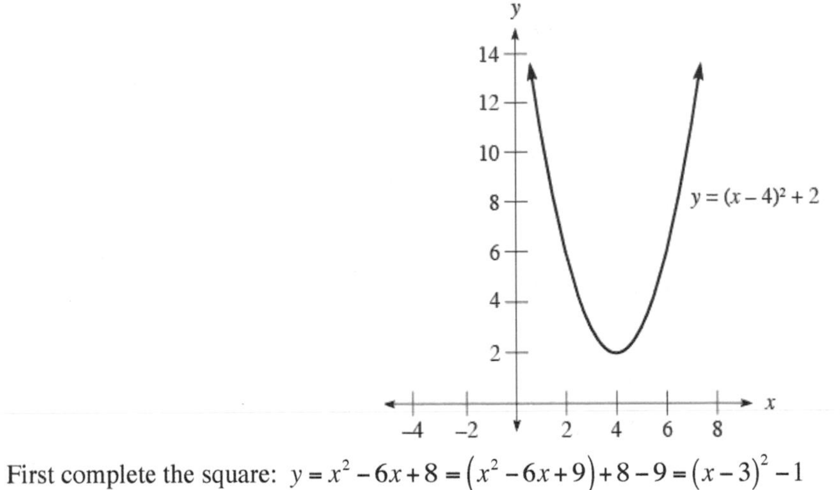

22. First complete the square: $y = x^2 - 6x + 8 = (x^2 - 6x + 9) + 8 - 9 = (x-3)^2 - 1$

The vertex is (3,–1). Sketching the graph:

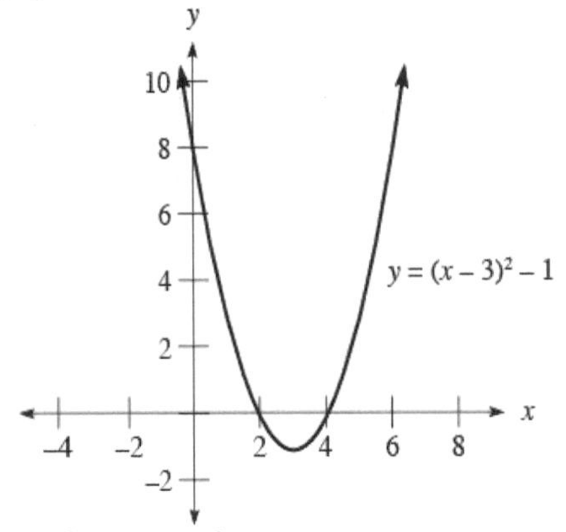

23. Let h represent the height. Using the Pythagorean theorem:

$$4^2 + h^2 = 5^2$$
$$16 + h^2 = 25$$
$$h^2 = 9$$
$$h = 3 \quad (h = -3 \text{ is impossible})$$

The height is 3 inches.

24. This graph matches C.

25. This graph matches B.

26. This graph matches A.

27. This graph matches D.